Lecture Notes in Mathematics 1700

Editors:
A. Dold, Heidelberg
F. Takens, Groningen
B. Teissier, Paris

T0224445

Springer
Berlin
Heidelberg
New York
Barcelona
Hong Kong
London
Milan
Paris
Singapore
Tokyo

Wojbor A. Woyczyński

Burgers-KPZ Turbulence

Göttingen Lectures

Springer

Author

Wojbor A. Woyczyński
Department of Statistics and
Center for Stochastic and Chaotic Processes
in Science and Technology
Case Western Reserve University
Cleveland, OH 44106, USA
e-mail: waw@po.cwru.edu

Cataloging-in-Publication Data applied for

Die Deutsche Bibliothek - CIP-Einheitsaufnahme

Woyczyński, Wojbor A.:
Burgers KPZ turbulence : Göttingen lectures / Wojbor A.
Woyczyński. - Berlin ; Heidelberg ; New York ; Barcelona ; Budapest
; Hong Kong ; London ; Milan ; Paris ; Santa Clara ; Singapore ;
Tokyo : Springer, 1998
  (Lecture notes in mathematics ; 1700)
  ISBN 3-540-65237-X

Mathematics Subject Classification (1991): 60H15, 60G60, 60K40, 70K40,
76L05, 83F05, 35Q53

ISSN 0075-8434
ISBN 3-540-65237-X Springer-Verlag Berlin Heidelberg New York

© Springer-Verlag Berlin Heidelberg 1998
Printed in Germany

Typesetting: Camera-ready T_EX output by the author
SPIN: 10650158   41/3143-543210 - Printed on acid-free paper

To the memory of my parents,

Otylia Sabina,
a dedicated teacher who studied mathematics under Dickstein and Sierpinski, and
Eugeniusz,
a mechanical engineer, who provided the high school student with a fierce in-house competition in solving mathematics and physics olympiads problems

Preface

These notes, woven around the subject of Burgers' turbulence/KPZ interface growth model—a study of the nonlinear parabolic equation $u_t + (u \cdot \nabla)u = \nu \Delta u$ with random initial data—are a written version of lectures given in the Summer of 1996 at the Georg-August-Universität Göttingen, Germany. The audience was principally the faculty and students of the Joint Graduate Program in Flow Stability and Turbulence of the Institutes of Fluid Flow, Mathematics, and Mathematical Stochastics, headed by Prof. Dr. Helmut Eckelman. A series of lectures given by the author on the same topic at the Nagoya University, Japan, during his visits there over the three-year period 1992-1994, also served as a source of material.

The volume should be taken for what it is, a relatively rough lecture notes, often sketchy and informal, in the form only slightly smoothed out (we are forever indebted to Donald Knuth—Case alumnus, too— for making our TEX manuscripts look better than they really are). However, compared to the handouts distributed to the audience, more details have been included. Hopefully, at some point in the future, a real book will emerge as a result of this exercise. Selection of the topics is highly idiosyncratic and reflects author's own interests; analysis is conducted mostly in the space-time domain with less attention paid to the frequency domain picture. However, the Bibliography contains more complete information about other work in the field which over the last decade enjoyed a sort-of revival, especially in the probability theory and stochastic analysis community, the latter in the midst of the golden period of the stochastic partial differential equations theory. Open problems are mentioned throughout the text.

The notes are addressed to a diverse audience, including mathematicians, statisticians, physicists, fluid dynamicists and engineers, and contain both rigorous and heuristic arguments. In many situations the material is presented at a "physical level of rigorousness". Usually, but not always, that means that we do not know how to prove things rigorously, and the area remains a challenge for mathematicians. Because of the multidisciplinary audience, the notes include a concise exposition of some classical and fairly elementary topics in probability theory such as Brownian motion, Wiener polynomial chaos, etc. From the mathematical perspective the subject matter of these lectures goes beyond the classical probability theory and faces the challenge of multidimensional stochastic flows constrained by physically motivated dynamics expressed in the form of nonlinear partial differential equations.

Thanks are due to Manfred Denker and Helmut Eckelman of the Göttingen University, and Tada-Hisa Funaki of the Nagoya University (currently at the University of Tokyo) for their hospitality. The author also takes this opportunity to express his brotherly love for (in the order of appearance on the stage) Yiming Hu, Donatas Surgailis, Stan Molchanov, Tada-Hisa Funaki, Sasha Saichev, Kolya Leonenko and Piotr Biler, with whom he had pleasure to collaborate on the subject of Burgers' turbulence over the last seven years. A significant portion of the material in this volume comes from our joint papers. Conversations and correspondence on Burgers' turbulence with Uriel Frisch, Murray Rosenblatt, Yasha Sinai, Sergei Shandarin and Jackson Herring are also acknowledged. Piotr Biler and Barbara Margolius carefully read through portions of the original manuscript and pointed out numerous inaccuracies and omissions. I thank them for their attention to detail.

Wojbor A. Woyczyński
Cleveland, Spring 1998

Contents

Lecture 1
Shock Waves and The Large Scale Structure (LSS) of the Universe

1.1 Nonlinear waves, shock formation, conservation laws

The simplest *nondispersive waves* (i.e. waves in media where the speed of propagation c is independent of the frequency of the wave) are *planar hyperbolic waves* described by the equation

$$u_t + cu_x = 0, \tag{1}$$

where $u = u(x, t)$, and where c is a constant. The obvious solution

$$u(x, t) = u_0(x - ct) \tag{2}$$

represents the distortionless propagation of the initial field $u_0(x) = u(x, 0)$. It's straightforward nonlinear analogue is a *hyperbolic conservation law* expressed by the equation

$$u_t + c(u)u_x = 0, \tag{3}$$

with the initial condition $u_0(x) = u(x, 0)$ (see e.g. Lax (1973)). Here, the speed of propagation $c(u)$ depends on the amplitude u. The characteristic equations for the above first-order partial differential equation take the form

$$\frac{dU}{dt} = 0, \quad \frac{dX}{dt} = c(U) \tag{4}$$

with the initial conditions

$$X(0) = y, \quad U(0) = u_0(y), \tag{5}$$

so that

$$X(y, t) = y + c(u_0(y))t, \quad U(y, t) = u_0(y), \tag{6}$$

which gives the solution

$$u = u_0(x - c(u)t) \tag{7}$$

in an implicit form. However, if $c' \not\equiv 0$, we encounter a nonuniqueness problem. Indeed, if

$$\frac{d}{dy} c(u_0(y)) < 0, \tag{8}$$

then, for two characteristics starting at y and $y + dy$, the difference

$$X(y + dy, t) - X(y, t) = \left(1 + \frac{d}{dy} c(u_0(y))t\right) dy, \tag{9}$$

and the two characteristics are bound to intersect for t large enough. The uniqueness can be guaranteed only in the interval

$$t \in \left(0, \left(\min \frac{d}{dy} c(u_0(y))\right)^{-1}\right). \tag{10}$$

The above analytic phenomenon is physically reflected in formation of *shock waves* (discontinuous solutions). One way to get around this difficulty is to take into account local interactions with the medium, such as the linear *viscous dissipation*, which leads to the nonlinear diffusion equation of the form

$$u_t + c(u)u_x = \nu u_{xx}. \tag{11}$$

It is known that if the viscosity coefficient $\nu \to 0$ then the solutions of (11) converge to the (generalized) solutions of (3) (see, e.g., DiPerna (1983)).

The special case of equation (3) with $c(u) = cu$ gives rise to the so-called *Riemann equation*

$$u_t + cuu_x = 0 \tag{12}$$

which describes the hydrodynamic flow of noninteracting particles moving along the x-axis with velocity u (see, e.g., Arnold (1988). Softening the shock fronts in the Riemann equation by addition of a linear dissipation term (*parabolic regularization*) leads to the (one-dimensional) *Burgers equation*

$$u_t + cuu_x = \nu u_{xx}, \tag{13}$$

which, coupled with the *random initial data* $u_0(x)$, is the main object of study in these notes. The term *Burgers turbulence* applied to this model is natural if one keeps in mind that the Burgers equation can be viewed as a special one-dimensional case of the Navier-Stokes equation

$$\vec{u}_t + (\vec{u} \cdot \vec{\nabla})\vec{u} = -\vec{\nabla}p + \nu \Delta \vec{u} + \vec{F}, \tag{14}$$

describing fluid flow, where the pressure p and external force field $\vec{F}$ terms were neglected. In such a flow, the velocity field appears to be random even without random initial conditions and "this contrast is the source of much of what is interesting in turbulence theory" (see Chorin (1975), p. 1, Frisch (1996)). The statistical approach has been the established tool in the study of turbulence for a long time (see Monin and Yaglom (1987)).

Describing the relationship of the Navier-Stokes equation to the Burgers' equation it is hard to improve on the following compact analysis penned some 25 years ago by Robert Kraichnan (1968).

> "The differences between Burgers' and Navier-Stokes' equations are as interesting as the similarities. The uu_x term [in (13)] conserves both $\int u(x,t)\,dx$ and $\int [u(x,t)]^2 dx$, as in the incompressible Navier-Stokes equation. In both cases, the advection term tends to produce regions of steepened velocity gradients, which implies a transfer of excitation from lower- to higher-wavenumber components of the velocity field. Perhaps the sharpest difference is that Burgers' equation appears to offer no counterpart to the hierarchy of instabilities which makes the small-scale structure of high Reynolds number [small ν] turbulence chaotic and unpredictable. If the initial Reynolds number is high, Burgers' equation leads to shock fronts which coalesce on collision so that, at later times when the Reynolds number is still high, an initially complicated u field is reduced to a sparse collection of shocks, with smooth and simple variation of u between fronts. The high-wavenumber excitation is then associated principally with the shocks themselves. Burgers' equation reduces initial chaos instead of increasing it [...] These similarities and differences make Burgers' equation a valuable vehicle

for exploring the limits of applicability of statistical approximations designed for Navier-Stokes turbulence. Interest is height-
ened because direct numerical integration of initial ensembles of
velocity fields forward in time is much more feasible for Burgers'
equation than for the Navier-Stokes equation."

Since its inception by J. Burgers in the late 1930s, the model held a
steady interest of the fluid dynamics and physics communities (see, e.g.,
Gotoh, Kraichnan (1993)). In view of the inelastic type of particles'
collisions, Burgers' equation (coupled with the continuity equation of
passive tracer transport) has been also studied as a model of evolution
of the self-gravitating matter. Thus, information about the time propagation of the initial fluctuations in the Burgers flow gives a theoretical
model for the observed large scale structure of the Universe in late nonlinear stages of the gravitational instability (see Shandarin, Zeldovich
(1989), Weinberg, Gunn (1990), Gurbatov, Malakhov, Saichev (1991),
and Albeverio, Molchanov, Surgalis (1995), Molchanov, Surgailis and
Woyczynski (1997)). We will return to this topic in the next section.

Over the last 10-15 years the mathematical community developed
a renewed interest in Burgers' turbulence and related models (KPZ,
forced Burgers' flows, anomalous nonlinear diffusion) ranging from the
study of *propagation of chaos* (see, e.g., Gutkin, Kac (1983), Sznitman
(1988), Funaki and Woyczynski (1998)), *asymmetric exclusion processes*
(Andjel, Bramson, Liggett (1988), Ferrari (1992)), *cellular automata*
(Boghosian, Levermore (1987), Brieger, Bonomi (1992)), scale renormalization (Rosenblatt (1987)), the *Hausdorff dimension* of the shocks
set (Sinai (1992), Janicki and Woyczynski (1996), Bertoin (1998)), to
maximum principles for moving average initial data (Hu, Woyczynski
(1994, 1995)), *parabolic and hyperbolic scaling limit behavior* (Funaki,
Surgailis and Woyczynski (1995), Molchanov, Surgailis and Woyczynski (1995), and Leonenko and Orsingher (1995)) and the *white noise*
forcing and initial data (Avellaneda and E (1995), Holden, Øksendal,
Ubøe and Zhang (1996), Bertini, Cancrini and Jona-Lasinio (1994)); a
large number of interesting problems remain unsolved.

In the first few lectures our main question is: How do the initial
random fluctuations of u propagate in Burgers' flow $u(x, t), x \in \mathbf{R}, t >$
0? The goal is to provide a rigorous *mathematical* study of the problem
for a precisely specified initial random data and the developments are
based on some relatively recent advances in the theory of random fields.
Here, the pioneering work was that of Bulinski and Molchanov (1991),
who also elucidated the importance of the initial shot noise type data.

Burgers' equation also often arises in the following generic situation: consider a flow of $u(t, x)$ on the real line, say, describing the density of a certain quantity per unit length, with the flux of this quantity through section at x described by another function $\phi(t, x)$. Assume that the flow is subject to the *conservation law*

$$\frac{\partial}{\partial t} \int_{x_0}^{x_1} u(t, x)\, dx + \phi(t, x_1) - \phi(t, x_0) = 0,$$

when $x_0 < x_1$. If we assume that the flux $\phi(t, x) = \Phi(u(t, x))$ depends on the local density only, then, as $x_0 \to x_1$, the above conservation law leads to a *quasilinear equation* equation of Riemann type

$$u_t - \Phi'(u)u_x = 0.$$

If the flux function is permitted to depend additionally on the gradient of the density u, say, $\phi(t, x) = \Phi(u(t, x)) - \nu u_x(t, x)$, then the above conservation law leads to the equation

$$u_t - \Phi'(u)u_x = \nu u_{xx},$$

of which the Burgers' equation is a special case.

1.2 The large scale structure of the Universe and the adhesion approximation

It is a well-known, albeit relatively recent, observational fact that matter in the Universe is distributed in cellular "pancake" structures, clusters and superclusters of galaxies, with giant voids between them (see, Figs. 1.2.1 and 1.2.2). The current wide-ranging Sloan Digital Survey is aimed at providing even finer data about the distribution of galaxies within π steradians (a quarter of the whole sky) to include all the point sources down to the 23rd magnitude and galaxies down to the 19th magnitude ($r' = 18$, where r' is the apparent magnitude in the spectral band with effective wavelength 6280 Å). It corresponds to about 600 Mpc of the effective depth. Meanwhile, over the last fifteen or so years, a major effort was undertaken by the astrophysicists (see the astrophysical literature quoted in references, from Zeldovich, Einasto, Shandarin (1981), through Gurbatov, Malakhov, Saichev (1991), to Bernardeau, Kofman (1995)) to provide a mathematical model of an evolution that, starting out with an essentially uniform distribution of matter following the Big Bang, with perhaps minute random quantum fluctuations, would lead to the presently observable rich structure with filaments, sheets and clusters of galaxies. At this late epoch of the formation of the large scale structure,

- the dark (nonluminous) matter dominates;
- it acts as collisionless dustlike particles;
- no pressure effects need to be taken into account, with the Newtonian gravity being the only force of consequence;
- the radiative and gas dynamics effects are short range.

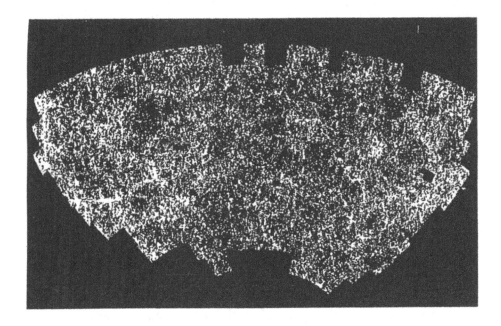

Figure 1.2.1. The distribution of galaxies in the APM galaxy survey. This uniform survey of ≥ 2 million galaxies, with $17 \leq b_J \leq 20.5$, constructed from machine scans of 185 OK Schmidt plates is shown in an equal area projection centered on the South Galactic pole. The small empty patches in the map are regions that were excluded around bright stars, nearby dwarf galaxies and globular clusters. From Maddox et al. (1990), van de Waygaert (1991).

Assuming the flat, expanding Universe, with the scale factor (rate of expansion)

$$a(t) = t^{2/3},$$

and the mean density

$$\bar{\rho} \propto a^{-3},$$

the evolution of the matter density $\rho = \rho(t, \vec{x})$, $\vec{x} \in \mathbf{R}^3$, is usually (see, e.g., Peebles (1980), Kofman et al. (1992)) described by the system of three coupled partial differential equations of hydrodynamic type

$$\frac{\partial \rho}{\partial t} + 3H\rho + \frac{1}{a}\nabla \cdot (\rho \vec{w}) = 0, \tag{1}$$

$$\frac{D\vec{w}}{Dt} + H\vec{w} = -\frac{1}{a}\nabla \varphi, \tag{2}$$

$$\nabla^2 \varphi = 4\pi G a^2 (\rho - \bar{\rho}), \tag{3}$$

SDSS Northern Survey, 6° slice

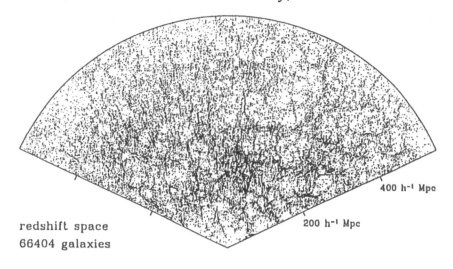

redshift space
66404 galaxies

400 h⁻¹ Mpc

200 h⁻¹ Mpc

Figure 1.2.2. A simulated slice, 6° by 130°, through the SDSS redshift survey of the north Galactic cap. Galaxies are plotted at the distance that would be inferred from their redshift, so cluster velocity dispersions create "fingers of God" that point towards the observer. The slice contains 66,404 galaxies, 6.6% of the number expected over the full area of the northern survey. This mock catalog is drawn from a large N-body simulation of a low density ($\Omega = 0.4, \Lambda = 0.6$) CDM model, performed by Changbom Park and Richard Gott; from Gunn and Weinberg (1994).

where $\vec{w}$ is the local velocity, φ is the gravitational potential, D/Dt stands for the usual Eulerian derivative, and H and G are, respectively, the Hubble and the gravitational constants. The three equations are, of course, the *continuity equation*, the *Euler equation* and the *Poisson equation*.

System (1-3) is not easy to analyze and several attempts have been made at simplifying it, while preserving the predictive ability of the reduced models. Introducing the velocity $\vec{v} = d\vec{x}/da$ in the coordinates co-moving with the expanding Universe (see, e.g., Sahni, Sathyaprakash, Shandarin (1994)), the Euler equation (2) is transformed into equation

$$\frac{\partial \vec{v}}{\partial a} + (\vec{v} \cdot \nabla)\vec{v} = -\frac{3}{2a}(\vec{v} + A\nabla\varphi), \tag{4}$$

with

$$A = \left(\frac{3}{2}H^2 a^3\right)^{-1} = \text{const},$$

where the right-hand side represents, in the Lagrangian approach, the force acting on the particle. It still involves a nonlocal operator so, in 1970, Zeldovich proposed a model where the nonlocal part was simply dropped. This gives a clear Lagrangian picture as (4) then becomes the classical Riemann equation, which can provide an explanation of the formation of pancake structures. Zeldovich' model has been adjusted and studied on the physical level of rigorousness by Gurbatov, Saichev and Shandarin (1984) (see also Shandarin and Zeldovich (1989), Weinberg and Gunn (1990), Bernardeau and Kofman (1995), Kofman and Raga (1992), Kofman et al. (1994)) who replaced the nonlocal term on the right-hand side of (4) by the Laplacian, to yield the Burgers equation

$$\frac{\partial \vec{v}}{\partial a} + (\vec{v} \cdot \nabla)\vec{v} = \nu\nabla^2\vec{v}, \qquad \vec{v} \in \mathbf{R}^3, \tag{5}$$

where the viscosity term is supposed to mimic the gravitational "adhesion". The constant ν should be small so that the viscosity effects do not affect the motion of matter outside clusters. This *adhesion model* of the large structure of the Universe demonstrates self-organization at large times and has the rough ability to reproduce formation of cellular structures in mass distribution. It has been extensively studied in the astrophysical literature and satisfactorily tested against high resolution (512 × 512) N-body simulations (Kofman et al. (1994), Mellott et al (1994)). In Sahni, Sathyaprakash, Shandarin (1994), extensive simulations and comparisons of different models were conducted. In particular, it made possible evaluation of values of the primordial gravitational potential at the centers of voids that would lead to the currently observable structures. Also, the void sizes were estimated.

The problem of evolution of the density fields $\rho(t, \vec{x})$ associated with Burgers' equations (5) (and its nonhomogeneous counterpart) can not be addressed at this point with similar degree of mathematical rigorousness (although, see the one-dimensional work by E, Rykov and Sinai (1996)), but can be studied via an approximate model equation (Saichev and Woyczynski (1994, 1995)) and by a statistical analysis of computer simulations (Janicki, Surgailis and Woyczynski (1995)). We shall discuss this work in one of the later lectures.

1.3 KPZ equation of interface growth and other physical models leading to Burgers' equation

Suppose that a surface is grown via the ballistic deposition process (such as some chemical vapor deposition processes for growing crystals or some sedimentation processes) which is schematically shown in Fig. 1.3.1. The only constraint is that the new particles are added in the direction perpendicular to the existing surface.

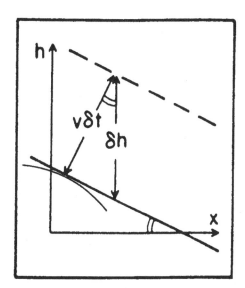

Fig. 1.3.1. Schematic picture of an interface growth via the ballistic deposition process. Function $h(x,t)$ deribes the evolution of the interface elevation.

With the function $h(x,t)$ describing evolution of the interface elevation the normal vector

$$\bar{n} = (-\nabla_x h, 1) \Big/ \sqrt{1 + (\nabla_x h)^2} \qquad (1)$$

where $(.,.)$ denotes the scalar product, so that the elevation increment

$$\delta h = v\frac{1}{\sqrt{(1+(\nabla_x h)^2}}\cdot \delta t \approx \left(v - \frac{v}{2}(\nabla_x h)^2\right)\delta t, \tag{2}$$

where v stands for the velocity of particles being deposited. Taking the limit $\delta t \rightarrow 0$, transforming to comoving coordinate frame, and taking into account the classical Brownian surface (height) diffusion (e.g., caused by surface tension) one arrives at the nonlinear partial differential equation for interface elevation

$$\frac{\partial h}{\partial t} + \frac{\lambda}{2}(\nabla h)^2 = \nu\nabla^2 h$$

where λ and ν are constants, which after the substitution $\vec{u} = \vec{\nabla}h$, gives the following Burgers equation for the velocity $\vec{u}$ of the interfacial growth

$$\frac{\partial u}{\partial t} + \lambda(v, \nabla v) = \nu\nabla^2 u. \tag{3}$$

In this context, equation (3) is called the *KPZ equation* (see, Kardar, Parisi and Zhang (1986), and Barábasi and Stanley (1995)).

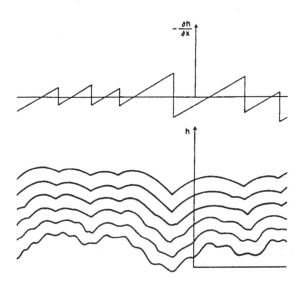

Fig. 1.3.2. *Top:* In the zero viscosity limit $\nu \rightarrow 0$, the solutions $\vec{u}(x,t)$ of Burgers' equation (4) assume a characteristic sawtooth shape with propagating shock fronts. *Bottom:* The corresponding interface profile $h(x,t)$, a solution of the KPZ equation (3), shows the familiar grained morphology.

We will see in Lecture 3 that , in the zero viscosity (weak) limit $\nu \to$ 0, the solutions $u(x, t)$ of Burgers' equation (4) assume a characteristic sawtooth shape with propagating shock fronts shown in Fig. 1.3.2 (Top). The corresponding interface profile $h(t, x)$, a solution of the KPZ equation (3), shows the familiar grained morphology of ballistically grown interfaces shown in Fig. 1.3.2 (Bottom).

In some situations the usual Brownian surface height diffusion accounted for in the Burgers-KPZ equation (2.3) is, however, supplemented by trapping effects and anomalous surface transport which, as argued by several authors (see, e.g., Zaslavsky (1995)), displays scaling and properties akin to the Lévy flights. In this context, we proposed to replace the usual Burgers equation (2.3) by what we call the *fractal Burgers equation* (see Biler and Woyczynski (1998), Funaki and Woyczynski (1998), Mann and Woyczynski (1997))

$$\frac{\partial u}{\partial t} + \lambda(v, \nabla v) = -\mu(-\nabla^2)^{\alpha/2}u, \qquad (2.4)$$

$0 < \alpha \leq 2$, perhaps augmented by the usual diffusion term. The fractal Burgers equation is, however, quite difficult to handle both theoretically and numerically as the fractional Laplacian $-(\nabla^2)^{\alpha/2}$ (defined most conveniently via the Fourier transform, see, e.g., Saichev and Woyczynski (1997)) is a nonlocal singular operator. We shall discuss these issues in Lecture 8.

Other physical examples, such as polymer chains, queueing systems, financial mathematics, and traffic problems, involving Burgers' dynamics abound and can be found in the Bibliography. Some of them will be explicitly mentioned in the subsequent lectures.

Lecture 2
Hydrodynamic Limits, Nonlinear Diffusions, and Propagation of Chaos

2.1 Random walks and linear diffusions

Consider a symmetric *random walk* on the one-dimensional lattice. Starting from the origin, a particle moves one step to the right or left with equal probabilities $1/2$, see Fig. 2.1.1.

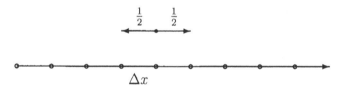

Fig. 2.1.1. One-dimensional random walk on a lattice with step-size Δx.

The consecutive steps are independent and are taken at times

$$t_k = k\Delta t, \qquad k = 1, 2, \ldots,$$

with the step size (lattice distance) Δx, so that the set of possible positions of the particle is

$$x_k = k\Delta x, \qquad k = 1, 2, \ldots.$$

Let $X(t, x)$ indicate whether the site x is occupied ($X = 1$), or unoccupied ($X = 0$) at time t. Denote by

$$u(t_k, x_k) = \mathbf{P}\{X(t_k, x_k) = 1\}, \tag{1}$$

the probability that at time t_k the particle is at site x_k. Then, clearly,

$$u(t_{k+1}, x_k) = \frac{1}{2}u(t_k, x_{k-1}) + \frac{1}{2}u(t_k, x_{k+1}). \tag{2}$$

This equation system can be rewritten as a system of difference equations for function u:

$$u(t_{k+1}, x_k) - u(t_k, x_k) = \frac{1}{2}\Big(u(t_k, x_{k-1}) + u(t_k, x_{x+1}) - 2u(t_k, x_k)\Big), \tag{3}$$

$k = 1, 2, \ldots$, with the initial condition $u(0, x) = \delta(x)$. Instead of directly solving system (3) it is easier to notice that with *parabolic scaling*

$$\Delta t = (\Delta x)^2$$

we can rewrite (3) in the form

$$\frac{u(t_{k+1}, x_k) - u(t_k, x_k)}{\Delta t} = \frac{1}{2}\frac{u(t_k, x_{k-1}) + u(t_k, x_{k+1}) - 2u(t_k, x_k)}{(\Delta x)^2}, \tag{3}$$

$k = 1, 2, \ldots$, which, in the *hydrodynamic limit* $\Delta t = (\Delta x)^2 \rightarrow 0$, becomes the usual *linear diffusion equation*

$$\frac{\partial u}{\partial t} = \frac{1}{2}\frac{\partial^2 u}{\partial x^2}. \tag{4}$$

If one permits *asymmetric transition probabilities*, $(1 - \bar{\alpha})/2$ for the particle moving to the right, and $(1 + \bar{\alpha})/2$ for the particle moving to the left, $-1 \leq \bar{\alpha} \leq 1$, see Fig. 2.1.2,

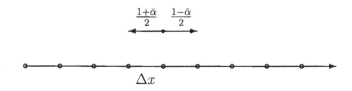

Fig. 2.1.2. One-dimensional asymmetric random walk on a lattice with step-size Δx.

then the hydrodynamic limit $\Delta t, \Delta x \rightarrow 0$, taken under the scaling conditions

$$c = \bar{\alpha}\frac{\Delta x}{\Delta t}, \qquad \nu = \frac{1}{2}\frac{(\Delta x)^2}{\Delta t}, \tag{5}$$

where $c \in \mathbf{R}, \nu > 0$, are constants, leads to the *linear diffusion equation with a drift*:

$$\frac{\partial u}{\partial t} + c\frac{\partial u}{\partial x} = \nu\frac{\partial^2 u}{\partial x^2}.$$

Physically, c is the *drift velocity* and ν is the *viscosity parameter* (*diffusion coefficient*) and (three-dimensional versions of) the above arguments go back to Einstein (1905) and Smoluchowski (1905), if not further back.

The above passage from the difference to differential equation has its analogue in terms of the random walk itself approaching the continuous-time *Brownian motion*.

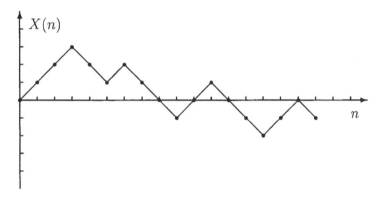

Fig. 2.1.3. A typical trajectory of random walk X_n.

Consider a sequence $\xi_i, i = 1, 2, \ldots$, of independent random variables with $\mathbf{P}\{\xi_i = \pm 1\} = 1/2$, so that $\mathbf{E}\xi_i = 0, \operatorname{Var}\xi_i = 1$. Then the position of the particle at time n is described by

$$X(n) = \xi_1 + \ldots + \xi_n,$$

with $\mathbf{E}X(n) = 0, \operatorname{Var}X(n) \nearrow \infty$ (see, Fig. 2.1.3). But

$$B_n = X(n)/\sqrt{n}$$

has variance 1 for all $n = 1, 2, \ldots$, and its Fourier transform

$$\phi_{B_n}(\lambda) = \mathbf{E}e^{i\lambda B_n} = \left(\mathbf{E}\exp\left[i\lambda\xi_1/\sqrt{n}\right]\right)^n = \cos^n(\lambda/\sqrt{n}).$$

As $n \to \infty$, applying twice the l'Hospital rule to variable n,

$$\lim_{n\to\infty} \log \phi_{B_n}(\lambda) = \lim_{n\to\infty} n\log\cos\frac{\lambda}{\sqrt{n}} = \lim_{n\to\infty} \frac{\log\cos(\lambda/\sqrt{n})}{1/n}$$

$$= \lim_{n \to \infty} \frac{-\sin(\lambda/\sqrt{n})(-\lambda/2n^{3/2})}{-(1/n^2)\cos(\lambda/\sqrt{n})}$$

$$= \lim_{n \to \infty} \frac{-\cos(\lambda/\sqrt{n})(\lambda^2/2n^{3/2})}{1/n^{3/2}} = -\frac{\lambda^2}{2},$$

which means that $\lim_{n \to \infty} \phi_{B_n}(\lambda) = \exp[-\lambda^2/2]$, so that, in law, the random variables B_n converge to a standard Gaussian random variable, say B_∞, i.e., for all $x \in \mathbf{R}$,

$$\mathbf{P}\{B_\infty \le x\} = \int_{-\infty}^{x} \frac{\exp[-y^2/2]}{\sqrt{2\pi}} \, dy. \tag{6}$$

This is, of course, an elementary version of the *Central Limit Theorem*. Interpolating linearly the parabolically rescaled in time and space random walk we get

$$B_n(t) = \frac{1}{\sqrt{n}} \sum_{i=1}^{\lfloor nt \rfloor} \xi_i = \frac{X(\lfloor nt \rfloor)}{\sqrt{n}}, \tag{7}$$

where $\lfloor x \rfloor$ is the greatest integer $\le x$, and the finite-dimensional distributions of processes $B_n(t), t \in \mathbf{R}$, converge to the finite-dimensional distributions of the Brownian motion $B(t)$, i.e., the Gaussian process with independent and stationary increments, mean $\mathbf{E}B(t) = 0$, and variance $\operatorname{Var} B(t) = t$, so that

$$\operatorname{Cov}(B(t), B(s)) = \min\{t, s\}. \tag{8}$$

This is the celebrated *Invariance Principle* (see, e.g., Billingsley (1986), Theorem 37.8).

The Brownian motion process has a *Wiener-Lévy representation*

$$B(t) = \sum_{i=1}^{\infty} \gamma_i \int_0^t \psi_i(s) \, ds, \qquad 0 \le t \le 1, \tag{9}$$

where $\gamma_i, i = 1, 2, \ldots$, are independent, identically distributed (iid) $N(0, 1)$ random variables, $\psi_i, i = 1, 2, \ldots$, is an orthonormal system in $L^2([0, 1])$, and the series converges uniformly with probability 1 (see, e.g., Kwapien and Woyczynski (1992)). Once the convergence of (9) is established it is easy to see that the series represents a Brownian motion. Indeed, since $\mathbf{E}\gamma_i\gamma_j = 0, i \ne j$, the covariance for such a series

$$
\begin{aligned}
\mathbf{E}B(t)B(s) &= \sum_{i=1}^{\infty} \mathbf{E}\gamma_i^2 \int_0^t \psi_i(z) \, dz \int_0^s \psi_i(z) \, dz \\
&= \int_0^1 \mathbf{1}_{[0,t]}(z)\mathbf{1}_{[0,s]}(z) \, dz = \min\{t, s\}
\end{aligned}
$$

in view of the *Parseval equality*. Since a Gaussian process (its finite-dimensional distributions) is completely characterized by its covariance function, in view of (8), the series (9) represents the Brownian motion.

In view of the Wiener-Lévy representation (9), the Brownian motion has, with probability 1, continuous sample paths. However, it is not differentiable in the classical sense as

$$\mathbf{E}\left[\frac{B(t+h) - B(t)}{h}\right]^2 = \frac{\mathbf{E}B^2(h)}{h^2} = \frac{1}{h} \nearrow \infty$$

as $h \nearrow 0$. Nevertheless, the *distributional derivative* exists, since for each ψ in the *Schwartz space* $\mathcal{S}(\mathbf{R})$, one can set

$$\int \dot{B}(t)\,\psi(t)\,dt = \int \psi(t)\,dB(t),$$

where the *stochastic integral* on the right is well defined; it is a Gaussian random variable with mean zero and variance $\int \psi^2(t)\,dt$. In other words, the mapping

$$L^2(\mathbf{R}) \ni \psi \mapsto \int \psi(t)\,dB(t) \in L^2_{Gauss}(\Omega)$$

is a linear isometry, and the basic rule of stochastic integration is $\mathbf{E}(dB(t))^2 = dt$!

The distributional process $\dot{B}$ is called *white noise*, or the delta-correlated generalized process since

$$\mathrm{Cov}\,(\dot{B}(t), \dot{B}(s)) = \delta(s - t).$$

Indeed, for any $\psi, \phi \in \mathcal{D}(\mathbf{R})$, the bilinear functional

$$\begin{aligned}
\mathrm{Cov}\,(\dot{B}(t), \dot{B}(s))[\psi, \phi] &= \mathbf{E}\left[\int \psi(t)\,dB(t) \int \phi(t)\,dB(t)\right] \\
&= \int \psi(t)\phi(t)\,dt = \int\int \psi(t)\phi(s)\delta(s - t)\,dt\,ds.
\end{aligned}$$

The evolving one-dimensional probability density function $\mathbf{P}\{B(t) \in dx\} = u(x, t)\,dx$ of the Brownian motion is

$$u(x, t) = \frac{\exp[-x^2/2t]}{\sqrt{2\pi t}},$$

and it satisfies the linear diffusion equation (4) which, in this context, is called the *Fokker-Planck-Kolmogorov equation*.

The Brownian motion $\boldsymbol{B}(t)$ in $\mathbf{R}^d$ is defined as a vector-valued process

$$\boldsymbol{B}(t) = (B^1(t), \ldots, B^d(t)),$$

where $B^1(t), \ldots, B^d(t)$, are independent real-valued Brownian motions. Its density $u(\boldsymbol{x}, t)$, $\mathbf{P}\{\boldsymbol{B}(t) \in d\boldsymbol{x}\} = u(\boldsymbol{x}, t) \, dx_1 \ldots dx_d$, (notice the spherical symmetry) satisfies the d-dimensional linear diffusion equation

$$\frac{\partial u}{\partial t} = \frac{1}{2}\Delta u,$$

where $\Delta = \partial^2/\partial x_1^2 + \ldots + \partial^2/\partial_d^2$ is the Laplacian in $\mathbf{R}^d$.

For a general linear diffusion, the Fokker-Planck-Kolmogorov equation takes the form

$$\frac{\partial u}{\partial t} = \mathcal{L}u$$

where $\mathcal{L}$ is a second-order elliptic operator with, perhaps, variable coefficients.

The above "crash course" in Brownian motion, stochastic integrals and linear diffusions is meant only as a preamble to our discussion of nonlinear diffusions in the next sections. The in-depth references are Stroock and Varadhan (1979), Ikeda and Watanabe (1981), Karatzas and Shreve (1988), Protter (1990), Kwapien and Woyczynski (1992), and Revuz and Yor (1994).

2.2 Hydrodynamic limit for asymmetric exclusion particle systems

The microscopic dynamics of the *exclusion particle system* (or, *exclusion cellular automaton*) is described as follows (see, Fig. 2.2.1):

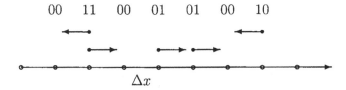

Fig. 2.2.1. Microscopic dynamics of the exclusion cellular automaton

Particles in the system move in time-steps on the integer lattice and at any particular discrete time their state is identified by their position, which is an integer, and the "velocity", which can be ± 1 per time step.

We demand that the *exclusion principle* be satisfied, that is no two particles can be in the same state at the same time or, equivalently, no two particles occupying the same site may be moving in the same direction. In each time step the system evolves in two substeps:

• *Collision Substep:* At its present lattice site $(t = l)$ each particle changes its direction randomly, provided the exclusion principle is not violated;

• *Advection Substep:* Particles move from their present lattice site to the neighboring site in their new direction.

It is convenient to describe the state of the system by a pair of functions

$$(L(k, l), R(k, l))$$

which, respectively, indicate the presence/absence of a particle moving to left at the site k at the time step l $(L(k, l) = 1/0)$, or moving to the right $(R(k, l) = 1/0)$. After the collision substep the state of the system will be denoted by

$$(L'(k, l), R'(k, l))$$

and the random dynamics is governed by an ensemble of independent random variables $\alpha(k, l) = \pm 1$ with the common mean $\bar{\alpha}$ so that the transitions are as follows:

$$
\begin{aligned}
(L, R) &= (0, 0) &\longmapsto& \quad (L', R') = (0, 0) \\
(L, R) &= (0, 1) &\longmapsto& \quad (L', R') = ((1 - \alpha)/2, (1 + \alpha)/2) \\
(L, R) &= (1, 0) &\longmapsto& \quad (L', R') = ((1 - \alpha)/2, (1 + \alpha)/2) \\
(L, R) &= (1, 1) &\longmapsto& \quad (L', R') = (1, 1)
\end{aligned}
$$

with $(L, R) = (L(k, l), R(k, l))$, $(L', R') = (L'(k, l), R'(k, l))$, $\alpha = \alpha(k, l)$. The rules are asymmetric as long as $\bar{\alpha} \neq 0$; if $\alpha = +1$ then only the pair $(1, 0)$ flips, and if $\alpha = -1$ then only the pair $(0, 1)$ flips. In an analytic form they can be written as follows:

$$L(k + 1, l + 1) = \frac{1 + \alpha(k, l)}{2}\Big(L(k, l) + R(k, l)\Big) - \alpha(k, l)L(k, l)R(k, l),$$

$$R(k - 1, l + 1) = \frac{1 - \alpha(k, l)}{2}\Big(L(k, l) + R(k, l)\Big) + \alpha(k, l)L(k, l)R(k, l).$$

Replacing the lattice step $k \mapsto k\Delta x$ and the time step $l \mapsto l\Delta t$ while imposing parabolic scaling $\Delta x^2 = \Delta t$, and averaging over the random ensemble $\{\alpha(k, l)\}$, we arrive at a difference equation for $T = L + R$

$$\frac{T(k\Delta x, (l+1)\Delta t) - T(k\Delta x, l\Delta t)}{\Delta t}$$

$$= \frac{\frac{1+\bar{\alpha}}{2}T((k-1)\Delta x, l\Delta t) + \frac{1-\bar{\alpha}}{2}T((k+1)\Delta x, l\Delta t) - T(k\Delta x, l\Delta t)}{\Delta x^2}$$

$$+\frac{\bar{\alpha}}{4}\left[\frac{T^2((k+1)\Delta x, l\Delta t) - T^2((k-1)\Delta x, l\Delta t)}{\Delta x^2}\right.$$

$$\left.-\frac{(L-R)^2((k+1)\Delta x, l\Delta t) - (L-R)^2((k-1)\Delta x, l\Delta t)}{\Delta x^2}\right].$$

which is an analogue of the equation (2.1.3).

Now, we take the hydrodynamic limit $\Delta x^2 = \Delta t \to 0$, making sure that

$$\frac{1}{2}\frac{\Delta x^2}{\Delta t} = \nu, \qquad \frac{\bar{\alpha}}{\Delta x}\frac{\Delta x^2}{\Delta t} = c$$

remain constant. This, of course, forces $\bar{\alpha} \to 0$, and the above difference equation becomes a nonlinear parabolic partial differential equation

$$\frac{\partial T}{\partial t} = \nu\frac{\partial^2 T}{\partial x^2} - c\left[\frac{\partial T}{\partial x} - \frac{1}{2}\frac{\partial T^2}{\partial x}\right]. \tag{1}$$

Changing variables $u(x, t) = -c(T(-x, t) - 1)$ results in the standard form of the Burgers equation

$$\frac{\partial u}{\partial t} + u\frac{\partial u}{\partial x} = \nu\frac{\partial^2 u}{\partial x^2}.$$

The above result is due to Boghosian and Levermore (1987) where one can find their more complete argument.

A related rigorous hydrodynamic limit result for the asymmetric *simple exclusion process* with continuous time which, however, gives the Riemann equation rather than the Burgers equation, can be described as follows:

Particles are distributed initially on the integer lattice $\mathbf{Z} = \{\ldots, -1, 0, 1, \ldots\}$ with at most one particle per site. The time is continuous and each particle waits an exponential time with parameter one and then moves one unit to the right with probability p if that site is unoccupied, and to the left with probability $1 - p$ with the same restriction. Otherwise, the particle does not move.

The existence of a unique Markov process $X(t, k), t \geq 0, k \in \mathbf{Z}$, (right-continuous-with-left-limits) corresponding to this description is guaranteed by the general existence theorem for interacting particle systems, see, e.g., Liggett (1985). More formally, the above rules can be written as an infinite system of stochastic differential equations in variable t:

$$
\begin{aligned}
dX(t,k) &= X(t-,k-1)[1-X(t-,k)]\,dP(t,k-1) \\
&\quad -X(t-,k)[1-X(t-,k+1)]\,dP(t,k) \\
&\quad +X(t-,k+1)[1-X(t-,k)]\,dQ(t,k+1) \\
&\quad -X(t-,k)[1-X(t-,k-1)]\,dQ(t,k),
\end{aligned}
$$

where $P(t,k), t \geq 0, k \in \mathbf{Z}$, and $Q(t,k), t \geq 0, k \in \mathbf{Z}$, are independent homogeneous Poisson processes with intensities p and $1-p$, respectively. They play the role of left and right "alarm clocks"; if one of them "rings", the particle moves in the corresponding direction.

To find the hydrodynamic limit for the simple exclusion process we need to look at *hyperbolically* rescaled quantities

$$
\begin{aligned}
X^h(t,x) &= \sum_{k \in \mathbf{Z}} X(t/h,k)\mathbf{1}_{[hk,h(k+1))}(x), \\
P^h(t,x) &= \sum_{k \in \mathbf{Z}} P(t/h,k)\mathbf{1}_{[hk,h(k+1))}(x), \\
Q^h(t,x) &= \sum_{k \in \mathbf{Z}} Q(t/h,k)\mathbf{1}_{[hk,h(k+1))}(x),
\end{aligned}
$$

where $\mathbf{1}_{[a,b)}$ is the indicator function of the interval $[a,b) \subset \mathbf{R}$, and $h > 0$.

With the notation

$$
F_{\pm h}u(x) = u(x)[1-u(x \pm h)],
$$

$$
D_{\pm h}u(x) = \pm[u(x \pm h) - u(x)]/h,
$$

for a real function $u(x)$, the above system of stochastic differential equations can be rewritten in terms of the rescaled process X^h as follows:

$$
\begin{aligned}
dX^h(t,x) &= -D_{-h}[F_h(X^h(t-,x))\,d(hP^h(t,x))] \\
&\quad +D_h[F_{-h}(X^h(t-,x))\,d(hQ^h(t,x))].
\end{aligned}
$$

Now, as $h \to 0$, with probability 1,

$$
hP^h(t,x) \to pt, \qquad hQ^h(t,x) \to (1-p)t.
$$

Also,

$$
D_{\pm h} \to \partial/\partial x, \qquad F_{\pm h}u \to u(1-u),
$$

so that, at the heuristic level, the random measure $X^h(t, x)\, dx$ converges weakly with probability 1 to $u(t, x)\, dx$, where $u(t, x)$ is a solution of the first-order nonlinear hyperbolic equation

$$\frac{\partial u}{\partial t} + (2p - 1)\frac{\partial u(1 - u)}{\partial x} = 0, \tag{2}$$

which, for $p \neq 1/2$ (that is, for an asymmetric exclusion process), is a linear transformation of the Riemann equation (1.1.12). For $p = 1/2$ the hyperbolic scaling is obviously wrong, and the correct parabolic scaling just leads to the linear diffusion equation (2.1.4).

The formal proof of the above hydrodynamic limit can be found in Benassi and Fouque (1987), and there exists a rich mathematical literature concerning hydrodynamic limits for similar interacting particle systems (and their queueing systems cousins), see, e.g., Andjel and Kipnis (1984), De Masi and Presutti (1991), Ferrari, Kipnis and Saada (1991), Kipnis (1986), Landim (1991), Spohn (1991), Varadhan (1993), and Seppäläinen (1996). An up-to-date survey, which also contains a description of unsolved problems in the area, is Liggett (1997). "The complex story of simple exlusion" processes is nicely told by Varadhan (1997).

2.3   Interacting and nonlinear diffusions, propagation of chaos

Initially, consider a general interacting diffusions in $\mathbf{R}^d$ described by a system of ordinary stochastic differential equations

$$\begin{cases} dX_t^{i,N} = dB_t^i + \frac{1}{N}\sum_{j=1}^{N} b(X_t^{i,N}, X_t^{j,N})dt, \\ i = 1, \dots, N, \end{cases} \tag{1}$$

where B_t^i are independent Brownian motions and the initial data $x_0^{i,N}$ are independent with common distribution v_0. Function $b(x, y)$ describes a potential of pair interactions. The infinitesimal generator for the above process is

$$L_N = \Delta_N + \frac{1}{N}\sum_{1 \leq i < j \leq N} b(x_i, x_j)\left(\frac{\partial}{\partial x_i} + \frac{\partial}{\partial x_j}\right), \tag{2}$$

and one can show that $X^{i,N} \to \bar{X}^i$ as $N \to \infty$ (cf. Sznitman (1986)). Also

$$\lim_{N \to \infty} E\left[\frac{1}{N}\sum_{i=1}^{N}\left(\frac{1}{N}\sum_{j=1}^{N} b(X_t^i, X_t^j) - \int b(X_t^i, y)v_t(dy)\right)^2\right] = 0, \tag{3}$$

where

$$v_t = \mathcal{L}(X_t). \tag{4}$$

Each $\bar{X}^i$ is an independent copy of the "nonlinear process" satisfying

$$\begin{cases} dX_t = dB_t + \int b(X_t, y)v_t(dy)dt, \\ \mathcal{L}(X_0) = v_0. \end{cases} \tag{5}$$

Take $f \in C_b^2(\mathbf{R}^d)$ and apply the Itô formula to (5) to obtain

$$\begin{aligned} f(X_t) &= f(X_0) + \int_0^t f'(X_s)dB_s \\ &\quad + \int_0^t \left(\frac{\Delta f(X_s)}{2} + \int_{\mathbf{R}^d} b(X_s, y)v_s(dy)\nabla f(X_s) \right) ds. \end{aligned}$$

Integrating by parts and taking expectations on both sides we get the equation

$$\int_0^t dE(X_s) = \int_0^t \left[\frac{1}{2}\Delta Ef(X_s) - \text{div}\left(\int b(X_{s,y})v_s(dy)Ef(X_s) \right) \right] ds,$$

which can be viewed as a "weak form" of the equation

$$\frac{\partial v}{\partial t} = \frac{1}{2}\Delta v - \text{div}\left(\int b(\cdot, y)v_t(dy)v \right). \tag{6}$$

In the special degenerate case, when $b(x,y) = \delta(x - y)$, one recovers (in $\mathbf{R}^1$) the Burgers equation

$$\frac{\partial v}{\partial t} = \frac{1}{2}\Delta v - \frac{\partial}{\partial x}(v^2).$$

The above result shows that, in a sense, the Burgers equation can be viewed as a Fokker-Kolmogorov-Planck equation for a "nonlinear" process X_t. A derivation of this result for an even more general fractal Burgers equation will be provided in Lecture 8.

The above limiting behavior of interacting diffusions can be reformulated as the "propagation of chaos" result which is due originally to Gutkin and Kac (1983). Consider interacting diffusions with singular interactions

$$L_N = \Delta_N + \frac{1}{N} \sum_{1 \le i < j \le N} \delta(x_i - x_j) \left(\frac{\partial}{\partial x_i} + \frac{\partial}{\partial x_j} \right),$$

with the corresponding evolution equation

$$\frac{\partial F_N}{\partial t} = L_N F_N,$$

and the initial product-form condition

$$F_N(0; x_1, x_2, \ldots, x_N) = v_0(x_1)v_0(x_2)\ldots v_0(x_N),$$

where

$$\int_{-\infty}^{\infty} v_0(x)dx = 1.$$

Then, it turns out that

$$\lim_{N\to\infty} \int_{\mathbf{R}^{N-1}} F_N(t; x_1, x_2, \ldots, x_N)dx_2\ldots dx_N = v(t, x),$$

where v satisfies the Burgers' equation with

$$v(0, x) = v_0(x).$$

Moreover, for each $l = 2, 3, \ldots,$

$$
\begin{aligned}
F_l(t; x_1, \ldots, x_l) &= \lim_{N\to\infty} \int_{\mathbf{R}^{N-l}} F_N(t; x_1, \ldots, x_N)dx_{l+1}\ldots dx_N \\
&= v(t, x_1)\ldots v(t, x_l),
\end{aligned}
$$

so that the original independence of the data is propagated and asymptotically preserved.

Several proofs of different variants of the "propagation of chaos" result exist; see, e.g., Calderoni and Pulvirenti (1983), Kotani and Osada (1985), Oelschläger (1985), Sznitman (1986). The original observation, due to Gutkin and Kac (1983), constructs an intertwining operators Q_n such that $L_N Q_N = Q_N \Delta_N$, so that

$$Q_N^{-1} e^{tL_N} v_0(x_1)\ldots v_0(x_N) = e^{t\Delta_N} Q_N^{-1} v_0(x_1)\ldots v_0(x_N).$$

Lecture 3
Hopf-Cole Formula and Its Asymptotic Analysis

3.1 Elementary, traveling wave, and self-similar solutions of Burgers' equation

Looking superficially at the structure of the Riemann equation

$$u_t + cuu_x = 0 \tag{1}$$

one immediately sees that if a polynomial in x is a solution of (1) then it has to be of degree one, and that if there is a power dependence in t then it has to be of degree minus one. The constants are readily calculated from the equation itself which gives an *elementary solution*

$$u(x, t) = \frac{x - x_0}{ct}, \tag{2}$$

which, of course, is no help in solving a general initial-value problems for the Riemann equation. However, it does give insight into the local structure of more general solutions of the equation. To develop more intuition for solutions of the Riemann equation let us take a look at the following example:

Example 1. Consider again equation (2.2.2)

$$u_t + (2p - 1)(u(1 - u))_x = 0, \qquad 0 \le p \le 1, p \ne 1/2, \tag{3}$$

which has been already encountered as the hydrodynamic limit for the simple asymmetric exclusion process in Lecture 2.2. A simple linear

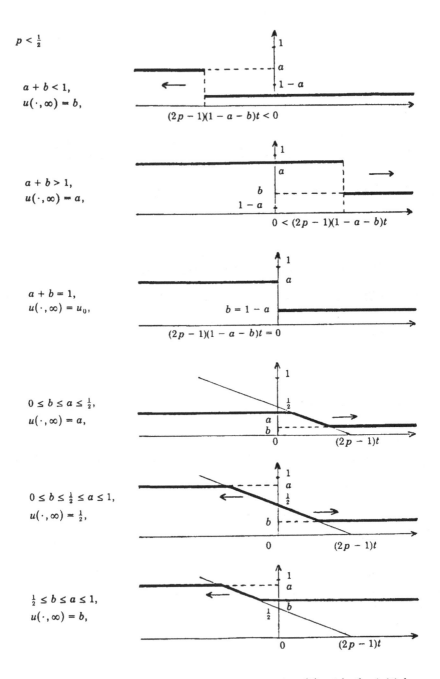

Fig. 3.1.1. Solution of the Riemann equation (3) with the initial condition (4). *Top:* In the case $p < 1/2$ the initial shock propagates; *Bottom:* In the case $p > 1/2$ the solution is continuous (from Benassi and Fouque (1987)).

substitution $\tau = t + (2p-1)x$ transforms equation (3) into the Riemann equation (1) with $c = -2(2p-1)$. For the initial condition

$$u_0(x) = a1_{(-\infty,0]} + b1_{(0,\infty)}, \tag{4}$$

and $p > 1/2$, the solution is continuous and has a nonconstant piece of the form (see Fig. 3.1.1)

$$\frac{1}{2}\left(1 - \frac{x}{(2p-1)t}\right).$$

Independently of the further discussion of the weak (viscosity) solutions of the Riemann equation we should mention at this point that a study of quasilinear hyperbolic equations of order one is a broad area and, for general results, the reader should consult, e.g., Oleinik (1957), Kružkov (1970).

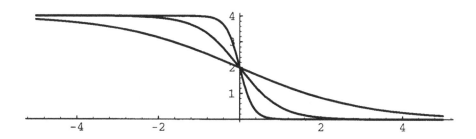

Fig. 3.1.2. A traveling wave solution $u(x,t) = W(x - ct)$ of the Burgers equation (5): $c = 2$, $\nu = 3, 1, 1/3$, $t = 0$, $W(\infty) = W'(\infty) = 0$. As the viscosity ν decreases the wave front becomes steeper.

For the Riemann equation (1), *traveling wave* solutions $u(x,t) = W(x - ct)$ are not very interesting. Indeed, a direct substitution gives $W \equiv 0$. However, for the Burgers equation

$$u_t + uu_x = \nu u_{xx} \tag{5}$$

a direct substitution gives the equation

$$-cW' + WW' = \nu W'',$$

which, after integration with $W(\infty) = 0, W'(\infty) = 0$, yields

$$-cW + W^2/2 = \nu W'. \tag{6}$$

If we are looking for the solutions with finite $W(-\infty)$, then it is natural to assume that $W'(y) \to 0$ as $y \to -\infty$. Then, clearly, $W(-\infty) = 2c$. In particular, such a solution has to be bounded.

Actually, equation (6) is separable and can be immediately solved explicitly to give a family of solutions

$$W(y) = c\left(1 + \tanh\left(-\frac{1}{2\nu}ct + a\right)\right), \tag{7}$$

with an arbitrary constant a. A special case corresponding to the propagation velocity $c = 2$ is pictured in Fig. 3.1.2. The soft shock of amplitude $W(\infty) - W(-\infty) = -2c = -4$ moves from left to right. As the viscosity ν decreases the wave front becomes steeper.

Another simple observation is that the Burgers equation (5) has a *self-similarity property*, that is, if a function u solves (5) then, for each $\lambda > 0$, the rescaled function

$$u_\lambda(t, x) = \lambda u(\lambda^2 t, \lambda x),$$

is also a solution of (5). The solutions satisfying the scaling invariance property

$$u_\lambda \equiv u, \qquad \forall \lambda > 0,$$

are called the *self-similar solutions*. It is expected that they describe large-time behavior of general solutions. Indeed, if

$$\lim_{\lambda \to \infty} \lambda u(\lambda^2 t, \lambda x) = V(t, x)$$

exists in an appropriate sense, then

$$\lim_{t \to \infty} t^{1/2}u(t, xt^{1/2}) = V(1, x)$$

(to see this, take $t = 1, \lambda = t^{1/2}$), and V satisfies the invariance property $V_\lambda \equiv V$. V is therefore a self-similar solution of (5) and since

$$V(t, x) = t^{-1/2}V(1, xt^{-1/2}),$$

it is completely determined by the function $\Phi(\xi) = V(1, \xi)$ of the single variable $\xi = xt^{-1/2}$. Note that, since $u(t, x) = t^{-1/2}\Phi(xt^{-1/2})$ is a solution of (5), a simple integration shows that $\Phi(\xi)$ satisfies the first order equation

$$\nu\Phi' + \frac{1}{2}\xi\Phi - \frac{1}{2}\Phi^2 = 0,$$

which, for $\Psi = \log \Phi$, gives the equation

$$\nu \Psi' + \frac{1}{2}\xi - \frac{1}{2}e^{\Psi} = 0.$$

Let us also observe that if

$$u_0(x) = \lim_{t \to 0} t^{-1/2} U(1, xt^{-1/2})$$

exists, then the initial condition u_0 is necessarily homogeneous of degree -1. If Burgers' equation is considered in dimension $d > 1$ ($x \in \mathbf{R}^d$), then such a $u_0 \not\equiv 0$ cannot have finite mass.

Self-similar properties of the Burgers equation will play a very important role in the study of *Burgers' turbulence*, that is a study of solutions of the Burgers equation with random initial conditions. This will be done in Lectures 4-7 and, for the fractal Burgers equation, in Lecture 8.

The *stationary solutions* $u(t, x) = u(x)$ for the Burgers equation (5) are also easy to find. Indeed, they have to satisfy the equation

$$u' = u^2/2$$

so that

$$u(x) = \frac{2}{C - x}.$$

3.2 The Hopf-Cole formula and exact solutions

An exact solution of the one-dimensional Burgers equation

$$u_t + u u_x = \nu u_{xx} \tag{1}$$

can be obtained via a transformation introduced in modern times independently by Hopf (1950) and Cole (1951), but which also can be found in the multi-volume treatise by Forsyth (1906). Let

$$u(x, t) = \frac{\partial}{\partial x} U(x, t),$$

where U is the potential of u. Integrating both sides of (1) with respect to x we obtain

$$\frac{\partial U}{\partial t} + \frac{1}{2}\left(\frac{\partial U}{\partial x}\right)^2 = \nu \frac{\partial^2 U}{\partial x^2}. \tag{2}$$

Now, if we make a logarithmic substitution

$$U(x, t) = -2\nu \log \varphi(x, t),\tag{3}$$

function φ satisfies the linear diffusion equation

$$\frac{\partial \varphi}{\partial t} = \nu \frac{\partial^2 \varphi}{\partial x^2}\tag{4}$$

with the initial condition

$$\varphi_0(x) = \varphi(x, 0) = \exp \frac{-U(x, 0)}{2\nu} = \exp \frac{-U_0(x)}{2\nu},\tag{5}$$

and

$$U_0(x) = \int_{-\infty}^{x} u_0(y) dy.\tag{6}$$

Since

$$u(x, t) = -2\nu \frac{\partial}{\partial x} \log \varphi(x, t),$$

the usual solution of the diffusion equation

$$
\begin{aligned}
\varphi(x, t) &= \frac{1}{\sqrt{4\pi\nu t}} \int_{-\infty}^{\infty} \varphi_0(y) \exp\left(-\frac{(x - y)^2}{4\nu t}\right) dy \\
&= \frac{1}{\sqrt{4\pi\nu t}} \int_{-\infty}^{\infty} \exp\left(-\left[U_0(y) - \frac{(x - y)^2}{2t}\right]/2\nu\right) dy
\end{aligned}
$$

gives the following explicit solution of the initial-value problem for the Burgers equation:

$$u(x, t) = \frac{\int_{-\infty}^{\infty} ((x - y)/t) \exp\left[-\Phi(x, y, t)/2\nu\right] dy}{\int_{-\infty}^{\infty} \exp\left[-\Phi(x, y, t)/2\nu\right] dy},\tag{7}$$

where

$$\Phi(x, y, t) = U_0(y) + \frac{(x - y)^2}{2t}.\tag{8}$$

Clearly, solution (7-8), although explicit, does not have a simple structure and the analysis of its properties, especially for random initial data, is quite nontrivial. Also, note that the Hopf-Cole solution can be viewed as a smoothed out solution (3.1.2) of the Riemann equation.

Example 1. (cf. Burgers (1974)) The structure of Hopf-Cole solutions can be elucidated by an asymptotic analysis of time evolution of the solution for the initial data of the form $u_0(x) = \delta_0(x)$. Then

$$\varphi(x, t) = 1 + \alpha - \text{erf} \frac{x}{2\sqrt{\nu t}}, \qquad \text{erf}(x) = \frac{2}{\sqrt{\pi}} \int_0^x e^{-t^2} dt,$$

and, approximately,

$$u(x,t) = 2\sqrt{\frac{\nu}{\pi t}} \frac{\exp(-x^2/4\nu t)}{1 + \alpha - \operatorname{erf}\,(x/2\sqrt{\nu t})}.$$

Writing the small parameter

$$\alpha = \frac{2\sqrt{\nu}}{a\sqrt{\pi}} \exp\left(-\frac{a^2}{4\nu}\right),$$

where a is a new constant, we get that

$$\text{for} \quad x \leq 0, \qquad u(x,t) \approx \sqrt{\frac{\nu}{t}} \exp\left(-\frac{x^2}{4\nu t}\right);$$

$$\text{for} \quad 0 \leq x < a\sqrt{t}, \qquad u(x,t) \approx \frac{x}{t};$$

$$\text{for} \quad x > a\sqrt{t}, \qquad u(x,t) \approx \frac{a}{\sqrt{t}} \exp\left(-\frac{x^2}{4\nu t} + \frac{a^2}{4\nu}\right).$$

In a small neighborhood of $a\sqrt{t})$, i.e. with $x = a\sqrt{t} + 4s\sqrt{t}$ where s is small,

$$u(x,t) \approx \frac{a}{2\sqrt{t}}\left(1 - \tanh\frac{as}{\nu}\right).$$

The shape of the solution is pictured in Fig. 3.2.1.

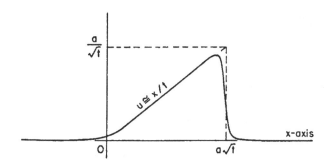

Fig. 3.2.1. The shape of a solution of Burgers' equation with point mass initial data $u_0(x) = \delta_0(x)$.

Notice that the momentum

$$\int_{-\infty}^{\infty} u\,ds = -2\nu \log \varphi|_{-\infty}^{\infty} = -2\nu \log \frac{\alpha}{\alpha + 2} \sim \frac{a^2}{2}$$

remains constant and independent of time.

The above discussion of the Hopf-Cole formula for the one-dimensional case can be carried over to higher dimensions in a straightforward manner only for potential fields $\boldsymbol{u}$. The multidimensional version of Burgers' equation is

$$\frac{\partial \boldsymbol{u}}{\partial t} + (\boldsymbol{u}, \nabla)\boldsymbol{u} = \nu \Delta \boldsymbol{u}, \tag{9}$$

where Δ is the Laplacian and ∇ is the gradient operator. As long as

$$\text{curl } \boldsymbol{u} = 0 \tag{10}$$

and

$$\boldsymbol{u}(\boldsymbol{x}, 0) = -2\nu \nabla U(\boldsymbol{x}), \tag{11}$$

the Hopf-Cole transformation

$$\boldsymbol{u}(\boldsymbol{x}, t) = -2\nu \nabla \log \varphi(\boldsymbol{x}, t) \tag{12}$$

works, and

$$\frac{\partial \varphi}{\partial t} = \nu \Delta \varphi \tag{13}$$

with the initial data

$$\varphi(\boldsymbol{x}, 0) = \exp[U(\boldsymbol{x})]. \tag{14}$$

The case of nonpotential field was discussed in special cases by Saichev and Woyczynski (1997) and we will mention it again later on.

In presence of an external potential field with potential V, equation (9) takes the form

$$\frac{\partial \boldsymbol{u}}{\partial t} + (\boldsymbol{u}, \nabla)\boldsymbol{u} = \nu \Delta \boldsymbol{u} - \nabla V, \tag{15}$$

and the Hopf-Cole transformation leads here to a parabolic equation of Schrödinger type

$$\frac{\partial \varphi}{\partial t} = \nu \Delta \varphi - V \varphi \tag{16}$$

for which localization questions are of interest for stationary and ergodic u_0 and V (see Carmona and Molchanov (1992)). The *forced Burgers turbulence* problems will be addressed in Lecture 7.

3.3 Asymptotic analysis of the Hopf-Cole formula in the inviscid limit

The following elegant asymptotic analysis of the Hopf-Cole formula is due to Burgers (1974) and is based on what is essentially the steepest descent argument (see, e.g., Saichev and Woyczynski (1997)). We begin with a smooth initial velocity potential U_0, a small (but positive) viscosity ν, and fixed x and t.

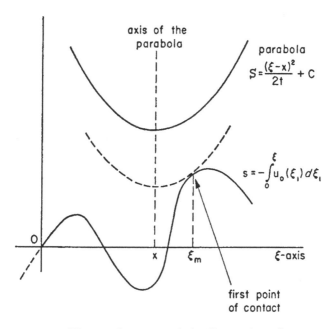

Figure 3.3.1. The osculating parabola illustration of asymptotic analysis of the Hopf-Cole solution of the Burgers' equation for fixed x and t, and for small viscosity ν. It is due to Burgers (1974).

The largest contribution to $\exp(-\Phi(y)/2\nu)$ in the integral formula (3.2.7) comes from the vicinity of the minimum $y_{\min} = \xi_m$ of $\Phi(y)$. There

$$v \approx \frac{x - \xi_m}{t}.$$

To find that minimum, we need to solve the equation $0 = d\Phi/dy$, which is equivalent to the condition

$$\frac{dU_0}{dy} = \frac{dS}{dy}$$

with

$$S(y) = \frac{(x-y)^2}{2t} + C,$$

as long as

$$0 < \frac{d^2\Phi}{dy^2} = \frac{du_0}{dy} + \frac{1}{t}.$$

Graphically, this process can be illustrated by considering a parabola $(x-y)^2/2t + C$. As the constant C decreases, the parabola is lowered to the point where it is tangent to the graph of the initial velocity potential $U_0(y)$ at the point $y = \xi_m$ (see Figure 3.3.1). For small values of t, that is for a sharp pointed parabola, the minimum $y = \xi_m$ is unique.

Next, let us take a look at the asymptotic behavior of the solution as a function of x, for a fixed and small t. As the osculating parabola slides along the x-axis it is possible that for some x's it will have two points of tangency, say ξ_1 and ξ_2, with the initial velocity potential curve U_0 (Fig. 3.3.2).

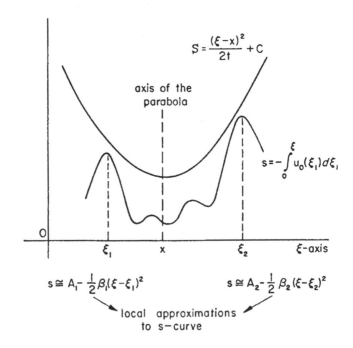

Fig. 3.3.2. Two points of tangency of the osculating parabola with the initial velocity potential curve results in creation of a steep wave front.

Now, the main contribution to $\exp(-\Phi(y)/2\nu)$ comes from the neighborhood of two points ξ_1 and ξ_2 in the vicinity of each, respectively,

$$U_0 \sim A_1 - \frac{\beta_1}{2}(y - \xi_1)^2, \qquad U_0 \sim A_2 - \frac{\beta_2}{2}(y - \xi_2)^2.$$

Then

$$\Phi(y) = \frac{(y-x)^2}{2t} + C - A_1 + \frac{\beta_1}{2}(y - \xi_1)^2$$

has a minimum at

$$\xi_1' = \frac{x + \beta_1 t \xi_1}{1 + \beta_1 t}$$

so that

$$\xi_1' - \xi_1 = \frac{x - \xi_1}{1 + \beta_1 t}.$$

Close to ξ_1, that is for small $\xi_1' - \xi_1$, the exponent is approximately

$$-\frac{1}{2\nu}\left(P_1 + \frac{1 + \beta_1 t}{2t}(y - \xi_1')^2\right)$$

with

$$P_1 = \frac{\beta_1(x - \xi_1)^2}{2(1 + \beta_1 t)} + C - A_1.$$

The linear factor

$$\frac{x - y}{t} = \frac{x - \xi_1'}{t} - \frac{y - \xi_1'}{t} = \frac{\beta_1(x - \xi_1)}{1 + \beta_1 t} - \frac{y - \xi_1'}{t}.$$

Hence, the integral in the denominator of solution (3.2.7) is approximately

$$\frac{2\sqrt{\pi \nu t}}{\sqrt{1 + \beta_1 t}} \exp\left(-\frac{P_1}{2\nu}\right),$$

and the one in the numerator is

$$\frac{\beta_1(x - \xi_1)}{1 + \beta_1 t} \frac{2\sqrt{\pi \nu t}}{\sqrt{1 + \beta_1 t}} \exp\left(-\frac{P_1}{2\nu}\right).$$

Contributions from P_2 can be accounted for in a similar fashion.

If $P_2 > 0$ and $(P_2 - P_1)/2\nu \gg 1$, the first contribution prevails and

$$u = u_1 = \frac{\beta_1(x - \xi_1)}{1 + \beta_1 t} = \frac{x - \xi_1'}{t}. \tag{1}$$

If $(P_1 - P_2)/2\nu \gg 1$, then

$$u = u_2 = \frac{\beta_2(x - \xi_2)}{1 + \beta_2 t} = \frac{x - \xi_2'}{t}. \tag{2}$$

Notice $\xi_1' \approx \xi_1$ and $\xi_2' = \xi_2$ for large t. Thus the solution are approximately linear there (compare formula (3.1.2)). Also, observe that if $\xi_2' > \xi_1'$ then $u_2 < u_1$!!

The changeover from one case to the other comes from a small change in x in the neighborhood of x_s which can be determined from condition

$$P_1 + \nu \log(1 + \beta_1 t) = P_2 + \nu \log(1 + \beta_2 t).$$

Hence, there

$$P_1 - P_2 + \nu \log \frac{(1 + \beta_1 t)}{(1 + \beta_2 t)} \approx (u_1 - u_2)(x - x_s) \tag{3}$$

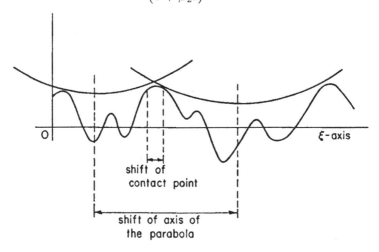

Fig. 3.3.3. Changeover from one double contact of the osculating parabola to another.

so that, for the velocity itself we get an approximation

$$u = \frac{u_1 + u_2 \exp((u_1 - u_2)(x - x_s)/2\nu)}{1 + \exp((u_1 - u_2)(x - x_s)/2\nu)}$$

$$\approx \frac{u_1 + u_2}{2} - \frac{u_1 - u_2}{2} \tanh \frac{(u_1 - u_2)(x - x_s)}{4\nu}.$$

The transition occurs when $x - x_s$ changes sign from negative to positive over the range of order $4\nu/u_1 - u_2$ (see Fig. 3.3.3).

The resulting asymptotic $(\nu \to 0)$ "sawtooth" velocity profile is shown on Fig. 3.3.4.

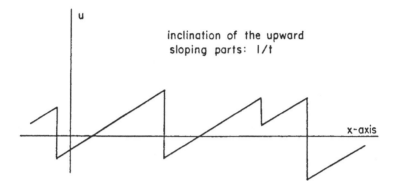

Fig. 3.3.4. Steep front in the velocity profile resulting from a change from one linear regime to another.

Finally, let t increase. This makes the osculating parabolas flatter and the process can lead to skipping some points of tangency with U_0 (see Fig. 3.3.5).

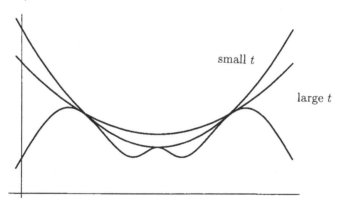

Fig. 3.3.5. As time t increases, the osculating parabolas become flatter, and skip some points of double tangence, resulting in larger shock catching up and merging with the smaller shocks.

Taking the time derivative of (3) (appr. $P_1 \sim P_2$) we get that

$$\frac{\beta_1(x_s - \xi_1)^2}{2(1 - \beta_1 t)} - \frac{\beta_2(x_s - \xi_2)^2}{2(1 - \beta_2 t)} = \text{const},$$

where x_s represents position of the front. Differentiating, and substituting (1), (2) for u_1, u_2 we get that

$$(u_1 - u_2)\frac{dx_s}{dt} = \frac{1}{2}(u_1^2 - u_2^2),$$

or equivalently, that

$$\frac{dx_s}{dt} = \frac{u_1 + u_2}{2}.$$

This gives the speed of advance of a steep front and also an explanation

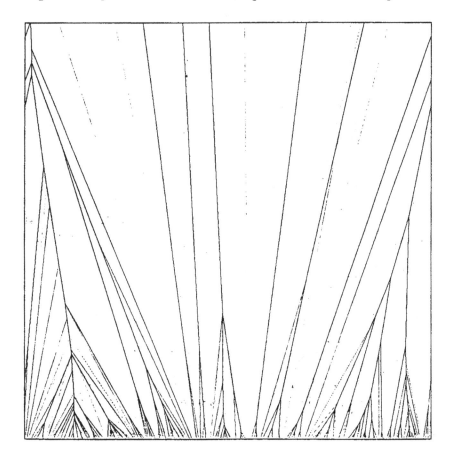

Fig. 3.3.6. A visualization of coalescing shock fronts in Burgers turbulence. The time t-axis is vertical and the space x-axis is horizontal. Each fixed t section shows the location of shock fronts at time t. The shock fronts are initially distributed randomly and densely on the x axis but, as the time progresses, the larger shocks "swallow" smaller shocks as they run into them. As time increases, the density of shocks decreases. (Courtesy A. Noullez)

of the phenomenon that makes this model acceptable as a model for density distribution of sticky particle dust in the adhesion approximation for the large scale structure of the Universe discussed in Lecture 1. Two different fronts can move at different speeds, but coalesce when they run into each other. This phenomenon is visualized in Fig. 3.3.6. In two and more spatial dimensions the inviscid limit structure of shock fronts results in a Voronoi type tessellation of the space. We shall discuss them in more detail in Lectures 6 and 7.

3.4 KdV equation and solitons

It is instructive to contrast the behavior of Burgers' shock waves with another nonlinear effect—the creation, propagation and interaction of solitary waves—the *solitons*. Although they appear as solutions of numerous nonlinear partial differential equations such as the cubic Schrödinger equation, and the Sine-Gordon equation, we will restrict our attention to the case of the Korteweg-deVries equation which has been studied since 1895. The discovery that they have two-soliton solutions is, however, a much more recent development and was made by Zabusky and Kruskal in 1965.

We will study the Korteweg-deVries equation in the form

$$\frac{\partial u}{\partial t} + \sigma u \frac{\partial u}{\partial x} + \frac{\partial^3 u}{\partial x^3} = 0, \tag{1}$$

so it is an evolution equation with the quadratic inertial nonlinear term identical to that in the Burgers equation. However the diffusion dissipation term in the Burgers equation has been replaced here by a third order dispersive term and that change has a dramatic impact on the behavior of the equation's solutions.

Equation (1) arises in several physical contexts. We just mention here the theory of water waves and the fact that the Korteweg-deVries equation governs the variation of the potential of the Schrödinger equation in such a way that the eigenvalues remain constant.

It turns out, that the substitution

$$\sigma u = 12(\partial^2/\partial x^2) \log v, \tag{2}$$

integration with respect to x, and multiplication by $\sigma v^4/12$, transform equation (1) into equation

$$v \frac{\partial Lv}{\partial x} - \frac{\partial v}{\partial x} Lv + 3 \left(\left(\frac{\partial^2 v}{\partial x^2} \right)^2 - \frac{\partial v}{\partial x} \frac{\partial^3 v}{\partial x^3} \right) = 0, \tag{3}$$

where the linear operator

$$L = \frac{\partial}{\partial t} + \frac{\partial^3}{\partial x^3}. \tag{4}$$

Inspecting the structure of this equation one can notice that equation

$$Lv = 0$$

has a solution

$$v = \exp[-\alpha(x - s) + \alpha^3 t] + \text{const}, \quad s = \theta_0/\alpha, \quad (5)$$

where θ_0 and α are parameters, which also annihilates the third term of equation (3). This serendipitous discovery gives a one soliton solution

$$u = \sigma^{-1} 3\alpha^2 \operatorname{sech}^2 \frac{\theta - \theta_0}{2} = \frac{12}{\sigma} \frac{\alpha^2 v}{(1 + v)^2}, \quad \theta = \alpha x - \alpha^3 t, \quad (6)$$

with the maximum of u corresponding to $v = 1$. It is easy to see that the maximum amplitude of u is $3\sigma^{-1}\alpha^2$ and that it is attained at $s + \alpha^2 t$, with the resulting wave velocity being α^2 (see Fig. 3.4.1).

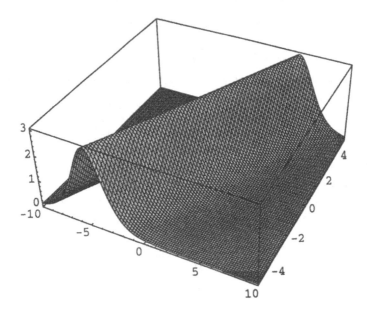

Fig. 3.4.1. Single soliton solution of the KdV equation with parameters $\alpha = \sigma = 1$, and times $-5 \leq t \leq 5$.

Since equation (3) is nonlinear, the superposition $v = v_1 + v_2$ of solutions

$$v_1 = \exp[-\alpha_1(x - s_1) + \alpha_1^3 t],$$
$$v_2 = \exp[-\alpha_2(x - s_2) + \alpha_2^3 t],$$

of the form (5) is not a solution, but a direct verification shows that

$$v = 1 + v_1 + v_2 + K v_1 v_2,$$

where

$$K = \frac{(\alpha_2 - \alpha_1)^2}{(\alpha_2 + \alpha_1)^2},$$

is an exact solution of (3) with the corresponding solution

$$u = \frac{12}{\sigma} \frac{\alpha_1^2 v_1 + \alpha_2^2 v_2 + 2(\alpha_2 - \alpha_1)^2 v_1 v_2 + K(\alpha_2^2 v_1^2 v_2 + \alpha_1^2 v_1 v_2^2)}{(1 + v_1 + v_2 + K v_1 v_2)^2}$$

of the Korteweg-deVries equation (1).

This is the famous two-soliton solution and we will provide its asymptotic analysis. For the sake of definiteness assume that

$$\alpha_2 > \alpha_1 > 0,$$

which implies that the second soliton u_2 has a larger amplitude and moves faster (to the right) than u_1.

As $t \to -\infty$, we have in the case $v_1 \approx 1, v_2 \ll 1$,

$$u \approx \frac{12}{\sigma} \frac{\alpha_1^2 v_1}{(1 + v_1)^2}$$

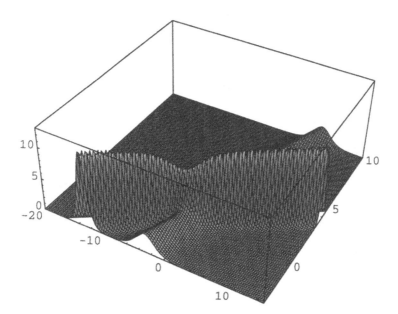

Fig. 3.4.2. Two-soliton solution of the KdV equation for parameters $\alpha_1 = \sigma = 1, \alpha_2 = 2, -3 \le t \le 10$.

and the solution looks like the first solitary wave u_1 cresting at $x = s_1 + \alpha_1^2 t$.

In the case $v_1 \gg 1, v_2 \approx 1$,

$$u \approx \frac{12}{\sigma} \frac{\alpha_1^2 K v_2}{(1 + K v_2)^2}$$

which can also be written as the soliton u_2 with a forward space shift $-\alpha_1^{-1} \log K$ or, in other words, with s_1 replaced by $s_2 + \alpha_2^{-1} \log K$. Equivalently, it describes a soliton cresting at

$$x = s_2 + \alpha_2^{-1} \log K + \alpha_2^2 t$$

In this case there is no-interaction region $v_1 \approx 1, v_2 \approx 1$.

For $t \to \infty$, the situation is symmetric to the one discussed above. At

$$x = s_1 + \alpha_1^{-1} \log K + \alpha_1^2 t$$

we have the shifted solitary wave u_1 $(v_1 \approx 1, v_2 \gg 1)$, and at

$$x = s_2 + \alpha_2^2 t$$

we have the solitary wave u_2.

So the stronger soliton u_2 which was behind u_1 for large negative times emerges ahead of it, without any shape change (except for a shift in space) at large positive times. So, what happened to the two solitons in the meantime?

In the intermediate region, more precisely in the neighborhood of

$$t = -\frac{s_2 - s_1}{\alpha_2^2 - \alpha_1^2}, \quad \text{and} \quad x = \frac{\alpha_2^2 s_1 - \alpha_1^2 s_2}{\alpha_2^2 - \alpha_1^2}$$

where $v_1 \approx 1, v_2 \approx 1$, the two solitons interact, merging into a single peak, and reemerge from this interaction in the reverse order and with phase shifts, but otherwise unscathed and unscattered (see Fig. 3.4.2).

Lecture 4
Statistical Description,
Parabolic Approximation

4.1   Statistical description in Burgers' turbulence

The rationale for studying the Burgers-KPZ equation with random initial data, which is traditionally called *Burgers' turbulence model*,

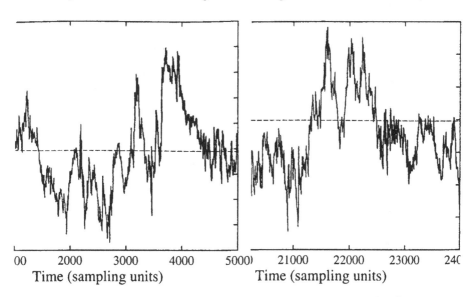

Time (sampling units)   Time (sampling units)

Fig. 4.1.1. *Left:* One second of a turbulent signal recorded by a hot-wire; *Right:* same signal, about four seconds later (Frisch (1996)).

is similar to the justification for the Navier-Stokes equations with random initial conditions as a model for the usual hydrodynamic turbulence. Why a probabilistic description of turbulence is necessary was crisply explained in a recent book by Uriel Frisch (1996). If one measures the velocity $v(t, x)$ in a turbulent flow for two different time intervals then the profiles look totally different and are not repeatable (Fig. 4.1.1). However, if one concentrates on the probability distribution of the measured turbulent signal one obtains a predictable object. (Fig. 4.1.2.).

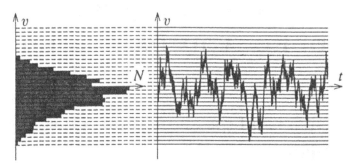

Fig. 4.1.2. The empirical distribution (histogram) of a turbulent signal remains stable.

For that reason, throughout the remainder of this lecture (and most of the following lectures) we will assume that the initial data $u(x, 0) = u_0(x), x \in \mathbf{R}^d$, is a random field on $\mathbf{R}^d$ with a given statistical description, and that the main object of study is the solution random field $u(x, t)$ on $\mathbf{R}_+ \times \mathbf{R}^d$, for which we are seeking the full statistical description in the form of all joint finite-dimensional distributions of

$$(u(x_1, t_1), \ldots, u(x_n, t_n)), \tag{1}$$

$$t_1, \ldots, t_n \in \mathbf{R}_+, \quad x_1, \ldots, x_n \in \mathbf{R}^d, \quad n = 1, 2, \ldots.$$

A more restricted program would seek the multipoint correlation functions

$$\rho^{(n)}(x_2, t_1; \ldots; x_n, t_n,) = \mathbf{E} \prod_{k=1}^{n} v(x_k, t_k), \tag{2}$$

or even simpler, but physically interesting, two-point spatial correlations $\rho^{(2)}(x_1, t; x_2, t)$ taken for the same time instant t. Recall that $\mathbf{E}X$ stands for the mathematical expectation of the random quantity X.

Since the solution of the initial-value problem for the Burgers equation is given by a fairly involved nonlinear Hopf-Cole functional of the initial data, one of the tools in the study of Burgers turbulence is the general theory of orthogonal expansions of nonlinear functionals of random fields which can be thought of as an infinite-dimensional analogue of the usual calculus power series expansions. It will be summarized in the next section using an approach borrowed from Kwapien and Woyczynski (1992).

4.2 Polynomial chaos and Wiener-Hermite expansions of nonlinear functionals

Let $\gamma_1, \gamma_2, \ldots$ be a canonical sequence of independent $N(0,1)$ Gaussian random variables and the sigma-field $\mathcal{G} = \sigma(\gamma_1, \gamma_2, \ldots)$. Denote by $\mathcal{K}_d^0$ the family of all real polynomial chaoses of degree d based on the sequence $\gamma_1, \gamma_2, \ldots$. Recall that if Q is a polynomial of degree d on $\mathbf{R}^m$, then the *polynomial chaos* $Q(\xi_1, \ldots, \xi_n)$ has a representation

$$Q(\xi_1, \ldots, \xi_n) = \sum_{k=0}^{d} \sum_{1 \leq i_1 \leq \ldots \leq i_k \leq n} x_{i_1, \ldots, i_k} \xi_{i_1} \cdot \ldots \cdot \xi_{i_k}. \tag{1}$$

In view of Proposition 6.5.1 of Kwapien and Woyczynski (1992), the topology of convergence in the second moment (or any other moment) coincides on $\mathcal{K}_d^0$ with the topology of convergence in probability. The main goal of this section is to identify the closure $\mathcal{K}_d$ of $\mathcal{K}_d^0$ in $\mathbf{L}^2(\Omega, \mathcal{G}, P)$, and to produce natural bases for both, $\mathcal{K}_d$ and the whole $\mathbf{L}^2(\Omega, \mathcal{G}, P)$. We begin our exposition by introducing the *Hermite polynomials*

$$h_n(x) = \frac{(-1)^n}{n!} e^{x^2/2} \frac{d^n}{dx^n} e^{-x^2/2},$$

for $n = 0, 1, 2, \ldots$ and $x \in \mathbf{R}$. It will follow from the results of this section that the sequence $(h_n/\sqrt{n!})$ is an orthonormal basis in $\mathbf{L}^2(\mathbf{R}, \frac{1}{\sqrt{2\pi}} e^{-x^2/2} dx)$ and, more generally, the set

$$\{h_{i_1}(x_1) \cdot \ldots \cdot h_{i_n}(x_m) : i_1, \ldots, i_n \in \mathbf{N}\}$$

forms an orthogonal basis in

$$\mathbf{L}^2\left(\mathbf{R}^m, (2\pi)^{-m/2} e^{-(x_1^2 + \ldots + x_m^2)/2} dx_1 \ldots dx_m\right).$$

Since

$$\exp\left(\frac{x^2}{2} - \frac{(x-t)^2}{2}\right) = \exp\left(\frac{x^2}{2}\right) \sum_{n=0}^{\infty} \frac{t^n}{n!} \frac{d^n}{dx^n} \exp\left(-\frac{(x-t)^2}{2}\right)\Bigg|_{t=0},$$

we see that, for every $t, x \in \mathbf{R}$,

$$\exp\left(tx - \frac{t^2}{2}\right) = \sum_{n=0}^{\infty} t^n h_n(x). \tag{2}$$

Clearly, $h_0(x) \equiv 1$. Moreover, the identity $(d/dx)\exp(tx - t^2/2) = t\exp(tx - t^2/2)$, implies that $\sum_{n=0}^{\infty} t^n h_{n-1}(x) = \sum_{n=0}^{\infty} t^n h_n'(x)$, which yields the following *Rodrigues Formula* for the Hermite polynomials: For each $x \in \mathbf{R}$ and $n = 1, 2, \ldots$,

$$h_n'(x) = h_{n-1}(x). \tag{3}$$

On the other hand, the identity $(d/dt)\exp(tx - t^2/2) = (x - t)\exp(tx - t^2/2)$, implies that

$$\sum_{n=0}^{\infty}(n+1)t^n h_{n+1}(x) = \sum_{n=0}^{\infty} t^n x h_n(x) - \sum_{n=0}^{\infty} t^n h_{n-1}(x),$$

which yields the following recurrence relation: For each $x \in \mathbf{R}$ and $n = 1, 2, \ldots$,

$$(n+1)h_{n+1}(x) = xh_n(x) - h_{n-1}(x). \tag{4}$$

The orthogonality property of the Hermite polynomials can be phrased in the following, more general, setting:

Proposition 1. *Let* $\mathbf{H}$ *be a Hilbert space with an orthonormal basis* (e_n). *For each* $h = \sum_{i=0}^{\infty} \lambda_i e_i \in \mathbf{H}$, *denote*

$$X_h = \sum_{i=0}^{\infty} \lambda_i \gamma_i. \tag{5}$$

Then, for each $n, k \in \mathbf{N}$, *and every* $g, h \in \mathbf{H}$ *such that* $\|g\| = \|h\| = 1$,

$$\mathbf{E}\Big(h_n(X_g)h_k(X_h)\Big) = \delta_n^k \frac{1}{n!} \langle g, h \rangle^n.$$

PROOF. The formula is obtained by identifying the coefficients of $s^n t^k$, $s, t \in \mathbf{R}$, on either end of the following chain of identities:

$$\mathbf{E}\left(\exp\left(tX_h - \frac{t^2}{2}\right)\exp\left(sX_g - \frac{s^2}{2}\right)\right)$$
$$= \mathbf{E}\exp\left(X_{sg+th} - \frac{s^2 + t^2}{2}\right) = \mathbf{E}\left(\frac{|sg + th|}{2} - \frac{s^2 + t^2}{2}\right)$$
$$= \sum_{n=0}^{\infty} \frac{(st)^n}{n!}\langle g, h\rangle^n = \mathbf{E}\left(\sum_{n=0}^{\infty} s^n h_n(X_g) \cdot \sum_{n=0}^{\infty} t^n h_n(X_h)\right).$$

∎

Definition 1. For each $d \geq 1$,

$$\mathcal{K}_d^0 := \{Q(\gamma_{i_1}, \ldots, \gamma_{i_k}) : i_1, \ldots, i_k \in \mathbf{N}, \ k \geq 1, \ \deg Q = d\}.$$

Each element of $\mathcal{K}_d^0$ is traditionally called a *Wiener polynomial chaos of order d*. The closure in $\mathbf{L}^2(\Omega, \mathcal{G}, P)$ of the set $\mathcal{K}_d^0$ will be denoted by $\mathcal{K}_d$.

The next theorem describes the *Wiener chaos decomposition* of $\mathbf{L}^2(\Omega, \mathcal{G}, P)$.

Theorem 1. *Let $\mathcal{H}_d^0$ be the linear span of the set $\{h_d(X_h) : h \in \mathbf{H}, \|h\| = 1\}$, and let $\mathcal{H}_d$ be the closure of $\mathcal{H}_d^0$ in $\mathbf{L}^2(\Omega, \mathcal{G}, P)$. Then*

$$\mathbf{L}^2(\Omega, \mathcal{G}, P) = \bigoplus_{d=0}^{\infty} \mathcal{H}_d,$$

and, for each $n = 1, 2, \ldots$,

$$\mathcal{K}_n = \bigoplus_{d=0}^{n} \mathcal{H}_d.$$

Moreover, for each $d = 1, 2, \ldots$, the set

$$H_d := \Big\{ \prod_{i \geq 1} \sqrt{\alpha_i} h_{\alpha_i}(\gamma_i) : a \in \Lambda, \ |a| = d \Big\},$$

forms a complete orthonormal system in $\mathcal{H}_d$. Here, Λ stands for the set of nonnegative integer sequences $a = (\alpha_i)$ with finitely many terms different from 0, and $|a| = \sum_{i=1}^{\infty} \alpha_i$.

PROOF. Proposition 1 gives the pairwise orthogonality of spaces $\mathcal{H}_d$ and that $\mathcal{H}_d \subset \mathcal{K}_n$ if $d \leq n$. Thus, it suffices to check that given $X \in \mathbf{L}^2(\Omega, \mathcal{F}, P)$, if $X \in \mathcal{H}_d^{\perp}$, for every $d = 1, 2, \ldots$, then $X = 0$. Clearly, for $h \in \mathbf{H}$ with $\|h\| = 1$, and $d = 1, 2, \ldots$, we have $E(X h_d(X_h)) = 0$, so that $E(X \exp(tX_h - t^2/2)) = 0$. Therefore, given $t_1, \ldots, t_d \in \mathbf{R}$, and $h_1, \ldots, h_d \in \mathbf{H}$, we get that

$$E\Big(X \exp\Big(\sum_{k=1}^{d} t_k X_{h_k}\Big)\Big) = 0.$$

Hence, by analytic extension, we obtain that

$$E\Big(X \exp\Big(-i \sum_{k=1}^{d} t_k \gamma_k\Big)\Big) = 0$$

for each d, and t_k's as above. Therefore,

$$EX \int_{\mathbf{R}^d} \exp\left(-i \sum_{k=1}^{d} t_k \gamma_k\right) \mu(dt_1, \ldots, dt_d) = 0$$

for each finite measure μ on $\mathbf{R}^d$, which proves that $X\psi(\gamma_1, \ldots, \gamma_d) = 0$ if ψ is the Fourier transform of a finite measure. Since the class of such functions is dense in

$$\mathbf{C}_0(\mathbf{R}^d) := \{\psi : \mathbf{R}^d \to \mathbf{R} : \psi \in \mathbf{C}(\mathbf{R}^d), \lim_{|x|\to\infty} \psi(x) = 0\},$$

we obtain that $X = 0$.

To prove the last statement of the theorem observe that Proposition 1 and the independence of X_{e_i}'s imply that

$$E\left(h_a(X_e)h_b(X_e)\right) = \prod_{i=1}^{\infty} \delta_{\alpha_i \beta_i} \frac{1}{\alpha_i!} = \delta_b^a \prod_{i=1}^{\infty} \frac{1}{\alpha_i!},$$

and if $|a| = d$, we have $h_a(X_e) \in \mathcal{K}_d^0$. Let $h = \sum_{i=1}^{\infty} \alpha_i e_i \in \mathbf{H}$ be such that $\|h\| = 1$. For each $N = 1, 2, \ldots$, set $h_N = \sum_{i=1}^{N} \beta_i e_i$ with $\beta_i = \alpha_i / \sum_{i=1}^{N} \alpha_i$. Then $\|h_N\| = 1$ and

$$\lim_{N\to\infty} h_d(X_{h_N}) = h_d(X_h).$$

Furthermore, for any $t \in \mathbf{R}$,

$$\begin{aligned}
\sum_{d=1}^{\infty} t^d h_d(X_{h_N}) &= \exp\left(-\frac{t^2}{2} + t \sum_{i=1}^{N} \beta_i X_{e_i}\right) \\
&= \prod_{i=1}^{N} \exp\left(-\frac{t^2 \beta_i^2}{2} + t\beta_i X_{e_i}\right) \\
&= \prod_{i=1}^{N} \left(\sum_{d=1}^{\infty} \beta_i^d t^d h_d(X_{e_i})\right) \\
&= \sum_{d=1}^{\infty} \left(\sum_{|a|=1} \beta^a h_a(X_e)\right),
\end{aligned}$$

which ends the proof of the theorem.   ∎

In the remainder of this section we will extend the above "discrete" polynomial chaos result to the case of "continuous" polynomial chaos.

Consider a Gaussian, orthogonally scattered random measure $\boldsymbol{w}$ on $(T = [0, 1], \mathcal{B})$ with the control measure

$$\mu(A) = E|\boldsymbol{w}(A)|^2,$$

$A \in \mathcal{B}$, which is assumed to be atomless. The principal example of such a measure is given by the formula

$$\boldsymbol{w}(A) = B(t_2) - B(t_1)$$

where $B(t)$ is the Brownian motion and $A = [t_1, t_2)$.

Given a partition $A_1 \cup \ldots \cup A_n = T$, and a $\mathcal{B} \times \ldots \times \mathcal{B}$-step real function f on T^d, $d \geq 1$, which vanishes on the "diagonals"

$$\Delta^d = \bigcup_{i \neq j} \{(t_1, \ldots, t_d) : t_i = t_j\},$$

and has the representation

$$f(t_1, \ldots, t_d) = \sum_{1 \leq i_1, \ldots, i_d \leq n} \alpha_{i_1, \ldots, i_d} I_{A_{i_1} \times \ldots \times A_{i_d}}(t_1, \ldots, t_d), \tag{6}$$

with $\alpha_{i_1, \ldots, i_d} = 0$ whenever two indices $i_1, \ldots, i_d$ are equal, the multiple integral

$$\int_{T^d} f(t_1, \ldots, t_d) \boldsymbol{w}(dt_1) \ldots \boldsymbol{w}(dt_d)$$

can be written in the form

$$\mathbf{I}_d(f) = \sum_{1 \leq i_1, \ldots, i_d \leq n} \alpha_{i_1, \ldots, i_d} \boldsymbol{w}(A_{i_1}) \cdot \ldots \cdot \boldsymbol{w}(A_{i_d}). \tag{7}$$

The mapping $\mathbf{I}_d$ is clearly a linear form in f and a d-linear form in values of the stochastic measure $\boldsymbol{w}$ on $A_1, \ldots, A_n$. It also enjoys the following three properties:

$1°$ The integal $\mathbf{I}_d(f)$ is invariant under the symmetrization of function f, i.e., if

$$\tilde{f}(t_1, \ldots, t_d) = \frac{1}{d!} \sum_{\sigma} f(t_{\sigma(1)}, \ldots, t_{\sigma(d)}),$$

where the summation extends over all the permutations of the set $\{1, \ldots d\}$, then

$$\mathbf{I}_d(f) = \mathbf{I}_d(\tilde{f}). \tag{8}$$

2° If f and g are two step functions (of the type considered above) on T^d and $T^{d'}$, respectively, and $d \neq d'$ then

$$E\Big(\mathbf{I}_d(f)\mathbf{I}_{d'}(g)\Big) = 0.$$

3° If f and g are two step functions (of the type considered above) on T^d then

$$E\Big(\mathbf{I}_d(f)\mathbf{I}_d(g)\Big) = d! \ \langle \tilde{f}, \tilde{g} \rangle,$$

where the inner product is taken in $\mathbf{L}^2(T^d, \mathcal{B}^d, \mu^d)$.

Property 1° follows directly from the definition as

$$
\begin{aligned}
\mathbf{I}_d(f) &= \sum_{1 \le i_1 < \ldots < i_d \le n} \Big(\sum_{\sigma} \alpha_{\sigma(i_1), \ldots \sigma(i_d)} w(A_{i_1}) \cdot \ldots \cdot w(A_{i_d}) \Big) \\
&= d! \sum_{1 \le i_1 < \ldots < i_d \le n} \Big(\frac{1}{d!} \sum_{\sigma} \alpha_{\sigma(i_1), \ldots \sigma(i_d)} \Big) w(A_{i_1}) \cdot \ldots \cdot w(A_{i_d}) \\
&= \mathbf{I}_d(\tilde{f}).
\end{aligned}
$$

The orthogonality $\mathbf{I}_d(f) \perp \mathbf{I}_{d'}(g)$ (Property 2°) for, say $d' < d$, can be demonstrated as follows. Any $\mathbf{I}_{d'}(g)$ is representable as a linear combination of expressions of the form $w(A_1') \cdot \ldots \cdot w(A_{d'}')$, where each of sets A_j' is contained in one of the sets from the representation (6), in which any product $w(A_{i_1}) \cdot \ldots \cdot w(A_{i_d})$ contains at least one factor $w(A_i)$ such that $A_i \cap A_j' = \emptyset, j = 1, \ldots, d'$. Consequently

$$w(A_1') \cdot \ldots \cdot w(A_{d'}') \perp w(A_{i_1}) \cdot \ldots \cdot w(A_{i_d}).$$

The covariance in Property 3° can be obtained by a straightforward computation. Without any loss of generality we can assume that

$$g(t_1, \ldots, t_d) = \sum_{1 \le i_1, \ldots, i_d \le n} \beta_{i_1, \ldots, i_d} I_{A_{i_1} \times \ldots \times A_{i_d}}(t_1, \ldots, t_d).$$

Then

$$
\begin{aligned}
&E\Big(\mathbf{I}_d(f)\mathbf{I}_d(g)\Big) \\
&= E\Big(\sum_{1 \le i_1 < \ldots < i_d \le n} \Big(\sum_{\sigma} \alpha_{\sigma(i_1), \ldots, \sigma(i_d)} w(A_{i_1}) \cdot \ldots \cdot w(A_{i_d}) \Big) \Big) \\
&\quad \times \Big(\sum_{1 \le i_1 < \ldots < i_d \le n} \Big(\sum_{\sigma} \beta_{\sigma(i_1), \ldots, \sigma(i_d)} w(A_{i_1}) \cdot \ldots \cdot w(A_{i_d}) \Big) \Big) \\
&= d! \sum_{1 \le i_1, \ldots, i_d \le n} \Big(\frac{1}{d!} \sum_{\sigma} \alpha_{\sigma(i_1), \ldots, \sigma(i_d)} \Big) \Big(\frac{1}{d!} \sum_{\sigma} \beta_{\sigma(i_1), \ldots, \sigma(i_d)} \Big) \\
&\quad \times \mu(A_{i_1}) \cdot \ldots \cdot \mu(A_{i_d}) = d! \ \langle \tilde{f}, \tilde{g} \rangle.
\end{aligned}
$$

The above properties give, in particular, that for any $\mathcal{B} \times \ldots \times \mathcal{B}$-step function f on T^d which vanishes on Δ^d,

$$E\mathbf{I}_d^2(f) = E\mathbf{I}_d^2(\tilde{f}) = d! \, \|\tilde{f}\|^2 \leq d! \, \|f\|^2,$$

where the norm is that of $\mathbf{L}^2(T^d, \mathcal{B}^d, \mu^d)$. Hence, the mapping $\mathbf{I}_d$ can be extended to a bounded linear operator

$$\mathbf{I}_d : \mathbf{L}^2(T^d, \mathcal{B}^d, \mu^d) \to \mathbf{L}^2(\Omega, \mathcal{G}, P),$$

where $\mathcal{G}$ is the σ-algebra generated by $\boldsymbol{w}(A), A \in \mathcal{B}$. Notice that $\mu^d(\Delta^d) = 0$ since μ was assumed to be atomless. Clearly, for $f \in \mathbf{L}^2(T^d, \mathcal{B}^d, \mu^d)$,

$$\mathbf{I}_d(f) = \int_{T^d} f(t_1, \ldots, t_d)\boldsymbol{w}(dt_1) \cdot \ldots \cdot \boldsymbol{w}(dt_d),$$

where the integral on the right hand side is the multiple Wiener stochastic integral, see Kwapien and Woyczynski (1992) for more details.

The relationship between the d-tuple Wiener integrals for different d's is explained in the following:

Theorem 3. *If $f \in \mathbf{L}^2(T^d, \mathcal{B}^d, \mu^d)$, $g \in \mathbf{L}^2(T, \mathcal{B}, \mu)$ and*

$$f \star_k g \, (t_1, \ldots, t_{k-1}, t_{k+1}, \ldots, t_d) := \int_T f(t_1, \ldots, t_d)g(t_k)\mu(dt_k)$$

then $f \star_k g \in \mathbf{L}^2(T^{d-1}, \mathcal{B}^{d-1}, \mu^{d-1})$ and, moreover,

$$\mathbf{I}_{d+1}\Big(f(t_1, \ldots, t_d)g(t_{d+1})\Big) = \mathbf{I}_d(f)\mathbf{I}_1(g) - \sum_{k=1}^{d} \mathbf{I}_{d-1}(f \star_k g).$$

PROOF. The first statement follows immediately from the Cauchy-Schwarz Inequality. To prove the identity, let — initially — f be a step function on T^d with the representation (6) and let $g = \sum_{i=1}^{n} \beta_i \mathbf{I}_{A_i}$. Set $\alpha = \|f\|_\infty$ and $\beta = \|g\|_\infty$. Without loss of generality we can assume that $\mu(A_i) < \epsilon$ for a given $\epsilon > 0$. Define

$$k_\epsilon(t_1, \ldots, t_d, t) = \begin{cases} \alpha_{i_1, \ldots, i_d}\beta_i & \text{if } i \notin \{i_1, \ldots, i_d\}, \text{ and} \\ & (t_1, \ldots, t_d, t) \in A_{i_1} \times \ldots \times A_{i_d} \times A_i, \\ 0 & \text{otherwise.} \end{cases}$$

Then, by a straightforward computation,

$$\mathbf{I}_d(f)\mathbf{I}(g)$$

$$= \left(\sum_{1 \le i_1,\dots,i_d \le n} \alpha_{i_1,\dots,i_d} w(A_{i_1}) \cdot \dots \cdot \dot{w}(A_{i_d}) \right) \left(\sum_{i=1}^{n} \beta_i w(A_i) \right)$$

$$= \sum_{\substack{1 \le i_1,\dots,i_d \le n \\ i \notin \{i_1,\dots i_d\}}} \alpha_{i_1,\dots,i_d} \beta_i w(A_{i_1}) \dots w(A_{i_d}) \cdot w(A_i)$$

$$+ \sum_{j=1}^{d} \sum_{1 \le i_1,\dots,i_d \le n} \alpha_{i_1,\dots,i_d} \beta_{i_j} w(A_{i_1}) \cdot \dots \cdot w(A_{i_{j-1}}) w^2(A_{i_j})$$

$$\times w(A_{i_{j+1}}) \cdot \dots \cdot w(A_{i_d})$$

$$= \mathbf{I}_{d+1}(k_\epsilon) + \sum_{j=1}^{d} \sum_{1 \le i_1,\dots,i_d \le n} \alpha_{i_1,\dots,i_d} \beta_{i_j} w(A_{i_1}) \cdot \dots \cdot w(A_{i_{j-1}}) \mu(A_{i_j})$$

$$\times w(A_{i_{j+1}}) \cdot \dots \cdot w(A_{i_d})$$

$$+ \sum_{j=1}^{d} \sum_{1 \le i_1,\dots i_d \le n} \alpha_{i_1,\dots,i_d} \beta_{i_j} w(A_{i_1}) \cdot \dots \cdot w(A_{i_{j-1}}) \left(w^2(A_{i_j}) - \mu(A_{i_j}) \right)$$

$$\times w(A_{i_{j+1}}) \cdot \dots \cdot w(A_{i_d})$$

$$= \mathbf{I}_{d+1}(k_\epsilon) + \sum_{j=1}^{d} I_{d-1}(f \star_j g) + \sum_{j=1}^{d} R_j.$$

In view of Property 3°,

$$\|\mathbf{I}_{d+1}(fg) - \mathbf{I}_{d+1}(k_\epsilon)\|_2^2 \;\le\; (d+1)! \, \|fg - k_\epsilon\|_{\mathbf{L}^2(\mu^{d+1})}$$

$$\le \; (d+1)! \, d \, \alpha^2 \beta^2 \epsilon \left(\sum_{i=1}^{n} \mu(A_i) \right)^d$$

$$= \; (d+1)! \, d \, \alpha^2 \beta^2 \epsilon \left(\mu(T) \right)^d,$$

and

$$\|R_j\|_2^2$$

$$= \sum_{1 \le i_1,\dots i_d \le n} a_{i_1,\dots,i_d}^2 b_{i_j}^2 \mu(A_{i_1}) \cdot \dots \cdot \mu(A_{i_{j-1}}) \left(2\mu(A_{i_j}) \right)^2 \mu(A_{i_{j+1}}) \cdot \dots \cdot \mu(A_{i_d})$$

$$\le 2\alpha^2 \beta^2 \epsilon \left(\mu(T) \right)^2.$$

Choosing $\epsilon \to 0$, we obtain the second statement of the theorem for step functions. Now the general case follows by the standard approximation argument in $\mathbf{L}^2(T^d, \mathcal{B}^d, \mu^d)$ and $\mathbf{L}^2(T, \beta, \mu)$. ∎

The next result explains the relationship of the multiple Wiener integrals to the Hermite polynomials, and hence the Wiener polynomial chaos discussed at the beginning of this section.

Theorem 4. *Let $h \in \mathbf{L}^2(T, \beta, \mu)$ be of norm 1, and let*

$$X_h := \int_T h(t)\boldsymbol{w}(dt).$$

If h_d is the Hermite polynomial of degree d, then

$$\int_{T^d} h(t_1) \cdot \ldots \cdot h(t_d)\boldsymbol{w}(dt_1) \cdot \ldots \cdot \boldsymbol{w}(dt_d) = d!\, h_d(X_h).$$

PROOF. The proof proceeds by induction in d. For $d = 1$ (and also for $d = 0$ if we agree that $\mathbf{I}_0(h) = 1$) the theorem holds true by the definition of X_h.

Now, assume that the above identity holds true for $i = 1, \ldots, d$. Then, by Theorem 3 and by the Recurrence Formula (4),

$$\mathbf{I}_{d+1}\Big(h(t_1)\ldots h(t_d)h(t_{d+1})\Big)$$

$$= \mathbf{I}_d\Big(h(t_1)\ldots h(t_d)\Big)\mathbf{I}_1\Big(h(t_{d+1})\Big)$$

$$- \sum_{j=1}^{d} \mathbf{I}_{d-1}\Big(\int_T h(t_1)\ldots h(t_{j-1})h^2(t_j)h(t_{j+1})\ldots h(t_d)\mu(dt_j)\Big)$$

$$= d!\, h_d(X_h)X_h - d\,(d-1)!\, h_{d+1}(X_h) = (d+1)!\, h_{d+1}(X_h).$$

This concludes the proof of Theorem 4. ∎

The final result of this section is a continuous-time analogue of the orthogonal decomposition from Theorem 1. It shows that the integrals $\mathbf{I}_d, d = 1, 2, \ldots$, form an orthogonal basis in $\mathbf{L}^2(\Omega, \mathcal{G}, P)$, where $\mathcal{G} = \sigma(\boldsymbol{w}(A), A \in \mathcal{B})$.

Theorem 5. *Let $d \geq 1$, and let $\mathcal{H}_d$ be the closed span in $\mathbf{L}^2(\Omega, \mathcal{G}, P)$ of $h_d(X_h)$, where $h \in \mathbf{L}^2(T, \mathcal{B}, \mu)$ is of norm 1. Then the mapping*

$$\mathbf{I}_d : \mathbf{L}^2(T^d, \mathcal{B}^d, \mu^d) \to \mathcal{H}_d$$

is one-to-one, and for any $X \in \mathbf{L}^2(\Omega, \mathcal{G}, P)$ there exist (unique) symmetric functions f_d on $T^d, d = 1, 2, \ldots$, such that

$$X = EX + \sum_{d=1}^{\infty} \mathbf{I}_d(f_d).$$

PROOF. Let $\mathbf{L}_s^2(T^d, \mathcal{B}^d, \mu^d)$ be a closed subset of $\mathbf{L}^2(T^d, \mathcal{B}^d, \mu^d)$ consisting of symmetric functions. For $f \in \mathbf{L}_s^2(T^d)$, by Property 3°,

$$E\Big(\mathbf{I}_d^2(f)\Big) = d! \, \|f\|_{\mathbf{L}^2(\mu^d)}^2$$

and

$$\mathbf{L}^2(\Omega, \mathcal{G}, P) \supset \overline{\mathbf{I}_d\Big(\mathbf{L}_s^2(T^d, \mathcal{B}^d, \mu^d)\Big)} = \mathbf{I}_d\Big(L_s^2(T^d, \mathcal{B}, \mu^d)\Big) \supset \mathcal{H}_d.$$

By Property 2°

$$\mathbf{I}_d\Big(L_s^2(\mu^d)\Big) \perp \bigcup_{j \neq d} \mathbf{I}_j\Big(L_s^2(\mu^j)\Big).$$

Therefore

$$\mathbf{I}_d\Big(\mathbf{L}_s^2(T^d)\Big) \perp \bigoplus_{j=d+1}^{\infty} \mathcal{H}_j,$$

so that

$$\mathbf{I}_d\Big(\mathbf{L}_s^2(\mu^d)\Big) \subset \mathcal{P}_d,$$

where $\mathcal{P}_d$ stands for the closure in $\mathbf{L}^2(\Omega, \mathcal{G}, P)$ of the set

$\{Q(X_{h_1}, \ldots, X_{h_k}) : k \geq 1, Q$ is a polynomial of degree $p \leq d$
in k variables and $h_i \in L^2(T, \mathcal{B}, \mu)\}.$

Hence (compare with the argument used in the proof of Theorem 1),

$$\mathbf{I}_d\Big(\mathbf{L}_s^2(T^d)\Big) \perp \bigoplus_{j=1}^{d-1} \mathcal{H}_j = \mathcal{P}_{d-1}.$$

This gives that $\mathbf{I}_d(\mathbf{L}^2(\mu^d)) = \mathcal{H}_d$, and an application of Theorem 4 and Properties 2° and 3° (again, compare with Theorem 1) concludes the proof. ∎

4.3   Parabolic scaling limits for regular initial data

Now we turn to asymptotic analysis of parabolically rescaled statistical solutions of the one-dimensional Burgers equation

$$\begin{cases} u_t + \frac{1}{2}(u^2)_x = u_{xx} \\ u(x, 0) = u_0(x) \, , \quad -\infty < x < \infty \end{cases} \tag{1}$$

Parabolic scaling is the most natural choice in the context of a parabolic equation, even nonlinear. This was made abundantly clear in Lecture 2, when hydrodynamic limit problems were discussed, and in Lecture 3, when we talked about self-similar solutions. Roughly speaking, one can think about parabolic approximation as a version of the central limit theorem for a *nonlinear stochastic flow* governed by the Burgers equation. The exposition in this section is based on Surgailis and Woyczynski (1993) and Funaki, Surgailis and Woyczynski (1995). Information about parabolic scaling limits for the 3-D equation can be found, for example, in Surgailis and Woyczynski (1994b).

The initial-value problem (1) admits the Hopf-Cole solution, which we will rewrite in the following form

$$u(x,t) = -\frac{\partial}{\partial x} \log \int_{\mathbf{R}} p(x,y,t) \exp(-U_0(y)) \, dy,$$

where

$$U_0(y) = \int_{-\infty}^{y} u_0(z) \, dz,$$

and

$$p(x,y,t) = \frac{1}{\sqrt{2\pi t}} \exp\left(-\frac{(x-y)^2}{2t}\right).$$

The critical observation is that with *parabolic rescaling*

$$x \longmapsto \sqrt{\beta}x, \qquad t \longmapsto \beta t, \tag{2}$$

one obtains that

$$u(\sqrt{\beta}x, \beta t) = -\frac{1}{\sqrt{\beta}} \frac{\int_R p_x(x,y,t) \exp(-U_0(\sqrt{\beta}y)) dy}{\int_R p(x,y,t) \exp(-U_0(\sqrt{\beta}y)) dy},$$

where the kernels p and p_x do not depend on β. To proceed further a general assumption has to be made to the effect that

$$E \exp(-U_0(y)) < \infty.$$

With the notation

$$V_\beta(y) = B(\beta)\left(e^{-U_0(\sqrt{\beta}y)} - A(\beta)\right), \tag{3}$$

where A, B are to be specified, one has the following basic result:

Theorem 1. *Assume that $V_\beta \to V$ as $\beta \to \infty$ in the Schwartz distribution space $\mathcal{S}'(\mathbf{R})$, where V is a generalized process in the following sense: for all $\phi \in \mathcal{S}(\mathbf{R})$*

$$\lim_{\beta \to \infty} E \exp(i \int_{\mathbf{R}} V_\beta(y)\phi(y)dy) = E \exp(i\langle V, \phi\rangle), \qquad (4)$$

and there exists an $a > 0$ such that, in P,

$$\lim_{\beta \to \infty} \int_{\mathbf{R}} e^{-U_0(\sqrt{\beta}y)} \phi(y)dy = a\langle 1, \phi\rangle. \qquad (5)$$

Then, as $\beta \to \infty$,

$$\beta^{1/2} B(\beta) u(\sqrt{\beta}x, \beta t) \to a^{-1}\langle V, p_x(x, ., t)\rangle$$

where the convergence is that of finite dimensional distributions of random fields in (x, t), $t > 0$, $x \in \mathbf{R}$.

PROOF. To show the convergence of finite dimensional distributions it suffices to prove that for each $\alpha_1, \ldots, \alpha_n, t_1, \ldots, t_n, x_1, \ldots, x_n$, the distribution of the linear combinations

$$w_\beta = \alpha_1 v_\beta(x_1, t_1) + \ldots + \alpha_n v_\beta(x_n, t_n)$$

converges weakly, as $\beta \to \infty$, to the distribution of the random variable

$$-a^{-1}\Big\langle V, \sum_{k=1}^n \alpha_k p_x(x_k, ., t_k) \Big\rangle. \qquad (6)$$

Observe that, because $\int_{\mathbf{R}} p_x(x, y, t)\, dy = 0$,

$$w_\beta = -a^{-1} \sum_{k=1}^n \alpha_k \Big(1 + \varepsilon(x_k, t_k)\Big) \int_{\mathbf{R}} p_x(x_k, y, t_k) V_\beta(y)dy,$$

where

$$\varepsilon(x, t) = a \left(\int_{\mathbf{R}} p(x, y, t) \exp(U_0(\beta y))dy \right)^{-1} - 1.$$

As $\beta \to \infty$, in view of assumption (5), we have that $\varepsilon(x_k, t_k) \to 0$ in probability. On the other hand,

$$\alpha_k p_x(x_k, ., t_k) \in \mathcal{S}(\mathbf{R}), \quad k = 1, \ldots, n,$$

so, in view of assumption (4), we obtain that in distribution

$$\lim_{\beta \to \infty} \sum_{k=1}^n \alpha_k \Big(1 + \varepsilon(x_k, t_k)\Big) \int_{\mathbf{R}} p_x(x_k, y, t_k) V_\beta(y)dy$$

$$= \langle V, \sum_{k=1}^{n} \alpha_k p_x(x_k, \cdot, t_k) \rangle. \qquad (7)$$

Hence, the distributions of w_β converge to the distribution of (6). ∎

Condition (5) really means that $\exp(-U_0)$ is in the domain of attraction of V in the generalized sense (see Dobrushin (1980)),

$$e^{-U_0} \in DA\{V; A(\beta), B(\beta)\},$$

in which case, necessarily

$$B(\beta) = \beta^\kappa L(\beta)$$

for some constant $\kappa \in \mathbf{R}$ and a slowly varying, locally bounded $L(\beta), \beta > 0$. In addition, if $A(\beta)$ is a constant independent of β, the limiting generalized process V is self similar with parameter κ i.e. for all $\beta > 0, \phi \in \mathcal{S}$

$$\langle V, \beta^{\kappa-1}\phi(\beta^{-1}.) \rangle \sim \langle V, \phi \rangle.$$

So now the question is, what kind of initial velocity potential processes U_0 satisfy assumptions (4-5))? Several cases have been investigated and we will describe some of them below.

Strictly stationary processes. In this case one can rely on the classical Ibragimov-Linnik (1965) central limit theory. In particular, in terms of the strong mixing coefficient

$$\alpha_x(x) = \sup_{A \in \mathcal{F}_{y_1}^-, B \in \mathcal{F}_{x+y}^+} |P(A \cap B) - P(A) \cdot P(B)|, \qquad (8)$$

where

$$\mathcal{F}_x^- = \sigma\{U_0(y) : y < x\}, \quad \mathcal{F}_x^+ = \sigma\{U_0(y) : y \geq x\},$$

and the covariance function

$$R(x) = E\left(e^{-U_0(0)} - Ee^{-U_0(0)}\right)\left(e^{-U_0(x)} - Ee^{-U_0(x)}\right),$$

Surgailis and Woyczynski (1993) obtained the following result:

Theorem 2. *Assume that* $X = \{X(y), y \in \mathbf{R}\}$ *is a strictly stationary process with mean zero, such that for some* $\delta > 0$, *the moment* $E|X(0)|^{2+\delta} < \infty$ *and*

$$\int_0^\infty \left(\alpha_X(x)\right)^{\delta/(2+\delta)} dx < \infty. \qquad (9)$$

Then

$$\sigma_X^2 = \int_{\mathbf{R}} R_X(x)\, dx < \infty, \tag{10}$$

where the integral converges absolutely, and

$$X \in \mathbf{DA}\{\sigma_X W; 0, \beta^{1/2}\}.$$

In other words, for any $\phi \in \mathcal{S}(\mathbf{R})$, $z \in \mathbf{R}$,

$$\lim_{\beta \to \infty} P\left(\beta^{1/2} \int_{\mathbf{R}} X(\beta y)\phi(y)\, dy < z\right) = P\left(\sigma_X \int_{\mathbf{R}} \phi(y)\, dW(y) < z\right), \tag{11}$$

where W is the Brownian motion process.

PROOF. The absolute convergence of the integral (10) can be shown in a way similar to the proof of (18.5.12) in Ibragimov and Linnik (1965) using (9) and the estimate of the covariance function by the mixing coefficient given in that book. By inspection of the proof of Theorem 18.5.3 in Ibragimov and Linnik (1965), and also of the corresponding (multidimensional) limit Theorem 7.3.1 in Ethier and Kurtz (1986), we obtain that, under the assumption of the theorem

$$\left\{\frac{1}{\sqrt{\beta}} \int_0^{\beta x} X(y)\, dy, x \ge 0\right\} \Rightarrow \{\sigma_X W(x), x \ge 0\}, \tag{12}$$

where $\Rightarrow$ denotes the convergence of finite dimensional distributions. By (16.3.15), we easily get that for any $n = 1, 2, \ldots$, $M = 1, 2, \ldots$, and any

$$\phi_n(y) = \sum_{k=-M}^{M} \phi_k \mathbf{1}_{[k/n,(k+1)/n)}(y), \quad |k| \le M,$$

we have

$$\frac{1}{\sqrt{\beta}} \sum_{k=-M}^{M} \phi_k \int_{\beta k/n}^{\beta(k+1)/n} X(y)\, dy \Rightarrow \sigma_X \int_{\mathbf{R}} \phi_n(y)\, dW(y). \tag{13}$$

Indeed, substituting $(-M + k)/n = x_k$ so that $x_0 = 0, \ldots, x_{2M+1} = (M+1)/n$, and $\tilde{X}(y) = X(y - M/n)$, we can rewrite the left-hand side of (13) as

$$\frac{1}{\sqrt{\beta}} \sum_{k=0}^{2M+1} \phi_k \int_{\beta x_k}^{\beta x_{k+1}} \tilde{X}(y)\, dy.$$

As (12) obviously remains true for the shifted process $\tilde{X}(y)$ as well, (12) implies (13).

To finish the proof it suffices to show that, for any $\phi \in \mathcal{S}(\mathbf{R})$, the integral on the left hand side of (11) can be approximated in the mean square, uniformly in $\beta > 0$, by the corresponding integral with respect to a step function ϕ_n introduced above. That is obviously the case since for any $\epsilon > 0$ there exist n, M and $\phi_k, k = -M, \ldots, M$ such that

$$\sup_{y \in \mathbf{R}} |\phi(y) - \phi_n(y)| < \epsilon,$$

which implies that

$$\sup_{y \in \mathbf{R}} |\phi * \phi(y) - \phi_n * \phi_n(y)| < C\epsilon,$$

where C is a constant independent of ϵ. Now, in view of (10),

$$E\left(\beta^{1/2} \int_{\mathbf{R}} \left(\phi(y) - \phi_n(y)\right) X(\beta y) \, dy\right)^2$$

$$= \beta^{-1} \int_{\mathbf{R}} \int_{\mathbf{R}} R_X(x - y) \left(\phi(x/\beta) - \phi_n(x/\beta)\right) \left(\phi(y/\beta) - \phi_n(y/\beta)\right) dx \, dy$$

$$= \int_{\mathbf{R}} R_X(x) \left((\phi - \phi_n) * (\phi - \phi_n)\right)(x/\beta) \, dx \leq C\epsilon \int_{\mathbf{R}} |R_X(x)| \, dx < \infty.$$

∎

An application of Theorems 1 and 2 gives the following Gaussian scaling limit result for the random field solution u of the Burgers equation:

Corollary 1. *Let $U_0 = \{U_0(y), y \in \mathbf{R}\}$ be a strictly stationary process satisfying the mixing conditions of Theorem 2, and such that for some $\delta > 0$*

$$E \exp\left((2 + \delta)U_0(0)\right) < \infty. \tag{14}$$

Then, as $\beta \to \infty$, the two-parameter random field in (x, t),

$$\beta^{3/2} u(\beta x, \beta^2 t) \Rightarrow -a^{-1}\sigma \int_{\mathbf{R}} p_x(x, y, t) \, dW(y), \tag{15}$$

where $a = E \exp(U_0(0))$,

$$R(y) = \text{Cov}\left(e^{U_0(0)}, e^{U_0(y)}\right),$$

and

$$\sigma^2 = \int_{\mathbf{R}} R(y) \, dy. \tag{16}$$

In particular, for each $t > 0$ and $x \in \mathbf{R}$, as $\beta \to \infty$,

$$\beta^{3/2} u(\beta x, \beta^2 t) \Rightarrow N\left(0, 4^{-1}\pi^{-1/2}a^{-2}\sigma^2 t^{-3/2}\right). \qquad (17)$$

PROOF. Applying Theorem 2 with $X(y) = e^{U_0(y)} - Ee^{U_0(y)}$ we can easily see that all the assumptions of Theorem 2 are satisfied so that for any $\phi \in \mathcal{S}(\mathbf{R})$, in distribution,

$$\lim_{\beta \to \infty} \int_{\mathbf{R}} V_\beta(y)\phi(y)\,dy = \sigma \int_{\mathbf{R}} \phi(y)\,dW(y), \qquad (18)$$

where

$$V_\beta(y) = \beta^{1/2}\left(e^{U_0(\beta y)} - a\right).$$

However, (16.3.21) is equivalent with assumption (4) of Theorem 1. On the other hand, for any $\phi \in \mathcal{S}(\mathbf{R})$, in probability,

$$\int_{\mathbf{R}} e^{U_0(\beta y)}\phi(y)\,dy = \beta^{-1/2}\int_{\mathbf{R}} V_\beta(y)\phi(y)\,dy + a\int_{\mathbf{R}} \phi(y)\,dy \to a\langle 1, \phi\rangle,$$

because the first term in the above sum converges to 0 in view of (18). Thus, assumption (5) is also satisfied and Theorem 1 gives statement (16). Statement (17) follows from (15) and the following computation of the variance:

$$a^{-2}E\left(\int_{\mathbf{R}} p_x(x, y, t)\,dW(y)\right)^2 = a^{-2}\int_{\mathbf{R}} p_x^2(x, y, t)\,dy = (4\sqrt{\pi}a^2 t^{3/2})^{-1}.$$

∎

Shot noise processes. By a general shot noise process we mean a process of the form

$$U_0(x) = \sum_{k=1}^{\infty} \xi_k \varphi\left(\frac{x - x_k}{\theta_k}\right), \quad x \in \mathbf{R},$$

where φ is an integrable and smooth function on $\mathbf{R}$, θ is a positive random variable, $(\xi, \theta), (\xi_1, \theta_1), \ldots$ are independent, identically distributed random vectors and (x_i) is a point process independent of the sequence (ξ_i, θ_i). So, intuitively speaking, the initial velocity process can be pictured as a family of "bumps" of fixed shape φ with random heights ξ_i, random scaling θ_i and randomly distributed in space according to the point process (in higher dimension, a random field) x_i. The simplest example of such a point process would be a Poisson point process, but our main interest is in situations where interactions between different bumps are present.

Such a situation arises when the above shot noise process is driven by what we call a Gibbs-Cox process. The following results were obtained by Funaki, Surgailis and Woyczynski (1995).

Let us recall the notion of a Gibbs measure on $\mathbf{R}^d$ (see e.g. Ruelle (1969) and Spohn (1991)). The Gibbs measure itself is determined by the pair potential $\Phi = \Phi(x)$, $x \in \mathbf{R}^d$, and the fugacity $\lambda = \lambda(x)$, $x \in \mathbf{R}^d$, which plays the role of the intensity in classical Poisson processes. Some extra technical conditions on the potential Φ are needed and can be found in the above quoted paper. In the usual references it is assumed that the fugacity λ is a constant and does not vary with $x \in \mathbf{R}^d$. In our case it is essential that space-dependent fugacity is permitted, but we impose a technical condition that λ is a function bounded by constant determined by potential Φ.

Let us introduce some notation. By $\mathcal{X}(\Lambda)$, $\Lambda \subset \mathbf{R}^d$, we denote the space of all $\mathbf{Z}_+$-valued Radon measures on Λ, i.e., all locally finite configurations on Λ. Notice, that if Λ is bounded, the space $\mathcal{X}(\Lambda)$ can be identified with $\cup_{n=0}^{\infty}\Lambda^n$. Let $\mathcal{B}_\Lambda$, $\Lambda \subset \mathbf{R}^d$, be the σ-field of $\mathcal{X}(\mathbf{R}^d)$ generated by $\{\langle\psi, \nu\rangle; \psi \in C_0^{\infty}(\mathbf{R}^d), \text{ supp } \psi \subset \Lambda\}$, where $\nu \in \mathcal{X}(\mathbf{R}^d)$ and $\langle\psi, \nu\rangle = \int \psi(x)\, d\nu(x)$.

Then, probability measure $\mu = \mu_\lambda$ on the space $\mathcal{X}(\mathbf{R}^d)$ is called a Gibbs measure (associated with pair potential Φ and fugacity λ) if it satisfies the so-called Dobrushin-Lanford-Ruelle (DLR) equation. More precisely, μ is a Gibbs measure if, for all bounded $\Lambda \subset \mathbf{R}^d$, its conditional probabilities with respect to $\mathcal{B}_{\Lambda^c}$ satisfy the following equation:

$$\mu(dx_1 \cdots dx_n | \mathcal{B}_{\Lambda^c})(\{y_j\})$$

$$= Z_\Lambda^{-1} \frac{1}{n!} \exp\left[-\sum_{1 \le i < j \le n} \Phi(x_i - x_j) - \sum_{i=1}^{n}\sum_{j=1}^{\infty} \Phi(x_i - y_j) \right] \prod_{i=1}^{n} \lambda(x_i)dx_i,$$

for $(x_1, \cdots, x_n) \in \Lambda^n$, $n = 0, 1, 2, \cdots$, and μ-a.a. $\{y_j\} \in \mathcal{X}(\Lambda^c)$, where

$$Z_\Lambda = \sum_{n=0}^{\infty} \frac{1}{n!} \int_{\Lambda^n} \exp\left[-\sum_{1 \le i < j \le n} \Phi(x_i - x_j) - \sum_{i=1}^{n}\sum_{j=1}^{\infty} \Phi(x_i - y_j) \right] \prod_{i=1}^{n} \lambda(x_i)dx_i$$

is a normalization constant.

Under some technical conditions on Φ and λ, the Gibbs measure $\mu = \mu_\lambda$ is uniquely determined by potential Φ and fugacity λ. Furthermore, μ has the exponential L^2-mixing property, uniformly in λ.

Now, a *Gibbs-Cox random field* on $\mathbf{R}^d$ is a random field generated by the Gibbs measure μ_λ with fugacity $\lambda = \lambda(x)$ being an independent stationary ergodic random field with bounded realizations and again some extra technical conditions.

The basic example of the Gibbs-Cox process is the classic Cox (sometimes called doubly-stochastic Poisson) process with the random intensity measure

$$\Lambda(A) = \int_A \lambda(x)\,dx, \quad A \subset \mathbf{R},$$

generated by a nonnegative process $\lambda = \{\lambda(x), x \in \mathbf{R}\}$ so that, in other words, if

$$N(A) := \sum_{k=1}^{\infty} \mathbf{1}_A(x_k),$$

is the number of counts in a Borel set A then, for any mutually disjoint Borel sets $A_1, \ldots, A_n \subset \mathbf{R}$, and any non-negative integers $j_1, \ldots, j_n$,

$$P\Big(N(A_1) = j_1, \ldots, N(A_n) = j_n\Big) = E\Big(\prod_{k=1}^{n} \frac{(\Lambda(A_k))^{j_k}}{j_k!} e^{-\Lambda(A_k)}\Big).$$

This example allows no interactions and corresponds to the case of zero pair potential ($\Phi \equiv 0$) for the Gibbs-Cox process.

The next result, obtained in Funaki, Surgailis, Woyczynski (1993), gives the limit behavior of solutions of the Burgers equations under the intial velocity potential which is a shot noise process with compact support of the "bumps" driven by the Gibbs-Cox process. It relies on an analysis of the limit behavior of a special class of functionals F on the Gibbs-Cox processes. The special case of not interacting bumps driven by the Cox (doubly Poisson) process was studied by Surgailis and Woyczynski (1993a) and (1994), with the latter paper analyzing the case of bumps with noncompact support.

We shall assume that the fugacity process λ is an, independent of everything else, stationary process, and that ξ, θ and φ satisfy some boundedness assumptions which can be found in the above quote paper. As we shall see, even under these restrictions, the scaling limit distribution of $u(x,t)$ may be non-Gaussian. This effect is in contrast to the case of the Poisson process driven shot noise when λ is nonrandom, where only Gaussian limiting distributions arise (see Bulinskii (1990) and Bulinskii and Molchanov (1991)).

Theorem 3. *If there exists a generalized process $V^{(\lambda)}$ and normalizing constants $A(\beta), B(\beta) > 0$, $\beta > 0$, such that*

$$E^{P_\lambda}(e^{U_0}) \in \mathbf{DA}\{V^{(\lambda)}; A(\beta), B(\beta)\},$$

and such that

$$\lim_{\beta \to \infty} \frac{B(\beta)}{\sqrt{\beta}} = d < \infty,$$

then, as $\beta \to \infty$,

$$\beta B(\beta) u(\beta x, \beta^2 t) \Rightarrow -a^{-1} \langle V, p_x(x, \cdot, t) \rangle,$$

where u is a solution of the Burgers equation with the initial velocity potential U_0,

$$a = E e^{U_0(0)},$$

and

$$V = \sigma d \cdot W' + V^{(\lambda)},$$

where σ is a constant, and W' is a white noise process independent of $V^{(\lambda)}$, i.e., a generalized process with the characteristic functional

$$E e^{i\langle W', \phi \rangle} = \exp\left[-\frac{1}{2} \int_{\mathbf{R}} \phi^2(x) dx\right].$$

Next, we consider the special case where the initial velocity potential process is a shot noise driven by the usual Cox process with

$$\lambda(x) = \zeta^2(x) + b, \quad x \in \mathbf{R},$$

where $\zeta = \{\zeta(x), x \in \mathbf{R}\}$ is a stationary Gaussian process with mean $m \in \mathbf{R}$ and covariance $R_\zeta(x) = \mathrm{Cov}\left(\zeta(0), \zeta(x)\right)$. Clearly, in order to guarantee that $\lambda(x) \geq 0$ a.s., we must assume that $b \geq 0$. Moreover, we shall assume that

$$R_\zeta(x) \sim x^{-\alpha} \quad (x \to \infty),$$

where $0 < \alpha \leq +\infty$ (for $\alpha = +\infty$, the above relation means that $R_\zeta(x) = o(x^{-\beta})$ for any $\beta > 0$).

Our main goal here is to find conditions on parameters α, m, b and $\sigma_0^2 = R_\zeta(0)$ which will imply the existence of a nontrivial large scale limit, i.e., conditions guaranteeing that for some $V^{(\lambda)}$

$$e^{\lambda * \Psi} \in \mathbf{DA}\{V^{(\lambda)} : A(\beta), B(\beta)\}.$$

In particular, one can show (see Surgailis, Woyczynski (1993) and Funaki, Surgailis and Woyczynski (1995)) that in some cases:

- If $m \neq 0$ and $0 < \alpha < 1$, then

$$e^{\lambda * \Psi} \in \mathbf{DA}\{\tilde{\sigma}_\alpha W'_\alpha; \tilde{a}, \beta^{\alpha/2}\},$$

where W'_α is the α-fractional Gaussian noise i.e. the generalized Gaussian process with

$$
\begin{aligned}
Ee^{i\langle W'_\alpha, \psi \rangle} &= \exp\left\{-\frac{1}{2}\int_{\mathbf{R}^2} \psi(x)\psi(y)|x-y|^{-\alpha}dx\,dy\right\} \\
&= \exp\left\{-\frac{1}{2}\int_{\mathbf{R}} |\hat{\psi}(p)|^2 |p|^{\alpha-1}dp\right\}
\end{aligned}
$$

with $\hat{\psi}$ being the Fourier transform of $\psi \in \mathcal{S}$;

- If $m = 0$ and $0 < \alpha < 1/2$, then

$$e^{\lambda * \Psi} \in \mathbf{DA}\{\tilde{\sigma}_\alpha^{(2)} W_\alpha^{(2)}; \tilde{a}, \beta^\alpha\},$$

where $W_\alpha^{(2)} =: (W'_\alpha)^2$: is the second Wick polynomial of W_α i.e., the generalized process given by the double Ito-Wiener integral (see Section 4.2 and, e.g., Kwapien and Woyczynski (1992) and Major (1981) as general references on multiple stochastic integrals)

$$\langle W_\alpha^{(2)}, \psi \rangle = \int_{\mathbf{R}^2} \hat{\psi}(p_1 + p_2)|p_1|^{(\alpha-1)/2}|p_2|^{(\alpha-2)/2}\hat{W}(dp_1)\hat{W}(dp_2),$$

where $\hat{W}$ is the complex Gaussian white noise (the Fourier transform of W'). Other results in this direction have been obtained by Leonenko et al. (1994, 1995).

4.4 The maximum energy principle for unimodal data

In this section we describe results, due to Hu and Woyczynski (1994), which show that a certain rearrangement of coefficients in initial data of discrete moving average type maximizes variance (energy) of the parabolic scaling limit distribution of the random field solving Burgers' equation. The proof, which can be found in the above quoted papers, is in the spirit of domination principles developed in Kwapien and Woyczynski (1992) book. A continuous analogue of the result was obtained in another paper by Hu and Woyczynski (1995a) and will be discussed at the end of this section.

Discrete case. Consider Burgers' equation

$$u_t + uu_x = \nu u_{xx},$$

with the initial data in the form of a finite moving average process

$$u_0(x) = \sum_{i=1}^{N} c_i \xi_{\lfloor x \rfloor - i}, \tag{1}$$

where $\xi, \ldots \xi_{-1}, \xi_0, \xi_1, \ldots$ are independent, indentically distributed random variables, and $\lfloor x \rfloor$ is the integer part of x. It turns out that a particular reordering of the coefficient vector $\vec{c} = (c_1, c_2, , \ldots, c_N)$ in the above initial condition affects the solution random field $u(t, x\sqrt{t})$ for large values of t.

Assume that $\mathbf{E}e^{a\xi} < \infty$ for $0 \le a < \infty$. Also, without loss of generality, we assume that $\operatorname{Var} \xi = 1$ in (2.1), and $\nu = 1/2$ in (1.1). Let

$$\sigma(\vec{c}) = \exp\left(\sum_{i=1}^{N} h(c_i, c_i)\right) + 2\sum_{k=1}^{N-1} \exp\left(\sum_{i=1}^{N-k} h(c_i, c_{i+k})\right) - (2N-1), \tag{2}$$

where

$$h(a, b) = \log \frac{\mathbf{E}e^{(a+b)\xi}}{\mathbf{E}e^{a\xi}\mathbf{E}e^{b\xi}}, \tag{3}$$

and

$$p(x) = \lim_{t \to \infty} t^{1/2} \sum_{k=-\infty}^{\infty} \left(\int_{\frac{k}{\sqrt{t}}}^{\frac{k+1}{\sqrt{t}}} (y - x) \exp\left(-\frac{(y-x)^2}{2}\right) dy\right)^2. \tag{4}$$

In this situation one can prove, utilizing an estimate in the central limit theorem for dependent random variables due to Bulinski (1987), that if $u(t, x)$ is a solution random field of the Burgers equation, then, for each fixed x, as $t \to \infty$, the distribution of $t^{5/4}u(t, x\sqrt{t})$ weakly converges to $N(0, p(x)\sigma(\vec{c})/2\pi)$.

As a corollary one obtains a monotonicity property of the limiting variance σ in terms of the componentwise ordering of the initial coefficient vectors $\vec{c}$.

Theorem 1. If $\vec{c} = (c_1, \ldots, c_N) \ge 0$, then function $\sigma(\vec{c}) = \sigma(c_1, \ldots, c_N)$ is increasing in each variable.

A proof of the above result can be found also in Hu and Woyczynski (1994). Finally, when ξ is either a Gaussian, or a Poissonian random variable, one can prove an extremal rearrangement property of the variance $\sigma(\vec{c})$ of the limit field u with respect to the permutation transformation on $\vec{c}$. A majorization method, and the Schur (1927) convexity ideas are used in the proof which uses combinatorial ideas and which can also be found in the above quote paper. The result is as follows:

Theorem 2. *If ξ is either a Gaussian or a Poissonian random variable, and the coefficient vector $\vec{c} = (c_1, \ldots, c_N)$ satisfies condition*

$$0 \leq c_1 \leq c_N \leq c_2 \leq c_{N-1} \leq c_3 \leq c_{N-2} \leq c_4 \leq \ldots, \tag{5}$$

then

$$\sigma(\vec{c}) = \max_{\Pi} \sigma(\Pi\vec{c}), \tag{6}$$

where $\Pi\vec{c}$ is a vector obtained from $\vec{c}$ by permutation Π of its components.

It is an open question whether an analog of Theorem 2 remains true for more general ξ.

Continuous case. The case of continuous moving average initial data leads to a slightly different result which was obtained in Hu and Woyczynski (1995a). The asymptotically maximum variance (energy) solution corresponds to the symmetric-unimodal equimeasurable version of the initial moving average kernel.

Consider the Burgers initial-value problem

$$u_t + uu_x = (1/2)u_{xx}, \tag{7}$$

$t > 0, x \in \mathbf{R}, u = u(t, x), u(0, x) = u_0(x)$, and a continuous-time version of the above extremal problem with initial input of the form

$$U_0(x) = \int_{\mathbf{R}} h(x - y)dM(y), \tag{8}$$

where $h(y) \geq 0$ is a continuous function with compact support, and $M(y)$ is either a Brownian motion, or a Poisson process, defined on the whole real line. Although the continuous-time phenomenon is similar to the one encountered for discrete moving averages, it turns out that the passage from discrete to continuous initial data requires nontrivial modifications in the proofs. We do not know if our results extend to processes other than the Brownian motion and the Poisson process.

The following result parallels Theorem 1. Notice, however, that the scaling in the continuous case is different from the one obtained previously in the discrete case.

Theorem 3. *Let* $u = u(t, x)$ *be a solution of (7) with initial data (8). Then, for each* $x \in \mathbf{R}$,

$$t^{3/4} u(t, x\sqrt{t}) \to N\left(0, \frac{1}{\sqrt{2\pi}}\sigma(h)\right) \tag{9}$$

in probability as $t \to \infty$, *where,*
  i) if $M(y)$ *is a Brownian motion,*

$$\sigma(h) = \int_{\mathbf{R}} \left(\exp\left(\int_{\mathbf{R}} h(x+y)h(y)dy\right) - 1\right) dx; \tag{10}$$

ii) if $M(y)$ *is a Poisson process,*

$$\sigma(h) = \int_{\mathbf{R}} \left(\exp\left(\int_{\mathbf{R}} (e^{h(x+y)} - 1)(e^{h(y)} - 1)dy\right) - 1\right) dx. \tag{11}$$

The next results shows that the variance $\sigma(h)$ in (9) is an increasing functional of kernel h.

Theorem 4. *If* $h_1(x) \le h_2(x)$ *for all* $x \in \operatorname{supp}(h)$, *then* $\sigma(h_1) \le \sigma(h_2)$.

Finally, the main result indicates that the maximum rearrangement principle also has a parallel for continuous time moving averages.

Theorem 5. *Suppose* $h(\cdot) \ge 0$ *is a symmetric-unimodal, continuous function on* $\mathbf{R}$ *with compact support. Then*

$$\sigma(h) = \max_{g \in \mathcal{M}_h} \sigma(g), \tag{12}$$

where $\mathcal{M}_h$ *is the family of all continuous functions equimeasurable with* h.

Recall that nonnegative functions with compact support, f_1, f_2, defined on $\mathbf{R}$ are said to be *equimeasurable* if, for all $c > 0$,

$$m(\{y : f_1(y) \ge c\}) = m(\{y : f_2(y) \ge c\}).$$

$m(A)$ is the Lebesgue measure of set $A \in \mathbf{R}$. f is said to be *unimodal* if, for any $c > 0$, $\{y : f(y) \ge c\}$ is an interval.

Function f is said to be *symmetric-unimodal* if $f(\cdot)$ is a unimodal function, and if there exists t_0 such that, for all $x \in \mathbf{R}$, $f(t_0 - x) = f(t_0 + x)$.

The property of being symmetric-unimodal is preserved under convolutions. The following result is due to Wintner (1938). Indeed, suppose f_1 and f_2 are two symmetric-unimodal functions such that their convolution is well defined on $\mathbf{R}$. Then $g(y) = \int_{-\infty}^{\infty} f_1(y-x)f_2(x)dx$ is symmetric-unimodal.

Now, let

$$R_f(x) = \int f(x-y)f(-y)dy.$$

It is easy to check that $R_f(-x) = R_f(x)$. Since $R_f(x)$ is an even function, we can restrict our attention to $x \in [0, \infty)$. Another function $\tilde{R}_f(x)$, related to $R_f(x)$, will be used in the remainder of this section. It is defined as follows. Let

$$\tilde{R}_f(x) = \begin{cases} c, & \text{if } x = m([0,\infty) \cap \{y : R_f(y) > c\}); \\ 0, & \text{if } x \geq m([0,\infty) \cap \{y : R_f(y) > c\}). \end{cases}$$

One can check, that the $\tilde{R}_f(\cdot)$ is well defined on $[0, \infty)$, continuous and decreasing.

We'll sketch the proofs of Theorems 3-5 indicating the intermediate series of lemmas and propositions.

Proposition 1. *(Domination Property.) Suppose f is a continuous and symmetric-unimodal function with compact support. Then, for any $g \in \mathcal{M}_f$, and any $x > 0$,*

$$\int_0^x \tilde{R}_g(y)dy \leq \int_0^x R_f(y)dy, \tag{13}$$

and

$$\int_0^\infty \tilde{R}_g(y)dy = \int_0^\infty R_f(y)dy. \tag{14}$$

Explicit solutions of the Burgers equation can be written as follows:

$$u(t,x) = t^{-1}\frac{Z(t,x)}{I(t,x)},$$

where

$$Z(t,x) = \int_{\mathbf{R}}(x-y)\exp\left(U_0(y) - \frac{(x-y)^2}{2t}\right)dy,$$

$$I(t,x) = \int_{\mathbf{R}}\exp\left(U_0(y) - \frac{(x-y)^2}{2t}\right)dy,$$

and

$$U_0(y) \equiv \int_{-\infty}^{y} u_0(x)dx$$

is the initial velocity potential. As stated above, the initial velocity potential is assumed here to be of the form

$$U_0(x) = \int_{\mathbf{R}} h(x-y)dM(y), \tag{15}$$

where $h(y) \geq 0$ is a continuous function with compact support, and $M(y)$ is either a Brownian motion, or a Poisson process, defined on the whole real line.

Lemma 1. i) *If $M(y)$ is a Brownian motion then, in probability,*

$$\lim_{t\to\infty} t^{-1/2} I(t, x\sqrt{t}) = \sqrt{2\pi} \exp\left(\frac{1}{2}\int h(s)^2 ds\right).$$

ii) *If $M(y)$ is Poisson processes then, in probability,*

$$\lim_{t\to\infty} t^{-1/2} I(t, x\sqrt{t}) = \sqrt{2\pi} \exp\left(\int (e^{h(s)} - 1)ds\right).$$

Lemma 2. *Let Z be a solution of the Burgers equation. Then,*

$$\mathbf{E}\left(t^{-3/4} Z(t, x\sqrt{t})\right) = 0,$$

and

$$\mathbf{Var}\left(t^{-3/4} Z(t, x\sqrt{t})\right) = \sqrt{2\pi}\tilde{\sigma}(h),$$

where, in the case when $M(y)$ is a Brownian motion,

$$\tilde{\sigma}(h) = \exp\left(\int h^2(s)ds\right) \int \left(\exp\left(\int h(v-s)h(-s)ds\right) - 1\right)dv,$$

and, in the case when $M(y)$ is a Poisson process,

$$\tilde{\sigma}(h) = \exp\left(2\int (e^{h(s)}-1)ds\right) \int \left(\exp\left(\int (e^{h(v-s)}-1)(e^{h(-s)}-1)ds\right)-1\right)dv.$$

The proof of following lemma depends on the rate-of-convergence result in the central limit theorem for dependent random variables which is due to Bulinski (1987).

Lemma 3. *The distribution of $t^{3/4}Z(t, x\sqrt{t})$ weakly converges to* $N(0, \sqrt{2\pi}\tilde{\sigma}(h))$ *as $t \to \infty$.*

Suppose $R_1(x)$ and $R_2(x)$ are integrable functions with compact support on $[0, C]$ where $C > 0$. Let $\tilde{R}_1(x)$ and $\tilde{R}_2(x)$ be defined as before. The following result is due to Burkill (1964).

Proposition 2. *Suppose that $\tilde{R}_1(x)$ and $\tilde{R}_2(x)$ satisfy the following relation:*

$$\int_0^x \tilde{R}_1(y)dy \le \int_0^x \tilde{R}_1(y)dy, \qquad 0 \le x \le C,$$

and

$$\int_0^C \tilde{R}_1(y)dy = \int_0^C \tilde{R}_1(y)dy.$$

Then, for all convex continuous functions ϕ,

$$\int_0^C \phi\big(\tilde{R}_1(y)\big)dy \le \int_0^C \phi\big(\tilde{R}_1(y)\big)dy.$$

Now, the proofs of Theorems 3-5 can be given in quick succession.

PROOF OF THEOREM 3. Since

$$t^{3/4}u(t, x\sqrt{t}) = \frac{t^{-3/4}Z(t, x\sqrt{t})}{t^{-1/2}I(t, x\sqrt{t})},$$

by Lemmas 1 and 3,

$$\frac{t^{-3/4}Z(t, x\sqrt{t})}{t^{-1/2}I(t, x\sqrt{t})} \to N\left(0, \frac{1}{\sqrt{2\pi}}\sigma(h)\right). \qquad \blacksquare$$

PROOF OF THEOREM 4. Notice that for all $x, y \in \mathbf{R}$, we have that $h_1(y + x)h_1(x) \le h_2(y + x)h_2(x)$ in the case of the Brownian motion, and $(e^{h_1(y+x)} - 1)(e^{h_1(x)} - 1) \le (e^{h_2(y+x)} - 1)(e^{h_2(x)} - 1)$ in the case of the Poisson processes. Since $\phi(x) = e^x - 1$ is an increasing function, the proof is complete. $\blacksquare$

PROOF OF THEOREM 5. In the case when $M(y)$ is a Brownian motion, since $\phi(x) = e^x - 1$ is convex and $R_h(x) = \int h(x+y)h(y)dy$ is continuous and has compact support, Proposition 1 and Proposition 2 give us the desired result. In the case when $M(t)$ is a Poisson process, consider $f(x) = e^{h(x)} - 1$. It is easy to check that $f(x)$ is continuous, symmetric-unimodal and has compact support if and only if $h(x)$ is continuous, symmetric-unimodal with the same compact support. Also, $e^x - 1$ is a strictly increasing function. Since $h \geq 0$, we get that $f \geq 0$, too. If $g \in \mathcal{M}_h$, then by simple computation, we have that $e^g - 1 \in \mathcal{M}_f$. Therefore, as in the Brownian motion case, applying Proposition 1 and Proposition 2 to function $f(\cdot)$, we obtain the result in the Poisson process case.   ∎

4.5 Parabolic scaling limits for singular initial data

In this section we discuss large time asymptotics and scaling limits of the solution $\boldsymbol{u}(t, x)$ of the multidimensional Burgers equation

$$\partial \boldsymbol{u}/\partial t + (\boldsymbol{u}, \nabla)\,\boldsymbol{u} = \frac{1}{2}\Delta \boldsymbol{u}, \qquad t > 0, \tag{1}$$

$$\boldsymbol{u}(0, x) = \boldsymbol{u}_0(x), \qquad x \in \mathbf{R}^3,$$

which describes evolution of the velocity field

$$\boldsymbol{u}(t, x) = (v^{(1)}(t, x), v^{(2)}(t, x), v^{(3)}(t, x)) \in \mathbf{R}^3, \qquad (t, x) \in [0, \infty) \times \mathbf{R}^3.$$

in $\mathbf{R}^3$ with random Gaussian initial data $\boldsymbol{u}(0, x) = -\nabla \phi(x)$, $x \in \mathbf{R}^3$. In the case when the spectral density $\rho(p)$ of $\phi(x)$ is singular, the limiting behaviour of $\boldsymbol{u}(t, x)$ as $t \to \infty$ may be non-Gaussian (a second order Itô-Wiener integral). The impact of the structure of singularities of $\rho(p)$ on the limiting distribution is investigated. The results are taken from Surgailis and Woyczynski (1994) and extend an earlier work.

A solution of equation (1) in the class of potential fields $\boldsymbol{u}(t, x) = -\nabla \phi(t, x)$ is given by the explicit Hopf-Cole formula:

$$\boldsymbol{u}(t, x) = -\frac{\langle e^{\phi(\cdot)}, \nabla g(t, x, \cdot)\rangle}{\langle e^{\phi(\cdot)}, g(t, x, \cdot)\rangle}, \tag{2}$$

where $\phi(x) = \phi(0, x)$ is the initial potential, i.e.

$$\boldsymbol{u}_0(x) = -\nabla \phi(x),$$

$\langle f, h\rangle = \int_{\mathbf{R}^3} f(x)\, h(x)\, dx$, and

$$g(t, x, y) = (2\pi t)^{-3/2} \exp\{-|x - y|^2/2t\}$$

is the Gaussian (heat) kernel. When the initial potential $\phi(x) = \phi(x; \omega)$ is random, one is interested in statistical properties of the solution (2); in particular, its limiting distribution as $t \to \infty$. Here, we focus on the Gaussian model (i.e. $\phi(x), x \in \mathbf{R}^3$, is a stationary zero-mean Gaussian field) with *long range dependence*, whose covariance $B(x) = E\phi(0)\phi(x)$ decays slowly as $|x| \to \infty$, (or, equivalently, the spectral density $\rho(p)$, $p \in \mathbf{R}^3$, is singular), and on the role of the structure of *singular spectrum* in determining the limiting distribution of $u(t, x)$ as $t \to \infty$. The case when $\rho(p)$ is non-singular (bounded), or has a unique power law singularity at the origin, is relatively simple. In this case, $u(t, x)$ is asymptotically Gaussian; see Theorems 1 and 2 below. A non-Gaussian limit distribution given by a two-fold Itô-Wiener integral is observed when the singular spectrum consists of a finite number of discrete points away from the origin (Theorem 3). Theorem 4 discusses a model situation when the singular spectrum is 1-dimensional, in which case the limiting distribution is Gaussian again (although different from that encountered in Theorems 1 and 2). This fact justifies a conjecture that a non-Gaussian limiting behavior can be observed only in presence of discrete singularities.

By the Jensen inequality, one can verify that the right hand side of (2) has a finite second moment and, therefore, admits an Itô-Wiener decomposition into an infinite series of multiple integrals with respect to random spectral measure

$$Z(dp), \ E|Z(dp)|^2 = \rho(p)\,dp,$$

of the Gaussian process

$$\phi(x) = \int_{\mathbf{R}^3} e^{i(x,p)} Z(dp).$$

However, for the limit problem mentioned above, a much simpler expansion of $e^{\phi(x)}$ suffices:

$$e^{\phi(x)} = a \sum_{k=0}^{\infty} : \phi(x)^k : /k!, \tag{3}$$

where $a = Ee^{\phi(0)} = e^{\frac{1}{2}B(0)}$, $: \phi(0)^0 := 1$, and

$$: \phi(x)^k := \int_{(\mathbf{R}^3)^k} e^{i(x, p_1 + \ldots + p_k)}\, d^k Z \tag{4}$$

is the k-tuple ($k = 1, 2, \ldots$) Itô-Wiener integral (Hermite polynomial), see Section 4.2, with $d^k Z = Z(dp_1)\ldots Z(dp_k)$. Let $\mathcal{S} = \mathcal{S}(\mathbf{R}^3)$ be the Schwartz space of rapidly decreasing C^∞-functions, with the dual $\mathcal{S}'$, and let $\mathcal{S}_1 = \{\varphi \in \mathcal{S} : \int_{\mathbf{R}^3} \varphi(x)\,dx = 0\}$. It follows from (4) that, for any $\varphi \in \mathcal{S}_1$

$$\langle e^\phi, \varphi \rangle = a \sum_{k=1}^\infty \frac{1}{k!} \langle : \phi^k :, \varphi \rangle = a \sum_{k=1}^\infty \frac{1}{k!} \int_{(\mathbf{R}^3)^k} \hat{\varphi}(p_1 + ... + p_k)\,d^k Z, \quad (5)$$

where $\hat{\varphi}(p) = \int_{\mathbf{R}^3} \varphi(x)\,e^{i(x,p)}\,dx$, $p \in \mathbf{R}^3$, is the Fourier transform.

Scaling Limits. Using scaling properties of the Gaussian kernel $g(t, x, y)$, and the observation that the denumerator of (3) tends to a constant as $t \to \infty$ (by the law of large numbers), under mild conditions on $\phi(x)$, one can prove the following simple but useful statement. Write $\zeta_\lambda(a) \Rightarrow \zeta_\infty(a)$, $a \in A$, for the weak convergence of finite dimensional distributions as $\lambda \to \infty$; here $\zeta_\lambda(a)$, $a \in A, \lambda \in (0, \infty]$, is a family of random variables, and A an arbitrary set. The following result rephrases the basic theorem of Section 4.3.

Proposition 1. *Assume that the Gaussian process $\phi(x)$ is ergodic, and*

$$A^{-1}(\lambda)\,\langle e^\phi, \varphi(\lambda^{-1}\cdot) \rangle \Rightarrow \langle \zeta, \varphi \rangle, \qquad \varphi \in \mathcal{S}_1, \quad (6)$$

as $\lambda \to \infty$, where $A(\lambda) \to \infty$ ($\lambda \to \infty$) are normalizing constants, and $\langle \zeta, \varphi \rangle$, $\varphi \in \mathcal{S}_1$, is a generalized process over $\mathcal{S}_1$. Then

$$\lambda^4\,A^{-1}(\lambda)\,u(\lambda^2 t, \lambda x) \Rightarrow -e^{\frac{1}{2}B(0)}\,\langle \zeta, \nabla g(t, x, \cdot) \rangle, \qquad (t, x) \in \mathbf{R}_+ \times \mathbf{R}^3. \quad (7)$$

Proposition 1 reduces the problem about the limit distribution of $u(t, x)$ to a problem about scaling limit of the exponential process $e^{\phi(x)}$, $x \in \mathbf{R}^3$. General theory of scaling limits of (generalized) random fields was discussed by Dobrushin (1979,1980). Scaling limits of nonlinear functions of Gaussian processes were studied by many authors, including Dobrushin and Major (1979), Funaki, Surgailis and Woyczynski (1995), Major (1981), Taqqu (1979), Breuer and Major (1983), Giraitis and Surgailis (1985), etc. An application of these results and methods yields the following result of Albeverio, Molchanov and Surgailis (1994):

Theorem 1. *Assume that the covariance function $B(x)$ of the Gaussian process $\phi(x)$ satisfies condition*

$$\int_{\mathbf{R}^3} |B(x)|\,dx < \infty. \quad (8)$$

Then

$$\lambda^{-3/2} \langle e^{\phi}, \varphi(\lambda^{-1}\cdot) \rangle \Rightarrow e^{\frac{1}{2} B(0)} \sigma \langle W, \varphi \rangle, \qquad \varphi \in \mathcal{S}_1, \tag{9}$$

and

$$\lambda^{5/2} \boldsymbol{u}(\lambda^2 t, \lambda x) \Rightarrow -\sigma \langle W, \nabla g(t, x, \cdot) \rangle, \qquad (t, x) \in \mathbf{R}_+ \times \mathbf{R}^3,$$

where $\langle W, \varphi \rangle$, $\varphi \in \mathcal{S}_1$ is the Gaussian white noise with $E\langle W, \varphi \rangle^2 = \langle \varphi, \varphi \rangle$, and $\sigma^2 = \int_{\mathbf{R}^3}(e^{B(x)} - 1)\, dx$.

Note that the dependence on the initial Gaussian process in the limiting distribution of $\boldsymbol{u}(t, x)$ is reduced only to parameter σ (the "width" of the white noise). As $B^k(x), x \in \mathbf{R}^3$, is positive definite for any $k = 1, 2, ...$, so $\sigma^2 \geq \frac{1}{2} \int_{\mathbf{R}^3} B^2(x)\, dx > 0$. Equation (8) is a condition of weak dependence of the Gaussian process $\phi(x)$. It implies continuity and boundedness of the spectral density $\rho(p)$.

Spectra with Discrete Singularities. In the rest of the section we discuss non-integrable covariances $B(x)$, or singular (unbounded) spectral densities $\rho(p)$. The case of regularly decaying covariances was discussed in Albeverio, Molchanov and Surgailis (1994), and Bulinskii and Molchanov (1991), using techniques of Dobrushin and Major (1979), see also Section 4.3. In that case, the scaling limit of $e^{\phi(x)}$ is determined by the lowest non-constant term of the Hermite expansions (3) or (5), which is linear (Gaussian).

Theorem 2. *Let $\alpha \in (0, 3)$, and let*

$$B(x) = L(|x|)\, |x|^{-\alpha},$$

where $L(\cdot)$ is assumed to be a slowly varying at infinity function. Then

$$\lambda^{\frac{\alpha}{2}-3} L^{-\frac{1}{2}}(\lambda) \langle e^{\phi}, \varphi(\lambda^{-1}\cdot) \rangle \Rightarrow e^{\frac{1}{2} B(0)} c \langle W_{\alpha}, \varphi \rangle, \qquad \varphi \in \mathcal{S}_1,$$

and

$$\lambda^{\frac{\alpha}{2}+1} L^{-\frac{1}{2}}(\lambda)\, \boldsymbol{u}(\lambda^2 t, \lambda x) \Rightarrow -c \langle W_{\alpha}, \nabla g(t, x, \cdot) \rangle, \qquad (t, x) \in \mathbf{R}_+ \times \mathbf{R}^3,$$

where $\langle W_{\alpha}, \varphi \rangle$, $\varphi \in \mathcal{S}$ is a Gaussian α-fractional noise, with

$$E\langle W_{\alpha}, \varphi \rangle^2 = \int_{\mathbf{R}^3} |\hat{\varphi}(p)|^2\, |p|^{\alpha-3}\, dp, \qquad c^2 = 2^{\alpha}\, \pi^{3/2}\, \Gamma(\alpha/2)/\Gamma((3-\alpha)/2).$$

Bulinskii and Molchanov (1991) raised a question about non-Gaussian limits for solutions $\boldsymbol{u}(t, x)$ of the Burgers equation when the covariance $B(x)$ has a more complex asymptotics (decays non-regularly). This question is discussed in Theorem 3 below, whose proof uses some ideas of Rosenblatt (1987). For technical reasons we formulate the corresponding assumptions on the spectral density rather than on the covariance. Put

$$\rho(p) = \sum_{j=-n}^{n} c_j \rho_0(p - \theta_j) + \rho_r(p), \qquad (10)$$

where $c_j > 0$, $\theta_j \in \mathbf{R}^3 / \{0\}$, $c_{-j} = c_j$, $\theta_{-j} = -\theta_j$, $j = 1, ..., n$, $\rho_r \in L^1 \cap L^\infty$; and

$$\rho_0(p) = L(|p|^{-1}) |p|^{\alpha-3} 1_{\{|p|<r\}}, \qquad (11)$$

where $r > 0$, $\alpha \in (0,3)$, and $L(\cdot)$ satisfies the same conditions as in Theorem 2. In other words, the spectral density $\rho(p)$ is a sum of a regular part $\rho_r(p)$, and a singular part which has singularities at a finite number of points $\theta_j \neq 0$, $j = -n, ..., n$.

Introduce a complex-valued α-fractional noise

$$\langle W_\alpha, \varphi \rangle = \langle W'_\alpha, \varphi \rangle + i \langle W''_\alpha, \varphi \rangle, \quad \varphi \in \mathcal{S},$$

where W'_α, W''_α are independent real-valued α-fractional noises, i.e. for any $\varphi \in \mathcal{S}$,

$$E \exp\{i \langle W'_\alpha, \varphi \rangle\} = E \exp\{i \langle W''_\alpha, \varphi \rangle\} = \exp\{-\frac{1}{2} \int_{\mathbf{R}^3} |\hat{\varphi}(p)|^2 |p|^{\alpha-3} dp\}.$$

Then, for $0 < \alpha < 3/2$, one can define the Wick square modulus $: |W_\alpha|^2 :$ as a double Itô-Wiener integral

$$\langle : |W_\alpha|^2 :, \varphi \rangle = \int_{(\mathbf{R}^3)^2} \hat{\varphi}(p_1 + p_2) \, d^2 Z'_\alpha + \int_{(\mathbf{R}^3)^2} \hat{\varphi}(p_1 + p_2) \, d^2 Z''_\alpha, \qquad \varphi \in \mathcal{S},$$

where Z'_α, Z''_α are random spectral measures of W'_α, W''_α, respectively, i.e. two mutually independent copies of a complex-valued Gaussian random measure

$$Z_\alpha(dp) = \overline{Z_\alpha(-dp)}, \qquad E |Z_\alpha(dp)|^2 = |p|^{\alpha-3} dp.$$

Theorem 3. *Assume that the spectral density (10) of the Gaussian process $\phi(x)$ satisfies the above conditions, and that $0 < \alpha < 3/2$. Then, for any $\varphi \in \mathcal{S}_1$,*

$$\lambda^{\alpha-3} L^{-1}(\lambda) \langle e^\phi, \varphi(\lambda^{-1} \cdot) \rangle \Rightarrow \frac{1}{2} e^{\frac{1}{2} B(0)} \sum_{j=1}^{n} c_j \langle : |W_{\alpha,j}|^2 :, \varphi \rangle, \qquad (12)$$

and

$$\lambda^{\alpha+1} L^{-1}(\lambda) \, \boldsymbol{u}(\lambda^2 t, \lambda x) \Rightarrow - \sum_{j=1}^{n} c_j \, \langle : |W_{\alpha,j}|^2 :, \nabla g(t, x, \cdot) \rangle, \qquad (13)$$

$(t, x) \in \mathbf{R}_+ \times \mathbf{R}^3$, *where* $W_{\alpha,1}, ..., W_{\alpha,n}$ *are mutually independent complex-valued α-fractional Gaussian noises.*

PROOF. It is well-known that if $f \in L^1$ is continuous at $0 \in \mathbf{R}^3$ and $f(0) > 0$ then, for any $\varphi \in \mathcal{S}$,

$$\int_{\mathbf{R}^3} |\hat{\varphi}(\lambda p)|^2 \, f(p) \, dp \ \sim \ \lambda^{-3} f(0) \, \langle \varphi, \varphi \rangle, \qquad (\lambda \to \infty). \qquad (14)$$

Here, and below, $a(\lambda) \sim b(\lambda)$ means that $\lim a(\lambda)/b(\lambda) = 1$, while $a(\lambda) \propto b(\lambda)$ means that the limit $\lim a(\lambda)/b(\lambda) \in \mathbf{R}_+$. Also, let $f \in L^1$ have the representation

$$f(p) = \Lambda(|p|^{-1}) \, |p|^{\alpha-3} + f_r(p)$$

in the neighborhood of $0 \in \mathbf{R}^3$, where $\alpha \in (0, 3)$, $\Lambda(\cdot)$ is a slowly varying at infinity function, and $f_r \in L^{\infty}$. Then, for any $\varphi \in \mathcal{S}$, as $\lambda \to \infty$,

$$\int_{\mathbf{R}^3} |\hat{\varphi}(\lambda p)|^2 \, f(p) \, dp \ \sim \ \lambda^{-\alpha} \Lambda(\lambda) \int_{\mathbf{R}^3} |\hat{\varphi}(p)|^2 \, |p|^{\alpha-3} \, dp + O(\lambda^{-3}). \quad (15)$$

Of course, the second term on the right-hand side of (15) is negligible unless $\Lambda(\cdot) \equiv 0$.

By (5), and by orthogonality of the Itô-Wiener decomposition,

$$E \, \langle e^{\phi}, \varphi(\lambda^{-1} \cdot) \rangle^2 \ = \ a^2 \sum_{k=1}^{\infty} \frac{d_k(\lambda)}{k!}, \qquad (16)$$

where

$$d_k(\lambda) \ = \ E \langle : \phi^k :, \varphi(\lambda^{-1} \cdot) \rangle^2 / k! \ = \ \lambda^6 \int_{\mathbf{R}^3} |\hat{\varphi}(\lambda p)|^2 \, \rho^{*k}(p) \, dp, \qquad (17)$$

and where

$$\rho^{*k}(p) \ = \ \int_{(\mathbf{R}^3)^{k-1}} \rho(p_1) ... \rho(p_{k-1}) \, \rho(p - p_1 - ... - p_{k-1}) \, dp_1 ... dp_{k-1},$$

$p \in \mathbf{R}^3$, is the k-fold convolution, $k = 2, 3, ...$; $\rho^{*1}(p) = \rho(p)$. Put

$$A^2(\lambda) = \lambda^{6-2\lambda} \, L^2(\lambda).$$

We claim that

$$d_1(\lambda) = O(\lambda^3), \tag{18}$$

$$d_2(\lambda) \propto A^2(\lambda), \tag{19}$$

and

$$\sum_{k=3}^{\infty} \frac{d_k(\lambda)}{k!} = o(A^2(\lambda)), \tag{20}$$

i.e. that the main contribution to $\langle e^\phi, \varphi(\lambda^{-1}\cdot)\rangle$ comes from the second term of the expansion. Here, (18) follows from (15) (with $\Lambda(\cdot) \equiv 0$) and from the fact that $\rho(p)$ of (10) is bounded near 0. Next,

$$\rho^{*2}(p) = 2\sum_{j=1}^{n} c_j^2\, \rho_0^{*2}(p) + f_r(p), \tag{21}$$

where

$$f_r(p) = \sum_{i\neq j} c_i c_j\, \rho_0^{*2}(p + \theta_i - \theta_j) + 2(\rho_s * \rho_r)(p) + \rho_r^{*2}(p),$$

and $\rho_s(p) = \rho(p) - \rho_r(p) = \sum_{j=-n}^{n} c_j \rho_0(p - \theta_j) \in L^1$. Here, $f_r \in L^1 \cap L^\infty$, while

$$\begin{aligned}
\rho_0^{*2}(p) &= \frac{L^2(|p|^{-1})}{|p|^{3-2\alpha}} \int_{\{|v|<r|p|^{-1}\}\cap\{|o-v|<r|p|^{-1}\}} \\
&\quad \times \frac{L(|p|^{-1}|v|^{-1})}{L(|p|^{-1})} \frac{L(|p|^{-1}|o-v|^{-1})}{L(|p|^{-1}}) \frac{dv}{|v|^{3-\alpha}|o-v|^{3-\alpha}} \\
&\sim\; L^2(|p|^{-1})\, |p|^{2\alpha-3} \int_{\mathbf{R}^3} |v|^{\alpha-3}|o-v|^{\alpha-3}\, dv, \qquad (p\to 0), \ (22)
\end{aligned}$$

where $o = p/|p|$, $|o| = 1$, and the last integral does not depend on o. Hence, (19) follows from (15) and (17).

To prove (20), it is convenient to use another representation of $d_k(\lambda)$:

$$\begin{aligned}
d_k(\lambda) &= \int_{\mathbf{R}^3}\int_{\mathbf{R}^3} \varphi(x/\lambda)\, \varphi(y/\lambda)\, B^k(x-y)\, dx\, dy \\
&= \int\int_{|x-y|\le N} \cdots\; +\; \int\int_{|x-y|>N} \cdots\; =:\; d_k'(\lambda) + d_k''(\lambda).
\end{aligned}$$

Here, for any $N < \infty$,

$$|d_k'(\lambda)| \;\le\; CN^3 B^k(0) \int_{\mathbf{R}^3} |\varphi(x/\lambda)|\, dx \;\le\; C\lambda^3 B^k(0),$$

where we designate by C, possibly different, constants independent of λ, k. On the other hand, since $B(x) \to 0$ ($|x| \to \infty$), for any $\epsilon > 0$ there is an $N < \infty$ such that $|\varphi(x)| \leq C(1+|x|^2)^{-2} =: \varphi_1(x)$. Hence, for $k \geq 3$,

$$
\begin{aligned}
|d_k''(\lambda)| &\leq \epsilon^{k-2} \int_{\mathbf{R}^3} \int_{\mathbf{R}^3} \varphi_1(x/\lambda)\varphi_1(y/\lambda)B^2(x-y)\,dx\,dy \\
&\leq C\epsilon^{k-2}\lambda^6 \int_{\mathbf{R}^3} e^{-2\lambda|p|}\rho^{*2}(p)\,dp \\
&\leq C\epsilon^{k-2} A^2(\lambda);
\end{aligned}
$$

see (17) and (19). Therefore, $\sum_{k=3}^{\infty} d_k(\lambda)/k! \leq C\epsilon A^2(\lambda) + O(\lambda^3)$, which implies (20), $\epsilon > 0$ being arbitrary small.

According to (18)-(20),

$$
\langle e^\phi, \varphi(\lambda^{-1}\cdot)\rangle = \frac{1}{2}a\langle : \phi^2 :, \varphi(\lambda^{-1}(\cdot))\rangle + R_\lambda, \tag{23}
$$

where $R_\lambda = o_P(A(\lambda))$. Put $Q = \cup_{j=1}^n Q_j$, $Q^c = (\mathbf{R}^3)^2/Q$, where $Q_j = Q_j^1 \cup Q_j^2$,

$$
Q_j^1 = \{(p_1, p_2) \in (\mathbf{R}^3)^2 : |p_1 - \theta_j| < \epsilon, |p_2 + \theta_j| < \epsilon\},
$$

$$
Q_j^2 = \{(p_1, p_2) \in (\mathbf{R}^3)^2 : |p_1 + \theta_j| < \epsilon, |p_2 - \theta_j| < \epsilon\},
$$

and where $\epsilon > 0$ is chosen so that $Q_i \cap Q_j = \emptyset$ ($i \neq j$), $0 \notin Q_j$, $i, j = 1, ..., n$. Then,

$$
\begin{aligned}
\langle : \phi^2 :, \varphi(\lambda^{-1}\cdot)\rangle &= \lambda^3 \int_{\mathbf{R}^3} \hat{\varphi}(\lambda(p_1 + p_2))\,d^2Z \\
&= \lambda^3 \int \hat{\varphi}(\lambda(p_1 + p_2))\,\mathbf{1}_Q(p_1, p_2)\,d^2Z \\
&\quad + \lambda^3 \int \hat{\varphi}(\lambda(p_1 + p_2))\,\mathbf{1}_{Q^c}(p_1, p_2)\,d^2Z \\
&\equiv I_\lambda' + I_\lambda''.
\end{aligned}
$$

Here, I_λ' is the main term and I_λ'' is the remainder term, which follows from (19) and from the fact that

$$
\begin{aligned}
E|I_\lambda''|^2 &= 2\lambda^6 \int |\hat{\varphi}(\lambda(p_1 + p_2))|^2\,\mathbf{1}_{Q^c}(p_1, p_2)\,\rho(p_1)\,\rho(p_2)\,dp_1\,dp_2 \\
&= 2\lambda^6 \sum_{j=-n}^n c_j^2 \int |\hat{\varphi}(\lambda(p_1 + p_2))|^2\,\mathbf{1}_{Q^c}(p_1, p_2) \\
&\quad \times \rho_0(p_1 - \theta_j)\rho_0(p_2 + \theta_j)\,dp_1\,dp_2 + O(\lambda^3) = O(\lambda^3),
\end{aligned}
$$

see the proof of (19). Next,

$$I'_\lambda \overset{d}{=} \lambda^3 \int \hat{\varphi}(\lambda(p_1 + p_2)) \, 1_Q(p_1, p_2) \, (\rho(p_1) \, \rho(p_2))^{1/2} \, d^2V, \qquad (24)$$

where $V(dp) = \overline{V(-dp)}$ is a complex-valued Gaussian white noise, $E|V(dp)|^2 = dp$, and $\overset{d}{=}$ stands for the equality in distribution. Introduce the shifted Gaussian random measures

$$V_{1,j}(du) = V(d(u + \theta_j)) \, 1_{\{|u| < \epsilon\}},$$

$$V_{2,j}(du) = V(d(u - \theta_j)) \, 1_{\{|u| < \epsilon\}},$$

$j = 1, ..., n$, each of them being independently scattered on $\mathbf{R}^3$, and satisfying

$$E|V_{1,j}(du)|^2 = E|V_{2,j}(du)|^2 = du, \qquad \overline{V_{1,j}(du)} = V_{2,j}(-du),$$

for $|u| < \epsilon$. Moreover, pairs $(V_{1,j}, V_{2,j})$, $j = 1, ..., n$, are independent because $Q_j, -Q_j$, $j = 1, ..., n$, are disjoint. It will be convenient to extend the noises $V_{1,j}, V_{2,j}$, $j = 1, ..., n$, to the whole space $\mathbf{R}^3$, with the preservation of the above mentioned properties.

Now, I'_λ of (24) can be rewritten as

$$I'_\lambda \overset{d}{=} 2\lambda^3 \sum_{j=1}^n \int_{Q_0} \hat{\varphi}(\lambda(u_1 + u_2)) \, (\rho(u_1 + \theta_j)\rho(u_2 - \theta_j))^{1/2} \, V_{1,j}(du_1) \, V_{2,j}(du_2),$$

$$(25)$$

where $Q_0 = \{(u_1, u_2) \in \mathbf{R}^3 : |u_1| < \epsilon, |u_2| < \epsilon\}$. Put

$$W_{1,j}(du) = \frac{1}{\sqrt{2}}(V_{1,j}(du) + V_{2,j}(du)),$$

$$W_{2,j}(du) = \frac{1}{\sqrt{2}}(V_{2,j}(du) - V_{1,j}(du)).$$

Then

$$W_{k,j}(du) = \overline{W_{k,j}(-du)}, \qquad k = 1, 2, \ j = 1, ..., n, \qquad (26)$$

are independent copies of a complex-valued Gaussian white noise; $E|W_{k,j}(du)|^2 = du$. The right hand side of (25) can be written as a sum of stochastic integrals with respect to $W_{k,j}$:

$$I'_\lambda \overset{d}{=} J'_\lambda + J''_\lambda, \qquad (27)$$

$$J'_\lambda = \lambda^3 \sum_{j=1}^{n} \sum_{k=1}^{2} \int_{Q_0} \hat{\varphi}(\lambda(u_1 + u_2)) \left(\rho(u + \theta_j)\, \rho(u_2 - \theta_j)\right)^{1/2} d^2 W_{k,j}, \quad (28)$$

$$J''_\lambda = i\lambda^3 \sum_{j=1}^{n} \int_{Q_0} \hat{\varphi}(\lambda(u_1 + u_2))\, f_j(u_1, u_2)\, W_{1,j}(du_1)\, W_{2,j}(du_2), \quad (29)$$

and

$$f_j(u_1, u_2) = (\rho(u_1 - \theta_j)\rho(u_2 + \theta_j))^{1/2} - (\rho(u_1 + \theta_j)\rho(u_2 - \theta_j))^{1/2}.$$

Note, that

$$E|J''_\lambda|^2 \;=\; O(\lambda^3) \;=\; o(A^2(\lambda)), \quad (30)$$

i.e., J''_λ is negligible in the limit $\lambda \to \infty$. Indeed,

$$E|J''_\lambda|^2 = \lambda^6 \sum_{j=1}^{n} \int_{Q_0} |\hat{\varphi}(\lambda(u_1 + u_2))|^2\, f_j^2(u_1, u_2)\, d^2 u. \quad (31)$$

From decomposition (10), one obtains

$$\rho(u \pm \theta_j) = c_j\, \rho_0(u) + \rho_{j,\pm}(u), \qquad (|u| < \epsilon), \quad (32)$$

where $\rho_{j,\pm}(u)$ are bounded. Hence $f_j^2(u_1, u_2) \leq C(\rho_0(u_1) + \rho(u_2))$, and, therefore, the right-hand side of (31) does not exceed

$$C\lambda^6 \int |\hat{\varphi}(\lambda u)|^2\, \rho_0 * 1_{\{|u| < \epsilon\}} du = O(\lambda^3),$$

(see (15)), which proves (30).

To prove statement (12) of the theorem, it suffices to show the convergence of J'_λ of (28) or, more exactly, that

$$A^{-1}(\lambda)\, J'_\lambda \;\Rightarrow\; \sum_{j=1}^{n} \sum_{k=1}^{2} c_j \int \hat{\varphi}(p_1 + p_2)\, |p_1|^{(\alpha-3)/2}\, |p_2|^{(\alpha-3)/2}\, d^2 W_{k,j}. \quad (33)$$

Put

$$h^{(j)}(p_1, p_2) \;=\; c_j \hat{\varphi}(p_1 + p_2)\, |p_1|^{(\alpha-3)/2}\, |p_2|^{(\alpha-3)/2},$$

$$h_\lambda^{(j)}(p_1, p_2) \;=\; A^{-1}(\lambda)\, \hat{\varphi}(p_1 + p_2)\, (\rho(\lambda^{-1}p_1 + \theta_j)\, \rho(\lambda^{-1}p_2 - \theta_j))^{1/2}\, 1_{Q_0}(p_1, p_2).$$

Using self-similarity $W_{k,j}(du/\lambda) \stackrel{d}{=} \lambda^{-3/2} W_{k,j}(du)$ of the white noise, and the change of variables formula for the Itô-Wiener integrals (see, e.g., Dobrushin (1979), Kwapien and Woyczynski (1992)), one obtains

$$A^{-1}(\lambda) J'_\lambda \stackrel{d}{=} \sum_{j=1}^{n} \sum_{k=1}^{2} \int h_\lambda^{(j)}(p_1, p_2)\, d^2 W_{j,k}. \quad (34)$$

Hence, (33) follows from the fact that

$$\int_{(\mathbf{R}^3)^2} |h_\lambda^{(j)}(p_1, p_2) - h^{(j)}(p_1, p_2)|^2 \, dp_1 \, dp_2 \;\to\; 0, \qquad (\lambda \to 0). \qquad (35)$$

In view of (32), one can rewrite $h^{(j)}\lambda(p_1, p_2)$ as

$$h_\lambda^{(j)}(p_1, p_2) \;=\; h^{(j)}(p_1, p_2) \prod_{k=1}^{2} \frac{L^{1/2}(\lambda|p_k|^{-1})}{L^{1/2}(\lambda)} \left(1 + \delta_{j,k}(p, \lambda)\right) 1_{\lambda Q_0}(p_1, p_2), \tag{36}$$

where

$$\delta_{j,1}(p, \lambda) = \frac{\rho_{j,-}(p/\lambda)}{c_j \rho_0(p/\lambda)} \quad \text{and} \quad \delta_{j,2}(p, \lambda) = \frac{\rho_{j,+}(p/\lambda)}{c_j \rho_0(p/\lambda)}$$

vanish as $\lambda \to \infty$, for any $p \in \mathbf{R}^3$, $|p| < \lambda\epsilon$, and are bounded in $\{|p| < \lambda\epsilon\}$. Therefore, the product on the right hand side of (36) tends to 1 as $\lambda \to \infty$, for any $(p_1, p_2) \in (\mathbf{R}^3)^2$. As $L(\cdot)$ is slowly varying, for any $\delta > 0$ there exists a $C < \infty$ such that $L(\lambda|p|^{-1})/L(\lambda) < C|p|^{-\delta}$ for all $0 < |p| < \lambda$. Hence,

$$|h_\lambda^{(j)}(p_1, p_2)| \leq C|\hat\varphi(p_1 + p_2)| \prod_{k=1}^{2} |p_k|^{-\delta+(\alpha-3)/2} \in L^2((\mathbf{R}^3)^2),$$

provided $2\delta < \alpha$, i.e., (35) follows from the Lebesgue dominated convergence theorem.

This proves (12) and Theorem 3 as well, as (13) follows from (12) and Theorem 1. ∎

Spectra Concentrated on Lines. In general, the structure of the singular part of spectrum of a Gaussian process $\phi(x)$ can be much more complicated than that considered in Theorem 3; e.g., singularities can form a 1- or 2-dimensional manifold in $\mathbf{R}^3$ rather than a discrete set. In the case when the singular spectrum is separated from 0, the main contribution to the scaling limit of e^ϕ again may be given by the quadratic term of the Hermite expansion, eventually resulting in a non-Gaussian asymptotics. Below, we discuss a model situation when the singular spectrum is concentrated on two symmetric lines in $\mathbf{R}^3$. Depending on whether the lines go through the origin (in which case they coincide) or not, the leading term of the Hermite expansion is linear or quadratic. However, in both cases the scaling limit is Gaussian, although its covariance and the normalizing constants are different.

This leads to a conjecture that a non-Gaussian behavior can be observed only in the presence of isolated singularities of the form described in Theorem 3.

We shall assume below that the spectral density has the form

$$\rho(p) = \big(\rho_0(q - \theta) + \rho_0(q + \theta)\big)\rho_{r,1}(p^1) + \rho_r(p), \qquad (37)$$

where $p = (p^1, q) \in \mathbf{R}^3$, $p^1 \in \mathbf{R}$, $q \in \mathbf{R}^2$, $\theta \in \mathbf{R}^2$, $0 \le \rho_{r,1} \in L^1(\mathbf{R}) \cap L^\infty(\mathbf{R})$, $\rho_r \in L^1(\mathbf{R}^3) \cap L^\infty(\mathbf{R}^3)$, and

$$\rho_0(q) = L(|q|^{-1})\,|q|^{\alpha-2}\,1_{\{|q|<r\}}, \qquad q \in \mathbf{R}^2,$$

where $L(\cdot)$ is a slowly varying at infinity function, $r \in \mathbf{R}_+$, and $0 < \alpha < 2$. In other words, the spectral density is singular near the lines $q = \pm\theta$, and is bounded elsewhere.

Theorem 4. *Let the spectral density (37) satisfy conditions listed above. Then:*

(i) If $0 < \alpha < 1$ and $\theta \neq 0$, then

$$\lambda^{\alpha-\frac{5}{2}}L^{-1}(\lambda)\langle e^\phi, \varphi(\lambda^{-1}\cdot)\rangle \;\Rightarrow\; \sigma_1\, e^{\frac{1}{2}B(0)}\,\langle W'_\alpha, \varphi\rangle, \qquad \varphi \in \mathcal{S}_1, \qquad (38)$$

and

$$\lambda^{\alpha+\frac{3}{2}}L^{-1}(\lambda)\,\boldsymbol{u}(\lambda^2 t, \lambda x) \;\Rightarrow\; -\sigma_1\,\langle W'_\alpha, \nabla g(t, x, \cdot)\rangle, \qquad (t, x) \in \mathbf{R}_+ \times \mathbf{R}^3,$$

where W'_α is a Gaussian process with mean zero and covariance

$$E\langle W'_\alpha, \varphi\rangle^2 = \int_{\mathbf{R}^3} |\hat{\varphi}(p)|^2\,|q|^{2\alpha-2}\,dp,$$

and where $p = (p^1, q)$, $p^1 \in \mathbf{R}$, $q \in \mathbf{R}^2$, and

$$\sigma_1^2 = \|\rho_{r,1}\|_{L^2}^2 \int_{\mathbf{R}^2} |q|^{2-\alpha}|0 - q|^{2-\alpha}\,dq.$$

(ii) If $0 < \alpha < 2$, $\theta = 0$ and $\rho_{r,1}$ is continuous at $0 \in \mathbf{R}$, then

$$\lambda^{(\alpha-5)/2}L^{-1/2}(\lambda)\langle e^\phi, \varphi(\lambda^{-1}\cdot)\rangle \;\Rightarrow\; \sigma_2 e^{\frac{1}{2}B(0)}\,\langle W'_{\alpha/2}, \varphi\rangle, \qquad (39)$$

$\varphi \in \mathcal{S}_1$, and

$$\lambda^{(3+\alpha)/2}L^{-1/2}(\lambda)\,\boldsymbol{u}(\lambda^2 t, \lambda x) \;\Rightarrow\; -\sigma_1\langle W'_{\alpha/2}, \nabla g(t, x, \cdot)\rangle,$$

$(t, x) \in \mathbf{R}_+ \times \mathbf{R}^3$, where $\sigma_1^2 = 2\rho_{r,1}(0)$.

(iii) If $1 < \alpha < 2$, $\theta \neq 0$ and $\rho(p)$ is continuous at $0 \in \mathbf{R}^3$, then

$$\lambda^{-3/2}\langle e^\phi, \varphi(\lambda^{-1}\cdot)\rangle \;\Rightarrow\; \sigma_3\, e^{\frac{1}{2}B(0)}\,\langle W, \varphi\rangle, \tag{40}$$

$\varphi \in \mathcal{S}_1$, *and*

$$\lambda^{5/2}u(\lambda^2 t, \lambda x) \;\Rightarrow\; -\sigma_3\langle W, \nabla g(t, x, \cdot)\rangle,$$

$(t, x) \in \mathbf{R}_+ \times \mathbf{R}^3$, *where W is a Gaussian white noise, $E\langle W, \varphi\rangle^2 = \langle \hat\varphi, \hat\varphi\rangle$, and*

$$\sigma_3^2 = \sum_{k=1}^{\infty} \frac{\rho^{*k}(0)}{k!}.$$

PROOF. (i) Let $d_k(\lambda)$ be defined by (17). Then, again, relations (18)-(20) hold with $A(\lambda) = \lambda^{5/2-\alpha}L(\lambda)$, and the proof is analogous to the one given above. In particular,

$$d_2(\lambda) = \lambda^6 \int_{\mathbf{R}^3} |\hat\varphi(\lambda p)|^2\, \rho^{*2}(p)\, dp, \tag{41}$$

where

$$\rho^{*2}(p) = 2\rho_{r,1}^{*2}(p^1)\, \rho_0^{*2}(q) + f_r(p), \qquad p = (p^1, q),$$

and $f_r \in L^1 \cap L^\infty$, $\rho_{r,1}^{*2}(p^1) \to \rho_{r,1}^{*2}(0) = \|\rho_{r,1}\|_{L^2}^2 > 0$, $(p^1 \to 0)$. Finally,

$$\rho_0^{*2}(q) \sim C_1 L^2(|q|^{-1})\, |q|^{2\alpha-2}, \qquad (q \to 0), \tag{42}$$

with $C_1 = \int_{\mathbf{R}^2} |u|^{\alpha-2}\, |o - u|^{\alpha-2}\, du < \infty$, independent of $|o| = 1$. Therefore, as in the proof of Theorem 3,

$$d_2(\lambda) \sim 2\sigma_1^2\, A^2(\lambda) \int_{\mathbf{R}^3} |\hat\varphi(p)|^2\, |q|^{2\alpha-2}\, dp, \tag{43}$$

with σ_1^2 defined in Theorem 4 (i), and

$$\langle e^\phi, \varphi(\lambda^{-1}\cdot)\rangle = \frac{1}{2}a\,\langle :\phi^2 :, \varphi(\lambda^{-1}\cdot)\rangle + R_\lambda,$$

where

$$\langle :\phi^2 :, \varphi(\lambda^{-1}\cdot)\rangle = \lambda^3 \int\int \hat\varphi(\lambda(p_1 + p_2))\, Z(dp_1)\, Z(dp_2),$$

and $R_\lambda = o_P(A(\lambda))$. Then, statement (38) of the theorem follows from the fact that

$$A^{-1}(\lambda)\langle :\phi^2 :, \varphi(\lambda^{-1}\cdot)\rangle \;\Rightarrow\; \mathcal{N}(0, 4\sigma^2(\varphi)),$$

where
$$\sigma^2(\varphi) = \sigma_1^2 \int |\hat{\varphi}(p)|^2 \, |q|^{2\alpha-2} \, dp.$$

In view of (43), to do this it suffices to show that the cumulants of order $n > 2$ vanish as $\lambda \to \infty$ (cf. p. 152 ff.), i.e., that for any $n = 3, 4, \ldots$

$$c_n(\lambda) := \mathrm{cum}_n \langle : \phi^2 :, \varphi(\lambda^{-1} \cdot) \rangle = o(A^2(\lambda)). \tag{44}$$

By a well-known formula for cumulants of the multiple Itô-Wiener integrals (Section 4.2),

$$c_n(\lambda) = 2^n \, n! \, \lambda^{3n} \int_{(\mathbf{R}^3)^n} \prod_{i=1}^n \hat{\varphi}(\lambda(p_i - p_{i+1})) \, \rho(p_i) \, dp_i, \qquad (p_{n+1} = p_1)$$

$$\tag{45}$$

$n = 2, 3, \ldots$ For any $\varphi \in S$ and any $N > 0$, there exists a $C < \infty$ such that

$$|\hat{\varphi}(p)| < C(1+|p^1|)^{-N} \, (1+|q|)^{-N} \equiv C \, h_1(p^1) \, h_0(q) \equiv C \, h(p), \ p = (p^1, q).$$

Then, from (45) we obtain that

$$|c_n(\lambda)| \leq C\lambda^{3n} \, (I_0(\lambda) \, I_{r,1}(\lambda) + I_r(\lambda)),$$

where

$$I_0(\lambda) = \int_{(\mathbf{R}^2)^n} \prod_{i=1}^n h_0(\lambda(q_i - q_{i+1})) \, \rho_0(q_i) \, dq_i, \qquad q_{n+1} = q_1,$$

$$I_{r,1}(\lambda) = \int_{\mathbf{R}^n} \prod_{i=1}^n h_1(\lambda(p_i^1 - p_{i+1}^1)) \, \rho_{r,1}(p_i^1) \, dp_i^1, \qquad p_{n+1}^1 = p_1^1,$$

$$I_r(\lambda) = \int_{(\mathbf{R}^3)^n} \prod_{i=1}^n h(\lambda(p_i - p_{i+1})) \, |\rho_r(p_1)| \, \rho(p_2) \ldots \rho(p_n) \, dp_1 \ldots dp_n,$$

$p_{n+1} = p_1$. First, let us estimate the last integral. By a repeated use of the Cauchy inequality,

$$I_r(\lambda) \leq \int_{\mathbf{R}^3} h^2(\lambda) \, (\rho_r * \rho)(p) \, dp \left(\int_{\mathbf{R}^3} h^2(\lambda p) \, \rho^{*2}(p) \, dp \right)^{(n-2)/2},$$

where, in view of the boundedness of convolution $\rho_r * \rho(p)$,

$$\int_{\mathbf{R}^3} h^2(\lambda p) \, \rho_r * \rho(p) \, dp = O(\lambda^{-3}),$$

(see (15)), while $\int h^\zeta \lambda p) \rho^{*2} \, dp = O(\lambda^{-6} A^2(\lambda))$, c.f. (41) and (43). Hence,

$$\lambda^{3n} I_r(\lambda) = O(\lambda^3 A^{n-2}(\lambda)) = o(A^n(\lambda)).$$

Similarly, using the boundedness of $\rho_{r,1}$,

$$I_{r,1}(\lambda) \leq C\lambda^{-n+1} h_{r,1}^{*n}(0) \int_{\mathbf{R}} \rho_{r,1}(u) \, du = O(\lambda^{-n+1}),$$

and

$$I_0(\lambda) \leq \left(\int_{\mathbf{R}^2} \int_{\mathbf{R}^2} h_0^2(\lambda(q_1 - q_2)) \, \rho_0(q_1) \, \rho_0(q_2) \, dq_1 \, dq_2 \right)^{n/2}$$

$$= \left(\int_{\mathbf{R}^2} h_0^2(\lambda q) \, \rho_0^{*2}(q) \, dq \right)^{n/2} = O(\lambda^{-n\alpha} L^n(\lambda)).$$

Hence, for $n > 2$,

$$\lambda^{3n} I_{r,1}(\lambda) I_0(\lambda) = O(\lambda^{2n+1-n\alpha} L^n(\lambda)) = o(A^n(\lambda)),$$

which proves (44) and Theorem 4 (i) as well.

(ii) Put $A^2(\lambda) = \lambda^{5-\alpha} L(\lambda)$. Then one can show, in a fashion similar to the one used above, that

$$d_1(\lambda) \sim \sigma_2^2 A^2(\lambda) \int_{\mathbf{R}^3} |\hat\varphi(p)|^2 \, |q|^{\alpha-2} \, dp, \tag{46}$$

$$\sum_{k=2}^\infty \frac{1}{k!} d_k(\lambda) = o(A^2(\lambda)),$$

so that

$$\langle e^\phi, \varphi(\lambda^{-1} \cdot) \rangle = a \langle \phi, \varphi(\lambda^{-1} \cdot) \rangle + R_\lambda, \tag{47}$$

with $R_\lambda = o_P(A(\lambda))$. Since the main term on the right hand side of (47) has Gaussian distribution, (39) follows from (46), or from the asymptotics of its variance.

(iii) Observe, that $\rho \in L^2 = L^2(\mathbf{R}^3)$. Hence, for any $k = 1, 2, \ldots$, $\rho^{*k}(p) \geq 0$ is continuous at $p = 0$. Morever, $\rho^{*2}(0) = \|\rho\|_2^2 > 0$, and, for $k = 2, 3, \ldots$,

$$\|\rho^{*k}\|_\infty \leq \bar\rho \|\rho^{*(k-1)}\|_\infty \leq \cdots \leq \bar\rho^{k-2} \|\rho\|_2^2, \tag{48}$$

where $\bar\rho = \|\rho\|_1$ and $\|\cdot\|_r \equiv \|\cdot\|_{L^r}$, $(1 \leq r \leq \infty)$; in particular, the series for σ_3^2 converges. Furthermore, using (15), (18) and the above observation, one obtains

$$E\langle :\phi^k:, \varphi(\lambda^{-1} \cdot) \rangle^2 = k! \, d_k(\lambda) \sim \lambda^3 \, k! \, \rho^{*k}(0) \, \|\hat\varphi\|_2^2,$$

$k = 1, 2, \ldots$ For $1 \leq N < \infty$, put

$$S_N(\lambda) = \lambda^{-3/2} a \sum_{k=1}^{N} \frac{1}{k!} \langle : \phi^k :, \varphi(\lambda^{-1} \cdot) \rangle.$$

Then

$$ES_N^2(\lambda) = \lambda^{-3} a^2 \sum_{k=1}^{N} \frac{1}{k!} d_k(\lambda) \sim a^2 \|\hat{\varphi}\|_2^2 \sum_{k=1}^{N} \frac{1}{k!} \rho^{*k}(0) \equiv a^2 \|\hat{\varphi}\|_2^2 \sigma_N^2.$$

$$\tag{49}$$

Then one can show, in a fashion similar to that used in the proof of (i), that for any $n > 2$, and any $k_1, \ldots, k_n \geq 1$,

$$\mathrm{cum}(\langle : \phi^{k_1} :, \varphi(\lambda^{-1} \cdot) \rangle, \ldots, \langle : \phi^{k_n} :, \varphi(\lambda^{-1} \cdot) \rangle) = o(\lambda^{3n/2}). \tag{50}$$

Therefore, from (49) and (50) we conclude that, for any $1 \leq N < \infty$,

$$S_N(\lambda) \Rightarrow \mathcal{N}(0, a^2 \sigma_N^2 \|\hat{\varphi}\|_2^2). \tag{51}$$

On the other hand, by (48),

$$\Delta_N(\lambda) := E(\lambda^{-3/2} \langle e^\phi, \varphi(\lambda^{-1} \cdot) \rangle - S_N(\lambda))^2$$

$$= a^2 \lambda^3 \sum_{k>N} \frac{1}{k!} d_k(\lambda) \leq C a^2 \|\rho\|_2^2 \|\hat{\varphi}\|_2^2 \sum_{k>N} \frac{1}{k!} \bar{\rho}^{k-2},$$

i.e., $\Delta_N(\lambda)$ can be made arbitrarily small by choosing N sufficiently large, uniformly in $\lambda > 0$. This, together with (51) and the fact that $\sigma_N^2 \to \sigma^2$ $(N \to \infty)$, proves the convergence in (40), and also the theorem. ∎

4.6 Spectral properties of scaling limits for singular initial data

In this section we again study the scaling limit of random fields which are solutions of a multidimensional Burgers equation under Gaussian initial condition with *long-range dependence*. However, this time we concentrate on their spectral properties and provide an explicit formula for the spectral density of the limiting homogeneous Gaussian field. The results are from Leonenko, Parkhomenko and Woyczynski (1996).

Consider the n-dimensional Burgers equation

$$u_t + (u, \nabla)u = \mu \Delta u, \qquad \mu > 0, \tag{1}$$

subject to initial random condition

$$u(0, x) = u_0(x) = \nabla \xi(x) \tag{2}$$

of the gradient form, which describes time-evolution of the velocity field

$$u(t, x) = [u_1(t, x), \dots, u_n(t, x)]', \qquad (t, x) \in [0, \infty) \times \mathbf{R}^n, \quad n \geq 1.$$

The potential $\xi(x), x \in \mathbf{R}^n$, is a scalar field, ∇ denotes the gradient operator on $\mathbf{R}^n$, and Δ stands for the n-dimensional Laplacian.

The spectral representation will be obtained in terms of stochastic integrals of the Gaussian random field which appears as the parabolic scaling limit of the Burgers equation solutions with a *singular initial data*. The latter condition means here that the spectral density is unbounded at zero, or equivalently, that the integral of the correlation function diverges.

The usual Hopf-Cole potential solutions of the initial-value problem (1-2) are given by the functional

$$u(t, x) = \frac{I(t, x)}{J(t, x)}, \tag{3}$$

where

$$I(t, x) = \int_{\mathbf{R}^n} \frac{x - y}{t} g(t, x - y) \exp\{-\xi(y)/(2\mu)\} dy,$$

$$\tag{4}$$

$$J(t, x) = \int_{\mathbf{R}^n} g(t, x - y) \exp\{-\xi(y)/(2\mu)\} dy,$$

and where

$$g(t, x - y) = (4\pi \mu t)^{-n/2} \exp\left\{-\frac{|x - y|^2}{4\mu t}\right\}, \qquad x, y \in \mathbf{R}^n, \quad t > 0, \tag{5}$$

is the Gaussian (heat) kernel.

In all of the results of this paper it is essential that the initial velocity potential $\xi(x)$ satisfies the following assumptions which we put under one umbrella as

Condition A. *The initial velocity potential* $\xi(x) = \xi(x,\omega)$ *is a zero-mean, measurable, mean-square differentiable, homogeneous and isotropic real Gaussian random field on* $\mathbf{R}^n \times \Omega$, *where* $(\Omega, \mathcal{F}, P)$ *is a complete probability space. In addition, its variance* $E\xi^2(x) = 1$, *and its covariance has a singularity at 0 and is of the form*

$$B(|x|) = E\xi(0)\xi(x) = \frac{L(|x|)}{|x|^\alpha}, \qquad 0 < \alpha < n, \quad x \in \mathbf{R}^n,$$

where function $L(t)$, $t > 0$, *is slowly varying for large values of* t, *and bounded on each finite interval. Recall, that* $L : (0,\infty) \mapsto (0,\infty)$ *is said to be slowly varying if, for all* $\lambda > 0$, $\lim_{t\to\infty} L(\lambda t)/L(t) = 1$.

Let $u = u(t,x)$, $(t,x) \in (0,\infty) \times \mathbf{R}^n$, be the solution of the initial-value problem (1-2) with random initial condition satisying condition A. The results of this section concern again the parabolic scaling limit for u, i.e., the limiting behavior of the random field $u(t, a\sqrt{t})$, $a \in \mathbf{R}^n$, when $t \to \infty$. The following basic result is another version of the basic limit theorems discussed earlier in this lecture.

Theorem 1. *Let* $u(t,x)$, $(t,x) \in [0,\infty) \times \mathbf{R}^n$, $n \geq 1$, *be the solution of the initial-value problem (1-2) with random initial data satisfying condition A. Then, the finite-dimensional distributions of the field*

$$X_t(a) = \frac{t^{1/2+\alpha/4}}{L^{1/2}(\sqrt{t})} u(t, a\sqrt{t}), \qquad a \in \mathbf{R}^n, \tag{6}$$

converge weakly, as $t \to \infty$, *to the finite-dimensional distributions of a homogeneous Gaussian random field* $X(a)$, $a \in \mathbf{R}^n$, *with* $EX(a) = 0$ *and the covariance function of the form*

$$R(a,b) = R(a-b) = \left(R_{ij}(a-b)\right)_{1\leq i,j\leq n} = EX(a)X(b)'$$

$$= \frac{1}{(2\mu)^{1+\alpha/2}} \left(\int_{\mathbf{R}^n}\int_{\mathbf{R}^n} \frac{w_i z_j}{|w - z - (a-b)/\sqrt{2\mu}|^\alpha} \phi_n(w)\phi_n(z)\, dw\, dz\right)_{1\leq i,j\leq n},$$

$$\tag{7}$$

where $0 < \alpha < n$, $w = (w_1, \dots, w_n)' \in \mathbf{R}^n$, $z = (z_1, \dots, z_n)' \in \mathbf{R}^n$, $a, b \in \mathbf{R}^n$,

$$\phi_n(w) = \frac{1}{(2\pi)^n} \exp\left\{-\frac{|w|^2}{2}\right\} = \prod_{j=1}^n \phi(w_j), \tag{8}$$

and

$$\phi(w) = \frac{1}{\sqrt{2\pi}} \exp\{-w^2/2\}, \qquad w \in \mathbf{R}^1. \tag{9}$$

The next theorem, the main result of the present section, provides an explicit spectral description of the limiting random field and takes advantage of the spectral representation of random fields in terms of stochastic integrals.

Theorem 2. *Let $u(t, x)$, $(t, x) \in [0, \infty) \times \mathbf{R}^n$, $n \geq 1$, be the solution of the initial-value problem (1-2) with random initial data satisfying condition A. If the random field $\xi(x)$, $x \in \mathbf{R}^n$, has the spectral density $f(\lambda) = f(|\lambda|)$, $\lambda \in \mathbf{R}^n$, i.e.,*

$$\xi(x) = \int_{\mathbf{R}^n} e^{i\langle \lambda, x \rangle} [f(\lambda)]^{1/2} W(d\lambda) \tag{10}$$

where $W(.)$ is the complex Gaussian white noise, and if that density is a decreasing function for $|\lambda| \geq \lambda_0 > 0$, then the assertion of Theorem 2.1 remains true and the limiting Gaussian field $X(a)$, $a \in \mathbf{R}^n$, has the following representation:

$$X(a) = -\frac{1}{i} \left[\frac{\alpha}{c_1(n, \alpha) c_2(n)} \right]^{1/2} \int_{\mathbf{R}^n} e^{i\langle \lambda, a \rangle} g(\lambda) \, W(d\lambda), \tag{11}$$

where

$$g(\lambda) = \lambda \frac{e^{-\mu|\lambda|^2}}{|\lambda|^{(n-\alpha)/2}}, \qquad \lambda \in \mathbf{R}^n, \tag{12}$$

$$c_1(n, \alpha) = 2^\alpha \Gamma \left(1 + \frac{\alpha}{2} \right) \Gamma \left(\frac{n}{2} \right) \Big/ \Gamma \left(\frac{n - \alpha}{2} \right), \quad c_2(n) = 2\pi^{n/2} / \Gamma(n/2), \tag{13}$$

the latter constant being the area of the unit sphere in $\mathbf{R}^n$. In particular, the random field $X(a)$, $a \in \mathbf{R}^n$, is a homogeneous Gaussian random field with mean zero and the covariance function R has a spectral representation

$$R(a - b) = \int_{\mathbf{R}^n} e^{\langle \lambda, a - b \rangle} q(\lambda) \, d\lambda \tag{14}$$

with the matrix-valued spectral density

$$q(\lambda) = \frac{\alpha}{c_1(n, \alpha) c_2(n)} \left(\lambda_r \lambda_j \frac{e^{-2\mu|\lambda|^2}}{|\lambda|^{n-\alpha}} \right)_{1 \leq r, j \leq n}, \tag{15}$$

$\lambda = (\lambda_1, \ldots, \lambda_n)' \in \mathbf{R}^n$, $0 < \alpha < n$.

Remark 1. Condition A implies, via the Tauberian Theorem (see, Leonenko and Olenko (1991)), the following asymptotics of the spectral density:

$$f(\lambda) = f(|\lambda|) \sim \alpha c_1^{-1}(n, \alpha) c_2^{-1}(n) L\left(\frac{1}{|\lambda|}\right) |\lambda|^{\alpha - n}, \qquad 0 < \alpha < n,$$

(16)

as $|\lambda| \to 0$, so that

$$f(|\lambda|) \uparrow \infty, \qquad |\lambda| \to 0. \tag{17}$$

This is the singularity property of the initial Gaussian data. Using (7), (12), and (15), we have

$$g(0) = O, \qquad q(0) = O, \tag{18}$$

where O is the matrix with all entries equal to zero, another display of the singularity of the initial data. So, the parabolically rescaled Hopf-Cole solution (3) of the Burgers equation transforms singularity (17) into singularity (18) and we should remember that for real data the statistical problems for random fields with these two types of singularities are very different.

PROOF OF THEOREM 2. The proof is based on the expansion involving Hermite polynomials

$$H_m(u) = (-1)^m e^{u^2/2} \frac{d^m}{du^m} e^{-u^2/2}, \qquad u \in \mathbf{R}, \quad m = 0, 1, \ldots,$$

which constitute a complete orthogonal system in the Hilbert space $L_2(\mathbf{R}^1, \phi(u)\,du)$, where the function $\phi(u)$ is the Gaussian density defined by (9) (see Section 4.2).

If $G : \mathbf{R}^1 \to \mathbf{R}^1$ is a function such that $EG^2(\xi(0)) < \infty$, for a random variable with the Gaussian density (9) (see condition A in Section 2) then, in $L_2(\mathbf{R}^1, \phi(u)\,du)$, we have the following expansion:

$$G(u) = \sum_{k=0}^{\infty} \frac{C_k}{k!} H_k(u), \qquad C_k = \int_{-\infty}^{\infty} G(u) H_k(u) \phi(u)\,du.$$

In particular, the coefficients of the Hermite expansion of the function

$$G(u) = \exp\left\{-\frac{u}{2\mu}\right\}, \qquad u \in \mathbf{R}^1,$$

are given by the formulas

$$C_0 = \exp\left\{\frac{1}{8\mu^2}\right\}, \qquad C_1 = -\frac{1}{2\mu} \exp\left\{\frac{1}{8\mu^2}\right\},$$

$$C_k = \frac{1}{(2\pi)^{1/2}} \int_{-\infty}^{\infty} \exp\left\{-\frac{u + u^2\mu}{2\mu}\right\} H_k(u)\, du, \qquad k = 3, 4, \ldots \quad (19)$$

In turn, these classical result imply the following expansion in the Hilbert space $L^2(\Omega)$ of random variables with finite second moments (in particular, under Condition A):

$$\exp\left\{-\frac{\xi(y)}{2\mu}\right\} = \sum_{k=0}^{\infty} C_k \frac{H_k(\xi(y))}{k!} \qquad (20)$$

where the C_k's are defined by (19).

Now, consider the random vectors

$$\eta_k(t, a) = \int_{\mathcal{D}_n(t)} \frac{a\sqrt{t} - y}{t} g(t, a\sqrt{t} - y) H_k(\xi(y))\, dy, \qquad k = 0, 1, \ldots,$$

where $\mathcal{D}_n(t) = \{x \in \mathbf{R}^n : |y| \le t\}$, $a, y \in \mathbf{R}^n$. Since (see, e.g., Leonenko, Orsingher and Rybasov (1994))

$$EH_k(\xi(y_1))H_j(\xi(y_2)) = k!\delta_k^j\big(B(|y_1 - y_2|)\big)^k, \qquad k, j \ge 0, \quad (21)$$

where δ_j^k is the Kronecker symbol, we have

$$E\eta_k(t, a)\eta_j(t, a)' = \delta_k^j E\eta_k(t, a)\eta_k(t, a)', \qquad k \ge 1, j \ge 1, a \in \mathbf{R}^n,$$

where

$$E\eta_k(t, a)\eta_k(t, a)' = \big(\psi_{k,i,j}^2(t)\big)_{1\le i,j\le n},$$

and

$$\psi_{k,i,j}^2(t) = k! \int_{\mathcal{D}_n(t)} \int_{\mathcal{D}_n(t)} \frac{a_i\sqrt{t} - y_{1i}}{t} \cdot \frac{a_j\sqrt{t} - y_{2j}}{t} \times$$

$$\times g(t, a\sqrt{t} - y_1) g(t, a\sqrt{t} - y_2) B^k(|y_1 - y_2|)\, dy_1\, dy_2,$$

where $y_1 = (y_{11}, \ldots, y_{1n})'$, $y_2 = (y_{21}, \ldots, y_{2n})'$, and B is the correlation function from Condition A. Changing the variables via the transformation

$$\frac{w_i^2}{2} = \frac{(a_i\sqrt{t} - y_{1i})^2}{4\mu t}, \qquad \frac{z_i^2}{2} = \frac{(a_i\sqrt{t} - y_{2i})^2}{4\mu t}, \qquad i = 1, \ldots, n,$$

and utilizing the basic properties of the slowly varying function L (see, e.g., Ivanov and Leonenko (1989), p. 56), we have, for $0 < \alpha < n/k$, $k \ge 1$, and $t \to \infty$,

$$\psi_{k,i,j}^2(t) = \frac{2\mu k!}{t} \int_{\Delta(t,a)} \int_{\Delta(t,a)} w_i z_j \phi_n(w)\phi_n(z) B^k(\sqrt{2\mu t}|w - z|)\, dz\, dw$$

$$= \frac{2\mu k! L^k(\sqrt{t})}{(2\mu)^{k\alpha/2} t^{1+(k\alpha)/2}} \int_{\Delta(t,a)} \int_{\Delta(t,a)} w_i z_j \phi_n(w) \phi_n(z) \frac{L^k(\sqrt{2\mu t}|w - z|)}{|w - z|^{k\alpha} L^k(\sqrt{t})} dz \, dw$$

$$= (2\mu)^{1-k\alpha/2} k! \kappa_{i,j}(k, \alpha) \frac{L^k(\sqrt{t})}{t^{1+(k\alpha)/2}} (1 + o(1)), \qquad k = 1, 2, \ldots, t \to \infty,$$
(22)

where

$$\Delta(t, a) = \left\{ y \in \mathbf{R}^n : \left| y + \frac{a}{\sqrt{2\mu}} \right| \le \sqrt{\frac{t}{2\mu}} \right\}, \qquad a \in \mathbf{R}^n, \ t > 0,$$

and

$$\kappa_{i,j}(k, \alpha) = \int_{\mathbf{R}^n} \int_{\mathbf{R}^n} \frac{w_i z_j \phi_n(w) \phi_n(z)}{|w - z|^{k\alpha}} dw \, dz, \qquad i, j, = 1, \ldots, n.$$

From (3-4), we have

$$e^{-1/(8\mu^2)} J(t, a\sqrt{t}) = e^{-1/(8\mu^2)} \int_{\mathbf{R}^n} g(t, a\sqrt{t} - y) e^{-\xi(y)/(2\mu)} \, dy \xrightarrow{P} 1, \quad (23)$$

From (4) and (20) we can write

$$I(t, a\sqrt{t}) = \int_{\mathbf{R}^n} \frac{a\sqrt{t} - y}{t} g((t, a\sqrt{t} - y) e^{-\xi(y)/(2\mu)} \, dy$$

$$= C_0 \eta_0(t, a) + C_1 \eta_1(t, a) + \sum_{k \ge 2} \frac{C_k}{k!} \eta_k(t, a) + R_t, \qquad (24)$$

where

$$R_t = \int_{\mathbf{R}^n \setminus \mathcal{D}_t} \frac{a\sqrt{t} - y}{t} g((t, a\sqrt{t} - y) e^{-\xi(y)/(2\mu)} \, dy.$$

We note that

$$\lim_{t \to \infty} C_0 \eta_0(t, a) = 0.$$

Random vectors

$$\frac{t^{1/2+\alpha/4}}{L^{1/2}(\sqrt{t})} \left[\sum_{k \ge 2} \frac{C_k}{k!} \eta_k(t, a) + R_t \right] \xrightarrow{P} 0, \qquad t \to \infty. \quad (25)$$

Hence, from (22-25) and by Slutsky's argument (see, for example, Leonenko et al. (1994b), Lemma 1),

$$\tilde{X}_t(a) = \frac{t^{1/2+\alpha/4}}{L^{1/2}(\sqrt{t})} e^{-1/(8\mu^2)} C_1 \eta_1(t, a) \xrightarrow{d} X(a),$$

and
$$X_t(a) \xrightarrow{d} X(a), \qquad a \in \mathbf{R}^n,$$

as $t \to \infty$, where $\xrightarrow{d}$ means convergence in distributions of random vectors and $X(a)$, $a \in \mathbf{R}^n$, is a vector homogeneous Gaussian field with zero mean and covariance matrix (7).

Let us show that the homogeneous field $X(a), a \in \mathbf{R}^n$, has the spectral representation (11) and spectral density (15).

Using the self-similar propertiy
$$W(d(c\lambda)) \overset{d}{=} \sqrt{c^n} W(d\lambda), \qquad c > 0,$$

of the Gaussian white noise we can write, in view of (10), that

$$\tilde{X}_t(a) \overset{d}{=} -\frac{t^{1/2+\alpha/4}}{2\mu L^{1/2}(\sqrt{t})} \int_{\mathbf{R}^n} \left[\int_{\mathcal{D}_t} \frac{a-y}{(4\pi\mu)^{n/2}} e^{-|a-y|^2/(4\mu)} e^{i\langle\lambda,y\rangle} \, dy \right] \times$$

$$\times \sqrt{f(|\lambda|/\sqrt{t})} t^{-n/4-1/2} W(d\lambda). \qquad (26)$$

Using the fact that

$$\nabla \int_{\mathbf{R}^n} \frac{\exp\{i\langle y, x\rangle - |x|^2/(2\sigma^2)\}}{(2\pi)^{n/2}\sigma^n} \, dx = \nabla \exp\{-\sigma^2|y|^2/2\},$$

after some transformations, we have (setting $\sigma^2 = 2\mu$)

$$\int_{\mathbf{R}^n} \frac{\exp\{i\langle\lambda, z\rangle - |z-a|^2/(4\mu)\}}{(4\pi\mu)^{n/2}} (a-z) \, dz = \frac{2\mu y}{i} e^{i\langle\lambda,a\rangle - \mu|\lambda|^2}. \qquad (27)$$

Then, with the help of (27), (11-12) and (26), we obtain the following estimate:

$$E|\tilde{X}_t(a) - X(a)|^2$$

$$= \frac{t^{1+\alpha/2}}{L(\sqrt{t})} E \left| -\int_{\mathbf{R}^n} \frac{\lambda \exp\{i\langle\lambda, a\rangle - \mu|\lambda|^2\}}{i t^{n/4+1/2}} \sqrt{f\left(\frac{|\lambda|}{\sqrt{t}}\right)} W(d\lambda) \right.$$

$$+ \int_{\mathbf{R}^n} \left[\int_{\mathbf{R}^n\backslash\mathcal{D}_t} \frac{(a-z) \exp\{i\langle\lambda, z\rangle - |a-z|^2/(4\mu)\}}{2\mu(4\pi\mu)^{1/2} t^{n/4+1/2}} dz \right] \sqrt{f\left(\frac{|\lambda|}{\sqrt{t}}\right)} W(d\lambda)$$

$$+ \int_{\mathbf{R}^n} \frac{\lambda \exp\{i\langle\lambda, a\rangle - \mu|\lambda|^2\} L^{1/2}(\sqrt{t})}{i t^{n/4+1/2}} \left[\frac{\alpha}{c_1(n,\alpha)c_2(n)} \right]^{1/2} \frac{W(d\lambda)}{|\lambda|^{(n-\alpha)/2}} \right|^2$$

$$\leq \frac{t^{1+\alpha/2}}{L(\sqrt{t})} \int_{\mathbf{R}^n} \left| \frac{\lambda \exp\{i\langle \lambda, a\rangle - \mu|\lambda|^2\}}{it^{n/4+1/2}} \sqrt{f\left(\frac{|\lambda|}{\sqrt{t}}\right)} \right.$$

$$- \frac{\lambda \exp\{i\langle \lambda, a\rangle - \mu|\lambda|^2\} L^{1/2}(\sqrt{t})}{it^{n/4+1/2}} \left[\frac{\alpha}{c_1(n,\alpha)c_2(n)} \right]^{1/2} \frac{1}{|\lambda|^{(n-\alpha)/2}}$$

$$- \left. \int_{\mathbf{R}^n\setminus\mathcal{D}_t} \frac{(a-z)\exp\{i\langle \lambda, z\rangle - |a-z|^2/(4\mu)\}}{2\mu(4\pi\mu)^{1/2}t^{n/4+1/2}} dz \sqrt{f\left(\frac{|\lambda|}{\sqrt{t}}\right)} \right|^2 d\lambda$$

$$\leq \left[\int_{\mathbf{R}^n} \left| \frac{t^{1/2+\alpha/4}}{L^{1/2}(\sqrt{t})} \frac{\lambda \exp\{i\langle \lambda, a\rangle - \mu|\lambda|^2\}}{it^{n/4+1/2}} \sqrt{f\left(\frac{|\lambda|}{\sqrt{t}}\right)} \right.\right.$$

$$- \left.\left. \left[\frac{\alpha}{c_1(n,\alpha)c_2(n)} \right]^{1/2} \frac{\lambda \exp\{i\langle \lambda, a\rangle - \mu|\lambda|^2\}}{|\lambda|^{(n-\alpha)/2}} \right|^2 d\lambda \right]$$

$$+ \left[\frac{t^{1+\alpha/2}}{L(\sqrt{t})2\mu} \int_{\mathbf{R}^n} \left| \int_{\mathbf{R}^n\setminus\mathcal{D}_t} \frac{(a-z)\exp\{i\langle \lambda, z\rangle - |a-z|^2/(4\mu)\}}{(4\pi\mu)^{n/2}t^{n/4+1}} dz \right|^2 \times \right.$$

$$\left. \times \sqrt{f\left(\frac{|\lambda|}{\sqrt{t}}\right)} d\lambda\, dz \right] = \Sigma_1(t) + \Sigma_2(t). \tag{28}$$

Function

$$h(\lambda) = |\lambda|^2 \exp\{2i\langle \lambda, a\rangle - 2\mu|\lambda|^2\}/|\lambda|^{n-\alpha}, \qquad \lambda \in \mathbf{R}^n,$$

is absolutely integrable. Using the Tauberian theorem (16) and the properties of slowly varying functions we obtain from (28) that

$$\lim_{t\to\infty} \Sigma_1(t) = \lim_{t\to\infty} \int_{\mathbf{R}^n} h(\lambda) Q_t(\lambda)\, d\lambda = 0,$$

where

$$Q_t(\lambda) \sim \frac{\alpha}{c_1(n,\alpha)c_2(n)} \left| \frac{L^{1/2}(\sqrt{t/|\lambda|})}{L^{1/2}(\sqrt{t})} - 1 \right|^2,$$

as $t \to \infty$.

For $\Sigma_2(t)$ we have the estimate

$$\Sigma_2(t) \leq \frac{1}{L(\sqrt{t})t^{(n-\alpha)/2}} \int_{\mathbf{R}^n} f(|\lambda|/\sqrt{t})\, d\lambda \left| \int_{\mathbf{R}^n} \frac{(a-z)e^{-|a-z|^2/(4\mu)}}{(4\pi\mu)^{n/2}} dz \right|^2$$

$$\leq \frac{\text{const}}{L(\sqrt{t})t^{(n-\alpha)/2}},$$

so that

$$\lim_{t\to\infty} \Sigma_2(t) = 0,$$

and

$$\lim_{t\to\infty} E|\tilde{X}_t(a) - X(a)|^2 = 0,$$

where $X(a), a \in \mathbf{R}^n$, is defined by (2.9-10).

Applying the Cramer-Wold arguments we conclude the proof of Theorem 2. The formula (7) also follows from (22). ∎

Lecture 5
Hyperbolic Approximation and Inviscid Limit

5.1 Hyperbolic scaling limit

In this section we again consider large time asymptotics of statistical solution $u(t, x)$ of the Burgers equation

$$u_t + uu_x = \nu u_{xx}, \tag{1}$$

$$t > 0, \quad x \in \mathbf{R}, \quad u = u(t, x), \quad u(0, x) = u_0(x)$$

but under a different scaling regime. The initial velocity potential

$$U_0(x) = \int_{-\infty}^{x} u_0(y)dy = \xi(x) = \xi_L(x) \sim \sigma_L \eta(x/L), \tag{2}$$

where $\eta(x)$ is assumed to be a stationary zero-mean Gaussian process, L—the "internal scale" parameter, and

$$\sigma_L = L^2(2\log L)^{1/2} \tag{3}.$$

We will show that as $L \to \infty$ the *hyperbolically scaled* random fields $u(L^2 t, L^2 x)$ converge in distribution to a random field with "sawtooth" trajectories defined by means of a planar Poisson process related to high fluctuations of $\xi(x)$. Such a scaling limit corresponds to *zero viscosity limit* solutions. The results of this section are taken from Molchanov, Surgailis and Woyczynski (1995), and Surgailis and Woyczynski (1995). For a degenerate shot noise process $\xi(x)$, see Albeverio, Molchanov, Surgailis (1994).

At the physical level of rigor, Kraichnan (1968), and Fournier, Frisch (1983), and Gurbatov, Malakhov, Saichev (1991) discussed asymptotics of $u(t, x)$ at high Reynolds numbers, in the case when the initial Gaussian data $\xi(x)$ are characterized by large "amplitude" $\sigma = (E(\xi(0))^2)^{1/2}$ and large "internal scale" $L = \sigma/\sigma' >> 1$, where $\sigma' = (E(\xi'(0))^2)^{1/2}$. They demonstrated (at the physical level of rigor) that "[...] *a strongly nonlinear regime of sawtooth waves* [...] *is set up,* [...] *and the field's statistical properties become self-preserving*" Also found one- and two-point distribution functions of the (limit) sawtooth velocity process.

The results rely on the Hopf-Cole explicit solution

$$u(t, x) = \frac{\int_{-\infty}^{\infty} [(x - y)/t] \exp[(2\mu)^{-1}(\xi(y) - (x - y)^2/2t)] dy}{\int_{-\infty}^{\infty} \exp[(2\mu)^{-1}(\xi(y) - (x - y)^2/2t)] dy}. \qquad (4)$$

In our rigorous set-up the basic tool is the *extremal theory* of stationary stochastic processes. We begin with a zero-mean stationary differentiable Gaussian $\xi_L(x) = \sigma_L \eta(x/L)$, with $\sigma_L = L^2\sqrt{2\log L}$. as defined in (2-3). Asymptotics of σ_L is dictated by the standard normalization constant in the extremal theory of Gaussian processes, and the scaling properties of the Hopf-Cole functional. Then

$$\xi'(x) = (\sigma_L/L)\eta(x/L)$$

and the "internal scale"

$$\frac{[E(\xi(x))^2]^{1/2}}{[E(\xi'(x))^2]^{1/2}} = L$$

is expressed by parameter L. Studying the solutions at large "internal scales" will mean letting $L \to \infty$.

Formal assumptions: Covariance function $r(x)$ of the process $\eta(x), x \in \mathbf{R}$, satisfies the following two conditions:

$$r(x) = o(1/\log x) \quad (x \to \infty), \qquad (5)$$

and

$$r(x) = 1 - \frac{1}{2!}\lambda_2 x^2 + \frac{1}{4!}\lambda_4 x^4 + o(x^4) \quad (x \to 0). \qquad (6).$$

Then, our main result can be formulated as follows.

Theorem 1. *Let $u(t, x)$ be the solution (4) of the Burgers equation (1) with the initial datum $\xi(x) = \xi_L(x), x \in \mathbf{R}$, of the form (2-3) and satisfying conditions (5) and (6). Then, as $L \to \infty$, the finite dimensional distributions of $u(L^2 t, L^2 x), (t, x) \in \mathbf{R}_+ \times \mathbf{R}$, tend to the corresponding distributions of the random field*

$$v(t, x) = \frac{x - y_{j^*(t,x)}}{t}. \qquad (7)$$

Here, $y_{j^*(t,x)} \equiv y_{j^*}$ is the ordinate of the point of a Poisson process $(u_j, y_j)_{j \in \mathbf{Z}}$ on $\mathbf{R}^2$, with intensity $e^{-u} du\,dy$, which maximizes $u_j - (x - y_j)^2/2t$, i.e.

$$u_{j^*} - \frac{(x - y_{j^*})^2}{2t} = \max_j \left(u_j - \frac{(x - y_j)^2}{2t} \right). \tag{8}$$

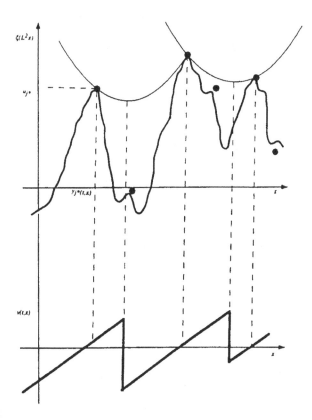

Fig. 5.1.1. Points (y_j, u_j) of the Poisson process (marked by •) correspond to high local maxima of the smooth Gaussian curve $\xi(x)$. Critical parabolas define discontinuity points and zeros of the limit velocity process $v(t, x)$.

Remarks: The qualitative and heuristic meaning of Theorem 1 which is illustrated in Fig. 5.1.1, can be found in the original Burgers asymptotic analysis of the Hopf-Cole formula which was discussed in Section 3.3. Also note that the limit random field $v(t, x)$ is independent of viscosity ν, its shape is what one sees for the Burgers equation in the zero viscosity limit. Indeed, the hyperbolic limit procedure is equivalent to taking the zero viscosity limit.

SKETCH OF THE PROOF. Consider only the convergence of 1-D distributions of $u(L^2 t, L^2 x)$ for $\mu = t = 1/2, x = 0$.

Put

$$H_L := u(L^2/2, 0),$$

$$a_L := \sigma_L/L^2 = \sqrt{2 \log L}, \tag{9}$$

$$b_L := \sqrt{2 \log L} + \frac{c_1}{\sqrt{2 \log L}}, \tag{10}$$

where $c_1 = \log(\sqrt{\lambda_2}/2\pi)$. According to (2) and (4),

$$H_L = -2 \frac{\int_{\mathbf{R}} y \exp[L^2(\eta_L(y) - y^2)] dy}{\int_{\mathbf{R}} \exp[L^2(\eta_L(y) - y^2)] dy} \tag{11}$$

where

$$\eta_L(y) = a_L(\eta(Ly) - b_L). \tag{12}$$

Let $y_j^{(\eta_L)}, u_j^{(\eta_L)} = \eta_L(y_j^{(\eta_L)})$ be positions and heights of local maxima of $\eta_L(x), x \in \mathbf{R}$, respectively. By (6), their number is a.s. finite on any finite interval. Let

$$u_{j*}^{(\eta_L)} - (y_{j*}^{(\eta_L)})^2 = \max_j \left(u_j^{(\eta_L)} - (y_j^{(\eta_L)})^2 \right). \tag{13}$$

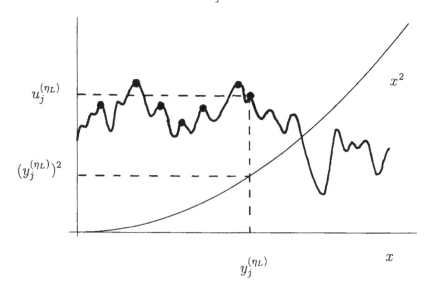

Fig. 5.1.2. The determination of the random point $y_j^{(\eta_L)}$ as the random point in the Poisson point process which minimizes the distance to parabola x^2.

Now, put

$$I(\Delta_{j^*}^{(\eta_L)}) = \int_{\Delta_{j^*}^{(\eta_L)}} \exp\left[L^2(\eta_L(y) - y^2)\right] dy, \tag{14}$$

where

$$\Delta_{j^*}^{(\eta_L)} = \left\{ y \in \mathbf{R} : |y - y_{j^*}^{(\eta_L)}| < 1/La_L \right\}. \tag{15}$$

Then, H_L of (11) can be written as

$$H_L = -2\frac{y_{j^*}^{(\eta_L)} + R_L + \rho_L}{1 + Q_L},$$

where

$$R_L = \int_{\mathbf{R}\backslash\Delta_{j^*}^{(\eta_L)}} y \exp\left[L^2(\eta_L(y) - y^2)\right] dy \;/\; I(\Delta_{j^*}^{(\eta_L)}), \tag{16}$$

$$Q_L = \int_{\mathbf{R}\backslash\Delta_{j^*}^{(\eta_L)}} \exp\left[L^2(\eta_L(y) - y^2)\right] dy \;/\; I(\Delta_{j^*}^{(\eta_L)}), \tag{17}$$

and

$$\rho_L = \int_{\Delta_{j^*}^{(\eta_L)}} \left(y - y_{j^*}^{(\eta_L)}\right) \exp\left[L^2(\eta_L(y) - y^2)\right] dy \;/\; I(\Delta_{j^*}^{(\eta_L)}). \tag{18}$$

Clearly, the convergence in distribution

$$H_L \Rightarrow v(1/2, 0) = -2y_{j^*}, \tag{19}$$

follows from the facts that

$$y_{j^*}^{(\eta_L)} \Rightarrow y_{j^*} \equiv y_{j^*(1/2,0)}, \tag{20}$$

$$R_L \Rightarrow 0, \tag{21}$$

$$Q_L \Rightarrow 0, \tag{22}$$

and from the trivial bound $|\rho_L| < 2/La_L \to 0 \; (L \to \infty)$.

• The proof requires a study of the Poisson convergence of functionals of a Gaussian trajectory near high local maxima.

• To prove (20) we need a criterion for convergence of the point process $(u_j^{(\eta_L)} y_j^{(\eta_L)})_{j \in \mathbf{Z}}$ in a topology matched to the Burgers equation.

Poisson convergence of local maxima. Denote by $\mathcal{M}$ the space of all locally finite point measures on $\mathbf{R}^2$, with the topology of vague convergence of measures, denoted by $\to$, by $\bar{\mathcal{M}}$ the space of all locally finite point measures on $\mathbf{R}^2$, taking values in $C[-1,1]$ with the sup norm $\|\,.\,\|$. The elements $\bar{\nu} \in \bar{\mathcal{M}}$ are identified with sequences

$$\bar{\nu} = (u_j, y_j, g_j)_{j \in \mathbf{Z}}, \tag{23}$$

where $(u_j, y_j) \in \mathbf{R}^2$ and $g_j \in C[-1,1]$, $j \in \mathbf{Z}$. Moreover,

$$\bar{\nu}_L \to \bar{\nu} \Leftrightarrow \nu_L \to \nu \text{ (in } \mathcal{M}) \quad \& \quad \|g_{j,L} - g_j\| \to 0 \tag{24}$$

for any $j \in \mathbf{Z}$. The space $\bar{\mathcal{M}}$, as well as $\mathcal{M}$, are complete, metrizable. The same notation $\Rightarrow$ is employed for the weak convergence of random elements from $\mathcal{M}$, $\bar{\mathcal{M}}$, and/or from a finite dimensional Euclidean space.

Introduce two point processes associated with the Gaussian process $\eta_L(x)$ of (12):

$$\nu^{(\eta_L)} = (u_j^{(\eta_L)}, y_j^{(\eta_L)})_{j \in \mathbf{Z}} \in \mathcal{M}$$

of local maxima, and

$$\bar{\nu}^{(\eta_L)} = (u_j^{(\eta_L)}, y_j^{(\eta_L)}, g_j^{(\eta_L)})_{j \in \mathbf{Z}} \in \bar{\mathcal{M}} \tag{25}$$

which includes the "germs" $g_j^{(\eta_L)}(\cdot) \in C[-1,1]$ of the sample path near local maxima, where, for $y \in [-1,1]$,

$$\begin{aligned} g_j^{(\eta_L)}(y) &= \eta_L(y_j^{(\eta_L)} + y/La_L) - \eta_L(y_j^{(\eta_L)}) \tag{26} \\ &= \eta_L(y_j^{(\eta_L)} + y/La_L) - u_j^{(\eta_L)}. \end{aligned}$$

Lemma 1. *The point process*

$$\bar{\nu}^{(\eta_L)} \Longrightarrow \bar{\nu}, \tag{27}$$

as $L \to \infty$, where $\bar{\nu} = (u_j, y_j, g_j)_{j \in \mathbf{Z}}$, with $\nu = (u_j, y_j)_{j \in \mathbf{Z}}$ being the Poisson process of Theorem 1, and

$$g_j(y) \equiv g(y) = -\frac{1}{2}\lambda_2 y^2, \quad y \in [-1,1] \tag{28}$$

being a deterministic parabola.

The proof of Lemma 1 is based on the Poisson limit theorem for maxima (see e.g. Leadbetter, Lindgren, Rootzen (1983), Theorem 9.5.2.) and the Slepian model process representation near a local maximum (due to Lindgren (1970)). It immediately yields the following lower bound for the exponential integral in (14).

Corollary 1. *For any compact $A \subset \mathbf{R}^2$, and any $\varepsilon, \delta > 0$, there exists an $L_0 < \infty$ such that, for every $L > L_0$,*

$$P\left[I(\triangle_{j^*}^{(\eta_L)}) < \frac{\exp[L^2(u_{j^*}^{(\eta_L)} - (y_{j^*}^{(\eta_L)})^2 - \delta)]}{\sqrt{e}L^2 a_L}, (u_{j^*}^{(\eta_L)}, y_{j^*}^{(\eta_L)}) \in A\right] < \varepsilon. \tag{29}$$

Burgers' topology on point measures. Fix $\alpha_0, \beta_0 \geq 0$. By definition, $\mathcal{B} = \mathcal{B}_{\alpha_0, \beta_0} \subset \mathcal{M}$ consists of all $\nu \in \mathcal{M}$ such that for any $\alpha > \alpha_0$, $\beta > \beta_0$,

$$I_{\alpha,\beta}(\nu) := \int_{\mathbf{R}^2} e^{\alpha u - \beta y^2} d\nu < \infty. \tag{30}$$

Definition 1. Let $\nu_L, \nu \in \mathcal{B}$. We shall say that $\nu_L \overset{\mathcal{B}}{\to} \nu$ as $L \to \infty$, if $\nu_L \to \nu$ and

$$I_{\alpha,\beta}(\nu_L) \to I_{\alpha,\beta}(\nu), \qquad L \to \infty, \tag{31}$$

for any $\alpha > \alpha_0$, $\beta > \beta_0$.

The convergence $\overset{\mathcal{B}}{\to}$ defines a topology in $\mathcal{B}$ which henceforth will be called the Burgers topology ($\mathcal{B}$-topology).

Compactness Criterion. *$\mathcal{B}$ is a complete separable metrizable space in the $\mathcal{B}$-topology. A Borel set $A \subset \mathcal{B}$ is compact in the $\mathcal{B}$-topology if, and only if, A is compact in the vague topology, and for any $\alpha > \alpha_0$, $\beta > \beta_0$,*

$$\sup_{\nu \in A} I_{\alpha,\beta}(\nu) < \infty. \tag{31}$$

Denote by $\mathbf{P}(\mathcal{M}), \mathbf{P}(\mathcal{B})$ the family of probability measures on $\mathcal{M}$, $\mathcal{B}$. The arrow $\Rightarrow$ and $\overset{\mathcal{B}}{\Rightarrow}$ stands for the weak convergence of probability measures on, or random elements in, $\mathcal{M}$ and $\mathcal{B}$, respectively.

Convergence Criterion. *Let $P_L, P \in \mathbf{P}(\mathcal{B})$. Then $P_L \overset{\mathcal{B}}{\Rightarrow} P$ if, and only if,*

    (i) $P_L \Rightarrow P$, and

    (ii) $P_L \circ I_{\alpha,\beta}^{-1} \Rightarrow P \circ I_{\alpha,\beta}^{-1}$, $\forall \alpha > \alpha_0, \forall \beta > \beta_0$.

The necessity of *(i)*, *(ii)* for $P_L \overset{\mathcal{B}}{\Rightarrow} P$ is easy. In particular, *(ii)* follows from the fact that $I_{\alpha,\beta}(\nu)$ is continuous on $\mathcal{B}$. To prove the converse part, it suffices to show that $\{P_L\}$ is tight in $\mathbf{P}(\mathcal{B})$. This takes little work.

Let us now return to the point processes from Lemma 1. Fix $\alpha_0 = 1$, $\beta_0 = 0$, so that

$$\mathcal{B} = \mathcal{B}_{1,0} = \{\nu \in \mathcal{M} : I_{\alpha,\beta}(\nu) < \infty, \quad \forall \alpha > 1, \beta > 0\}.$$

Proposition 1. *For any $L > 0$, the point processes $\nu^{(\eta_L)}$ and ν of Section 3 belong to $\mathcal{B}$ a.s. and , as $L \to \infty$,*

$$\nu^{(\eta_L)} \overset{\mathcal{B}}{\Longrightarrow} \nu. \tag{32}$$

In Proposition 2 we use Proposition 1 to prove the convergence (20) of the local maximum point $(u_{j^*}^{(\eta_L)}, y_{j^*}^{(\eta_L)})$.

Proposition 2. *As $L \to \infty$,*

$$(u_{j^*}^{(\eta_L)}, y_{j^*}^{(\eta_L)}) \Rightarrow (u_{j^*}, y_{j^*}). \tag{33}$$

In particular,

$$u_{j^*}^{(\eta_L)} - (y_{j^*}^{(\eta_L)})^2 \Rightarrow u_{j^*} - y_{j^*}^2. \tag{34}$$

Now, the proof of Theorem 1 can be completed by considering the set $\mathcal{B}_0 = \mathcal{B} \cap \mathcal{M}_0$, where $\mathcal{M}_0$ is the set of all *simple* point measures $\nu \in \mathcal{M}$. Each $\nu \in \mathcal{B}_0$ can be identified with its range, that is, a locally finite, countable set $\cup_{j \in \mathbf{Z}}\{(u_j, y_j)\}$.

Consider also the functional

$$h(\nu) = (u_{j^*}, y_{j^*}), \qquad \nu = (u_j, y_j)_{j \in \mathbf{Z}}, \tag{35}$$

with values in $\mathbf{R}^2$, where $j^* = j^*(1/2, 0)$. Then $h(\nu)$ is a.e. well defined and continuous on $\mathcal{B}_0$ in the $\mathcal{B}$-topology, with respect to the Poisson measure of Theorem 1. Moreover, for a.e. $\nu = (u_j, y_j)_{j \in \mathbf{Z}} \in \mathcal{B}_0$, the maximum on the right hand side of (8) is attained at a single point $(u_{j^*}, y_{j^*}) \in \nu$. Hence, if $\nu_L \overset{\mathcal{B}}{\to} \nu$, $\nu_L = (u_{j,L}, y_{j,L})_{j \in \mathbf{Z}}$, then $h(\nu_L) \to h(\nu)$, i.e., the functional $h(\nu)$ is a.e. $\mathcal{B}$-continuous. ∎

Remarks. The above rigorous result is available at this point only for one-dimensional Burgers flows. It is an important open problem for multidimensional flows and its solution depends on some yet unanswered questions in the theory of maxima of Gaussian random fields (see, e.g., Sun (1993), for recent progress in the area). Heuristically, and on the physical level of rigor (confirmed by computer simulations), the pictures is relatively clear. The shock fronts in the inviscid limit form an evolving *Voronoi tessellation* in space with high peaks of the initial potential random field serving as tessellation's seeds. The size of the average cell increases and larger cells "swallow" smaller cells. We will discuss this scenario in more detail in Lectures 6 and 7 in the context of passive tracer transport in Burgers' turbulence, and for forced Burgers' flows.

5.2   Densities and correlations of the limit velocity field

The fundamental Theorem 5.1.1 permits approximation of finite-dimensional distributions of the solution random field $u(t, x)$ by finite-dimensional distributions of the standard random field $v(x, t)$ with sawtooth-like trajectories. For the latter, an explicit calculation of the joint distributions of

$$(v(t_1, x_1), \ldots, v(t_n, x_n)) \tag{1}$$

is possible. Observe, that they are *not* absolutely continuous in $\mathbf{R}^n$ but rather a sum of absolutely continuous distributions on some k-dimensional hyperplanes of $\mathbf{R}^n$, $1 \le k \le n$, because $P[y_{j^*(t_i,x_i)} = y_{j^*(t_j,x_j)}] > 0$ for $i \ne j$.

Indeed, the distribution of (1), and n-point correlation function,

$$\rho^{(n)}(t_1, x_1, \ldots, t_n, x_n) = \mathbf{E}v(t_1, x_1) \ldots v(t_n, x_n)$$

can be obtained from the distribution

$$P^*(\, \cdot \, ; (t, x)_n) = \mathbf{P}[(y^*)_n \in \, \cdot \,] \tag{2}$$

of the random vector

$$(y^*)_n = (y_1^*, \ldots, y_n^*), \tag{3}$$

where

$$y_i^* = y_{j_i^*} \equiv y_{j^*(t_i, x_i)},$$

$i = 1, \ldots, n$, and we use the notation

$$(y)_n = (y_1, \ldots, y_n) \in \mathbf{R}^n,$$

$$(t, x)_n = ((t_1, x_1), \ldots, (t_n, x_n)) \in (\mathbf{R}_+ \times \mathbf{R})^n.$$

In particular,

$$\rho^{(n)}(t, x)_n = \int_{\mathbf{R}^n} \prod_{i=1}^n \frac{x_i - y_i}{t_i} P^*(d(y)_n; (t, x)_n). \tag{4}$$

We have

$$P^*(\, \cdot \, ; (t, x)_n) = \sum_{(A)_m} P^*_{(A)_m}(\, \cdot \, ; (t, x)_n), \tag{5}$$

where $(A)_m = (A_1, \ldots, A_m)$ — partitions of $\{1, \ldots, n\}$, and $P^*_{(A)_m}(d(y)_m; (t, x)_n)$ is a measure on $\mathbf{R}^m$ which can be identified with the distribution of $(y^*)_n$ on the m-dimensional hyperplane

$$y_i^* = y_k, \quad i \in A_k, \quad k = 1, \ldots, m. \tag{6}$$

Observation: The last event occurs if, and only if, for every $k = 1, \ldots, m$, and any Poisson point $(u_j, y_j), j \neq k$,

$$u_j < \bigwedge_{i \in A_k} g_{ik}(y_j), \tag{7}$$

where

$$g_{ik}(y) = u_k + \frac{1}{2t_i}((y - x_i)^2 - (y_k - x_i)^2) \tag{8}$$

is the parabola going through the point (u_k, y_k) and "centered" at $x_i, i \in A_k$.

From (7), for each partition $(A)_m$, measure $P^*_{(A)_m}(\ \cdot\ ; (t, x)_n)$ has density

$$p^*_{(A)_m}((y)_m; (t, x)_n) \tag{9}$$

$$= \int_{W(y)_m} \exp\Big[-\sum_{i=1}^m u_i - \int_{\mathbf{R}} \bigvee_{k=1}^m \bigvee_{i \in A_k} e^{-g_{ik}(z)} dz\Big] d(u)_m,$$

where

$$W(y)_m = \{(u)_m \in \mathbf{R}^m : u_j < \bigwedge_{i \in A_k} g_{ik}(y_j), \forall j \neq k, \ j, k = 1, \ldots, m.\} \tag{10}$$

For $n = 1, 2$, formulas (4), (9) can be made much more explicit.

$$p^*(y; (t, x)) = (1/\sqrt{2\pi t})e^{-(y-x)^2/2t},$$

i.e., $y_{j^*(t,x)}$ is Gaussian with parameters x, t, and consequently, $v(t, x)$ is Gaussian with parameters $0, 1/t$.

For $n = 2, (t, x)_2 = ((t_1, x_1), (t_2, x_2))$, there are two partitions of $\{1, 2\}$, namely, $(A)_1 = \{1, 2\}$, and $(A)_2 = (\{1\}, \{2\})$.

From (9), we have

$$p_1^*(y) \equiv p^*_{(A)_1}(y; (t, x)_2) = [B_1(0; y, y) + B_2(0; y, y)]^{-1}, \tag{11}$$

and

$$p_2^*(y_1, y_2) \equiv p^*_{(A)_2}(y_1, y_2; (t, x)_2)$$

$$= \int_{a_1}^{a_2} [e^{-v/2} B_1(v; y_1, y_2) + e^{v/2} B_2(v; y_1, y_2)]^{-2} dv, \tag{12}$$

where

$$a_i = \frac{1}{2t_i}((y_2 - x_i)^2 - (y_1 - x_i)^2),$$

$$B_i(v; y_1, y_2) = e^{(y_i - x_i)^2/2t_i} \int_{C_i(v; y_1, y_2)} e^{-(z - x_i)^2/2t_i} dz,$$

$$C_1(v; y_1, y_2) =$$

$$\left\{ z \in \mathbf{R} : \frac{(z - y_1)(z + y_1 + 2x_1)}{2t_1} - \frac{(z - y_2)(z + y_2 + 2x_2)}{2t_2} < v \right\}$$

and

$$C_2(v; y_1, y_2) = \mathbf{R} \setminus C_1(v; y_1, y_2).$$

The space-time covariance

$$\rho^{(2)}(t_1, x_1; t_2, x_2) = Ev(t_1, x_1)v(t_2, x_2)$$

For $t_1 < t_2$

$$\rho^{(2)}(t_1, x_1; t_2, x_2) = \frac{1}{t_1 t_2} \int_{-\infty}^{\infty} (z - x_1)(z - x_2) A^{-1}(z; x_2, x_2) dz$$

$$- \frac{t_2 - t_1}{t_1 t_2} \int_{-\infty}^{\infty} |z|(1 - e^{-2(x_2 - x_1)z/(t_2 - t_1)})^2 A^{-2}(z; x_1, x_2) dz,$$

where

$$A(z; x_1, x_2) = e^{(z - x_1)^2/2t_1} \int_{|y| < |z|} e^{-(y - x_1)^2/2t_1} dy$$

$$+ e^{(z - x_2)^2/2t_2} \int_{|y| > |z|} e^{-(y - x_2)^2/2t_2} dy.$$

For fixed time $(t_1 = t_2 \equiv t)$ covariance was obtained in Gurbatov, Malakhov, Saichev (1991) p. 181, and is somewhat simpler, namely

$$\rho^{(2)}(t, x_1; t, x_2) = \frac{1}{t} \frac{d}{dx}(x P_t(x)),$$

where

$$P_t(2x) = (1/\sqrt{2\pi t}) \int_{-\infty}^{\infty} [e^{(x+z)^2/2t} \Phi_t(x + z) + e^{(x-z)^2/2t} \Phi_t(x - z)]^{-1} dz,$$

is the probability that the points $x_1, x_2, x_2 - x_1 = 2x$, belong to the same line segment of continuity of the sawtooth process $v(t, x)$.

5.3   Statistics of shocks

In this section we evaluate the expected *density of shock fronts* in 1 - D
Burgers turbulence (5.1.1) for stationary initial velocity potential data.
It can decrease with time as $t^{-\alpha}$ ($1/2 \le \alpha < \infty$). A new interpretation
of α is given. Our principal objective is to use the results on limit distributions for the velocity fields in the Gaussian and shot noise initial
velocity potential models obtained in Section 5.1-2 and in Albeverio,
Molchanov, Surgailis (1994) (see also Surgailis and Woyczynski (1994))
to calculate the temporal rate of change of shock density. Such a rate
was of course of long standing interest for fluid dynamicists (see e.g.
Tatsuma and Kida (1972)) and we compare our method with their approach. We should also mention that, in the case of some nonstationary
and nonsmooth initial velocity potential, Sinai (1992) has calculated the
Hausdorff dimension of the set of shocks. We'll get back to this question later on. The material of this section is due to Hu and Woyczynski
(1996).

The following result complements the hyperbolic scaling limit result from Theorem 5.1.1 and deals with the initial conditions on the
opposite, dicrete end of the spectrum. It makes an assumption that
the initial velocity potential data $\xi(x)$ are of shot noise type, or more
precisely that

$$e^{\xi(x)} = \sum_i e^{\xi_i} \delta(x - x_i), \tag{1}$$

where $\{\xi_i\}$ is a sequence of independent, identically disctributed random variables, independent of the Poisson ensemble $\{x_i\}$. Then, the
Hopf-Cole formula takes the form

$$u(t, x) = -2\nu \frac{\sum_i [(x - x_i)/t] \exp\left(\xi_i - (x - x_i)^2/4\nu t\right)}{\sum_i \exp\left(\xi_i - (x - x_i)^2/4\nu t\right)}. \tag{2}$$

Additional technical assumptions are that

(i) Function $H_\xi(a) = \mathbf{P}[e^{\xi_i} > a]$ varies slowly as $a \to \infty$;

(ii) The integral $\int_{\mathbf{R}} H_\xi(e^{x^2}) dx < \infty$, and

$$\lim_{a \to \infty} \frac{H_\xi(a(\log a)^{-1/2})}{H_\xi(a)} = 1;$$

(iii) There exist a strictly increasing and continuous inverse function
$A(\lambda) = H_\xi^{-1}(1/\lambda)$, and a function $B(\lambda) > 0$, $\lambda \ge 1$, regularly varying

at infinity with exponent $\theta \in [0, 2/3)$ such that, for any $u \in \mathbf{R}$,

$$\lim_{\lambda \to \infty} \lambda H_\xi(A(\lambda) + uB(\lambda)) = G_\xi(u) \in [0, \infty] \tag{3}$$

exists.

Before we formulate the second hyperbolic limit result we also need another definition and the concept of ϕ-thinning of a point process $(x_i, \xi_i)_{i \in \mathbf{Z}}$ on $\mathbf{R}^2$, where $0 \le \phi = \phi(x) \le \infty$, $x \in R$, is an even Borel function. For simplicity, we proceed under assumption that $\nu = 1/2$.

For (x_i, ξ_i) from the definition (1) of the velocity potential define $x_{i^*} = x_{i^*}(x, t)$ and $\xi_{i^*} = \xi_{i^*}(x, t)$ so that

$$\xi_{i^*} - \frac{(x - x_{i^*})^2}{2t} = \max_j \left(\xi_j - \frac{(x - x_j)^2}{2t} \right). \tag{4}$$

Also, formally, the ϕ-thinning $(x_i^{(\phi)}, \xi_i^{(\phi)})_{i \in \mathbf{Z}}$ of an arbitrary point process $(x_i, \xi_i)_{i \in \mathbf{Z}}$ is another point process defined by the following condition: for any continuous function f on $\mathbf{R}^2$ with compact support

$$\sum_i f(x_i^{(\phi)}, \xi_i^{(\phi)}) := \sum_i f(x_i, \xi_i) \mathbf{1}\left[\xi_i \ge \xi_j - \phi(x_i - x_j), \forall j \ne i \right]. \tag{5}$$

Notice that, for $\phi(x) = \phi_t(x) = x^2/2t$ and x restricted to the set $\{x_i\}$, we also have

$$(x_i^{(\phi_t)}, \xi_i^{(\phi_t)}) = (x_{i^*}, \xi_{i^*}). \tag{6}$$

Under these conditions we have the following result due to Albeverio, Molchanov and Surgailis (1993):

Theorem 1. *If $v(t, x)$ is the solution of the Burgers equation with $\nu = 1/2$ with the initial data described by (5.1.1) then, for each $x \in \mathbf{R}$,*

$$\frac{v(t, x)}{t^{-1}(x - x_{i^*})} \to 1 \qquad (t \to \infty), \tag{7}$$

in probability. Moreover, there are constants $b_t = \Lambda(t)t^{1/(2-3\theta)}$ with slowly varying $\Lambda(t), \to \infty$, and $a_t = A(b_t^3)$, such that in distribution

$$\left\{ \left(\frac{x_i^{(\phi_t)}}{b_t}, \frac{(\xi_i^{(\phi_t)} - a_t)t}{b_t^2} \right) \right\} \to \{(x_i^{(\infty)}, \xi_i^{(\infty)})\}, \qquad (t \to \infty), \tag{8}$$

where $\{(x_i^{(\infty)}, \xi_i^{(\infty)})\}$ coincides in distribution with the ϕ_1-thinning of a Poisson point processes on $\mathbf{R} \times (G_\xi^-, +\infty)$, $G_\xi^- := \inf\{u \in \mathbf{R} : G_\xi(u) < \infty\}$, with intensity measure $-dxdG_\xi(u)$.

It is well-known (Leadbetter, Lindgren, Rootzen (1983)) that the limit function of $G_\xi(u)$ can only be of the following two types: either

$$G_\xi(u) = \begin{cases} 1/(1+cu)^\gamma, & \text{for } u > -1/c; \\ +\infty, & \text{for } u \leq -1/c, \end{cases}$$

where $c > 0$, $\gamma > 0$, or

$$G_\xi(u) = e^{-cu},$$

for all $u \in \mathbf{R}$, with $c > 0$.

Counting maxima of a point process in a parabolic set. In this subsection we will calculate the rate of change of expected values of counts of certain local maxima for a time-dependent point process. We will need them in the next subsection to obtain the rate of change of shock density in Burgers turbulence.

Proposition 1.   *Consider a planar Poisson point processes $(y_j, u_j)_{j \in \mathbf{Z}}$ with intensity $-dydG(u)$. Let $l > 0$, and define*

$$N(t, [-l, l] \times [a, b]) \tag{9}$$

$$= \sum_i \mathbf{1}\left[u_i \geq u_j - \frac{(y_i - y_j)^2}{2t}, \forall j \neq i, \text{ and } (y_k, u_k) \in [-l, l] \times [a, b]\right],$$

such that $\mathbf{E}N(t, [-l, l] \times [a, b]) < \infty$. Then

$$\lim_{a \to -\infty, b \to \infty} \lim_{l \to \infty} \frac{\mathbf{E}N(t_1, [-l, l] \times [a, b])}{\mathbf{E}N(t_2, [-l, l] \times [a, b])}$$

$$= \frac{\int_{\mathbf{R}} \exp\{-\int_{\mathbf{R}} G(u + x^2/2t_1) dx\} dG(u)}{\int_{\mathbf{R}} \exp\{-\int_{\mathbf{R}} G(u + x^2/2t_2) dx\} dG(u)}, \tag{10}$$

where

$$\mathbf{E}N(t, [-l, l] \times [a, b]) \tag{11}$$

$$= \sum_i \mathbf{P}\left(u_i \geq u_j - \frac{(y_i - y_j)^2}{2t}, \forall j \neq i, \text{ and } (y_k, u_k) \in [-l, l] \times [a, b]\right).$$

PROOF. Since the Poisson point process $(y_i, u_i)_{i \in \mathbf{Z}}$ is stationary along the y-axis, we get that

$$\lim_{l \to \infty} \frac{\mathbf{E}N(t_1, [-l, l] \times [a, b])}{\mathbf{E}N(t_2, [-l, l] \times [a, b])} \tag{12}$$

$$= \frac{\mathbf{P}\left(u_i \geq u_j - (y_i - y_j)^2/2t_1, \forall j \neq i, u_k \in [a, b]\right)}{\mathbf{P}\left(u_i \geq u_j - (y_i - y_j)^2/2t_2, \forall j \neq i, u_k \in [a, b]\right)},$$

and

$$\lim_{a \to -\infty, b \to \infty} \lim_{l \to \infty} \frac{EN(t_1, [-l, l] \times [a, b])}{EN(t_2, [-l, l] \times [a, b])}$$

$$= \frac{\mathbf{P}\left(u_i \geq u_j - (y_i - y_j)^2/2t_1, \forall j \neq i, \right)}{\mathbf{P}\left(u_i \geq u_j - (y_i - y_j)^2/2t_2, \forall j \neq i, \right)}.$$

In order to calculate the probabilities on the right-hand side we need to know the *n-point correlation functions* (Radon-Nikodym derivative)

$$\rho_n(y_1, \ldots, y_n) \tag{13}$$

$$= \frac{\mathbf{P}[(y_i, u_i) \in (y_1, y_1 + dy_1) \times \mathbf{R}, \ldots, (y_i, u_i) \in (y_n, y_n + dy_n) \times \mathbf{R}]}{dy_1 \ldots dy_n},$$

for the point process $(y_i, u_i)_{i \in \mathbf{Z}}$, where $y_i \neq y_j, (i \neq j)$, $i, j = 1, \ldots, n$, $n = 1, 2, \ldots$ These are expressed by the well known (see e.g. Ruelle (1969)) formula

$$\rho_n(y_1, \ldots, y_n) \tag{14}$$

$$= (-1)^n \int_{U(y)_n} \exp\left\{-\int_{\mathbf{R}} \bigvee_{i=1}^n G\left(u_j + \frac{(x - y_j)^2}{2t}\right) dx\right\} \prod_{i=1}^n dG(u_j),$$

where

$$U(y)_n$$

$$:= \left\{(u)_n \in (G^-, +\infty)^n : u_i \geq u_j - \frac{(y_i - y_j)^2}{2t}, i \neq j, i, j = 1, \ldots, n\right\},$$

where $G^- = \inf\{u \in \mathbf{R} : G(u) < \infty\}$.

Now, for $n = 1$, we have $U(y)_1 = (G^-, \infty)$, and if we suppose that $dG(u) = 0$ for $u \in (-\infty, G^-]$ then

$$\rho_1(y) = \rho_1(t, y) = -\int_{\mathbf{R}} \exp\left\{-\int_{\mathbf{R}} G\left(u + \frac{(x - y)^2}{2t}\right) dx\right\} dG(u) \tag{15}$$

$$= -\int_{\mathbf{R}} \exp\left\{-\int_{\mathbf{R}} G\left(u + \frac{x^2}{2t}\right) dx\right\} dG(u),$$

which does not depend on y. Therefore, we get that

$$\mathbf{P}\left(u_i \geq u_j - \frac{(y_i - y_j)^2}{2t}, \forall j \neq i, y_i \in (y, y + dy)\right) = \rho_1(t, y) dy, \tag{16}$$

which is also independent of y. Using the stationarity of the point process along the y-axis again, we get that

$$\frac{\mathbf{P}\Big(u_i \geq u_j - (y_i - y_j)^2/2t_1, \forall j \neq i, y_i \in (y, y+dy)\Big)}{\mathbf{P}\Big(u_i \geq u_j - (y_i - y_j)^2/2t_2, \forall j \neq i, y_i \in (y, y+dy)\Big)} = \frac{\rho_1(t_1, y)dy}{\rho_1(t_2, y)dy}.$$
(17)

Finally, using (12) and (15), we get the statement of Proposition 1. ∎

Let us calculate the above quantity in a couple of special cases of the intensity G which will be useful in the next subsection.

For $G = e^{-cu}$, we have

$$\lim_{a \to -\infty, b \to \infty} \lim_{l \to \infty} \frac{EN(t_1, [-l, l] \times [a, b])}{EN(t_2, [-l, l] \times [a, b])}$$
(18)

$$= \frac{\int_{\mathbf{R}} \exp[-cu - \int_{\mathbf{R}} e^{-c(u+y^2/t_1)} dy] du}{\int_{\mathbf{R}} \exp[-cu - \int_{\mathbf{R}} e^{-c(u+y^2/t_2)} dy] du} = \left(\frac{t_1}{t_2}\right)^{-1/2}.$$

For $G = (1 + cu)^{-\gamma}$, $u > -1/c$, and $G = \infty$, for $u \leq -1/c$, and assuming that $\gamma > 1/2$, we get, by obvious change of variables, that

$$\lim_{a \to -\infty, b \to \infty} \lim_{l \to \infty} \frac{EN(t_1, [-l, l] \times [a, b])}{EN(t_2, [-l, l] \times [a, b])}$$
(19)

$$= \lim_{\epsilon_1 \to 0} \lim_{\epsilon_2 \to 0} \frac{\int_{\epsilon_1}^{\infty} \exp\Big\{\int_{\{z+cy^2/t_1 > \epsilon_2\}} (z + cy^2/t_1)^{-\gamma} dy\Big\} dz^{-\gamma}}{\int_{\epsilon_1}^{\infty} \exp\Big\{\int_{\{z+cy^2/t_2 > \epsilon_2\}} (z + cy^2/t_2)^{-\gamma} dy\Big\} dz^{-\gamma}}$$

$$= \lim_{\epsilon_1 \to 0} \frac{\int_{\epsilon_1}^{\infty} \exp\Big\{-z^{-\gamma}\sqrt{zt_1}(\int_{-\infty}^{\infty} (1 + y^2)^{-\gamma}) dy)\Big\} dz^{-\gamma}}{\int_{\epsilon_1}^{\infty} \exp\Big\{-z^{-\gamma}\sqrt{zt_2}(\int_{-\infty}^{\infty} (1 + y^2)^{-\gamma}) dy)\Big\} dz^{-\gamma}} = \left(\frac{t_1}{t_2}\right)^{-\frac{\gamma}{(2\gamma-1)}}.$$

Remark 1. Since we assumed $\gamma > 1/2$ the exponent $\alpha = \gamma/(2\gamma - 1)$ varies between $1/2$ and ∞). In particular, if $\gamma = 2$ then $\alpha = 2/3$. Also, in the limiting case $\gamma = \infty$, we get that $\alpha = 1/2$, which corresponds to the value obtained for the exponential intensity of the point process.

Shock density. In this section we calculate the rate of change of the expected number of shocks in a fixed finite interval after the sawtooth structure formation (see Gurbatov, Malakhov, Saichev (1991)). We shall consider separately the two cases discussed respectively in Theorem 5.1.1 and Theorem 1, and will base our computation on the results of the above subsection concerning local maxima of planar Poisson point processes.

Gaussian velocity potential. Consider first the case of a Gaussian initial velocity potential described in Theorem 5.1.1. This case corresponds to the intensity $G(u) = e^{-u}$. In view of Theorem 5.1.1, one can see that the number of shocks in the interval $(-l, l)$ of the limit velocity field $v(t, x)$ from (5.1.7) is asymptotically equal to $N(t, [-l, l] \times [a, b])$ defined in formula (9). Therefore, in this case, the rate of change (as a function of time t) of the expected shock density $\lim_{a \to -\infty, b \to \infty} EN(t, [-l, l] \times [a, b])/2l$ is expressed by the formula

$$\lim_{a \to -\infty, b \to \infty} \lim_{l \to \infty} \frac{EN(t_1, [-l, l] \times [a, b])}{EN(t_2, [-l, l] \times [a, b])} = \left(\frac{t_1}{t_2}\right)^{-1/2}, \tag{20}$$

in view of results of the above subsection.

Shot noise type velocity potential. Here the situation is described by Theorem 1 and our notation is taken from it. Let us define

$$N(t, [-l, l] \times [a, b]) = \sum_i \mathbf{1}\left\{\xi_i^{(\phi_t)} \geq \xi_j^{(\phi_t)} - \frac{(x_i^{(\phi_t)} - x_j^{(\phi_t)})^2}{2t}, \right.$$

$$\left. \forall j \neq i \text{ and } (x_k^{(\phi_t)}, \xi_k^{(\phi_t)}) \in [-l, l] \times [a, b]\right\}. \tag{21}$$

Then

$$N(t, [-b_t l, b_t l] \times [a, b]) = \sum_i \mathbf{1}\left\{\xi_i^{(\phi_t)} \geq \xi_j^{(\phi_t)} - \frac{(x_i^{(\phi_t)} - x_j^{(\phi_t)})^2}{2t}, \right.$$

$$\left. \forall j \neq i \text{ and } (x_k^{(\phi_t)}, \xi_k^{(\phi_t)}) \in [-b_t l, b_t l] \times [a, b]\right\}. \tag{22}$$

Since condition

$$\xi_i^{(\phi_t)} \geq \xi_j^{(\phi_t)} - \frac{(x_i^{(\phi_t)} - x_j^{(\phi_t)})^2}{2t}, \qquad i \neq j, \tag{23}$$

is equivalent to the condition

$$\frac{(\xi_i^{(\phi_t)} - a_t)t}{b_t^2} \geq \frac{(\xi_j^{(\phi_t)} - a_t)t}{b_t^2} - \frac{(x_i^{(\phi_t)}/b_t - x_j^{(\phi_t)}/b_t)^2}{2}, \qquad i \neq j,$$

in view of Theorem 1, in the limit $t \to \infty$, the probability of event (23) is equal to the probability of event

$$\xi_i^{(\infty)} \geq \xi_j^{(\infty)} - \frac{(x_i^{(\infty)} - x_j^{(\infty)})^2}{2}, \qquad i \neq j. \tag{24}$$

Therefore, as $t \to \infty$,

$$\mathbf{E}N(t, [-b_t l, b_t l] \times [a, b]) \sim \sum_i \mathbf{P}\left\{\xi_i^{(\infty)} \geq \xi_j^{(\infty)} - \frac{(x_i^{(\infty)} - x_j^{(\infty)})^2}{2};\right.$$

$$\left. \forall j \neq i, \ (x_k^{(\infty)}, \xi_k^{(\infty)}) \in [-l, l] \times [a, b]\right\},$$

and, as $t_1 \to \infty$ and $t_1/t_2 \to c > 0$,

$$\lim_{l \to \infty} \frac{\mathbf{E}N(t_1, [-b_{t_1} l, b_{t_1} l] \times [a, b])}{\mathbf{E}N(t_2, [-b_{t_2} l, b_{t_2} l] \times [a, b])} = 1. \tag{25}$$

Since $(x_i^{(\infty)}, \xi_i^{(\infty)})$ is stationary on x-axis, we have

$$2b_t \mathbf{E}N(t, [-l, l] \times [a, b]) \sim \mathbf{E}N(t, [-b_t l, b_t l] \times [a, b]),$$

for large l and b_t. Notice, that in view of the assumptions in Theorem 1, $b_t = \Lambda(t) t^{1/(2-3\theta)}$, and

$$\lim_{a \to -\infty, b \to \infty} \lim_{l \to \infty} \frac{\mathbf{E}N(t_1, [-l, l] \times [a, b])}{\mathbf{E}N(t_2, [-l, l] \times [a, b])} \sim \left(\frac{b_{t_1}}{b_{t_2}}\right)^{-1} \sim \left(\frac{t_1}{t_2}\right)^{-\frac{1}{2-3\theta}} \tag{26}$$

as $t_1 \to \infty$ and $t_1/t_2 \to c$.

Remark 2. Fig. 3.3.6 shows time decay of shock density in one-dimensional Burgers turbulence. The horizontal line represents the space variable and the vertical axis the time variable. As the time increases, the shocks merge and their density decreases.

Remark 3. Tatsuma, Kida (1972), in their paper on statistical mechanics of Burgers turbulence, have obtained the rate of decay of the shock density of the form $(t_1/t_2)^{-\alpha}$. They also argued, on physical grounds, in favor of the choices of $\alpha = 1/2$ and $\alpha = 2/3$. The former can be realized in our Gaussian model. The latter, already appearing as a possibility in Remark 1, is now immediately recognizable as corresponding to the case of $\theta = 1/6$. As a matter of fact, since θ can range over the interval $[0, 2/3)$, the decay rate for the shock density in the shot noise model is of power type $(t_1/t_2)^{-\alpha}$ with α varying from $1/2$ to ∞.

5.4 Sinai's theorem—Hausdorff dimension of shock points

In this section we will consider random initial data that are not smooth, e.g., Brownian motion, the white noise, or the Lévy α-stable process. We begin with an illuminating result from Hopf (1950) who was one of the first authors to discuss properties of the nonrandom Burgers equation

$$u_t + uu_x = \nu u_{xx} \tag{1}$$

in the zero viscosity limit, i.e. when $\nu \to 0$.

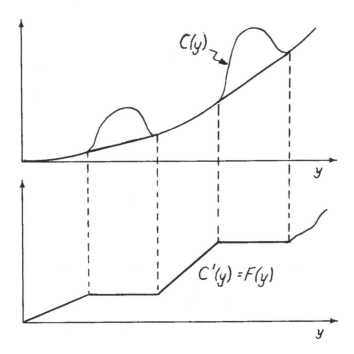

Fig. 5.4.1. Function $F(y)$ whose generalized inverse $F^{-1}(x)$ appears in Hopf's Theorem 1. It is the derivative of the convex envelope of the modified initial velocity potential W_0.

Consider a modified initial velocity potential

$$W_0(y) = \int_0^y \left(u(z,0) + z \right) dz. \tag{2}$$

Define $C(y)$ to be the largest convex function bounded from above by W_0. The graph of $C(y)$ consists of straight line intervals and a closed set lying outside them. The derivative

$$F(y) = \frac{dC(y)}{dy} \tag{3}$$

is constant where Cq is linear, see, Fig. 5.4.1. Denote by F^{-1} the (generalized multi-valued) inverse of F (an obvious connection of F^{-1} with the Legendre transform of the modified initial potential W_0 will be utilized later in this section). For a fixed time t (say, $t = 1$) the Hopf (1950) result can then be formulated as follows:

Theorem 1. *Let x be such that the line $xy + c$ is tangent to the graph of $W_0(y)$ at exactly one point $y(x) = F^{-1}(x)$. Then the limit of the Hopf-Cole solution as $\nu \to 0$ exists and*

$$\lim_{\nu \to 0} u(x, 1) = x - F^{-1}(x). \tag{4}$$

If x is such that $F^{-1}(x)$ is an interval, then the zero viscosity limit solution is discontinuous with jump (shock) size equal to the length of the interval $F^{-1}(x)$. More precisely,

$$\lim_{x' \to x - 0} \lim_{\nu \to 0} u(x, 1) = x - \min F^{-1}(x), \tag{5}$$

and

$$\lim_{x' \to x + 0} \lim_{\nu \to 0} u(x, 1) = x - \max F^{-1}(x). \tag{6}$$

Self-similar Gaussian initial data. She, Aurell and Frisch (1992) considered the case when the initial condition is a Gaussian process with mean 0 and scaling exponent $0 < H < 1$, i.e., a *fractional Brownian motion*. The case $H = 1/2$ corresponds to the Brownian motion. These types of initial conditions are not differentiable but possess self-similarity and power-law, based on which one can give heuristic argument to derive the power-law for the probability distribution of shock amplitudes, and extract the multifractal property of shock points. The latter refers to the singularity structure of the velocity field. In addition, they conjectured that, with probability 1, the set of Lagrangian shock points (to be defined below) has *Hausdorff dimension H*. Sinai (1992) gave an ingeneous proof of this conjecture in the case of Brownian motion:

Theorem 2. *If $u(x, 0)$ is a Brownian motion process for $x \geq 0$ and $u(x, 0) = 0$ for $x < 0$ then, with probability 1, the union of intervals* *where F is constant is of full measure on the halfline y > 0. Moreover, its complement has the Hausdorff dimension equal to 1/2.*

Sinai's proof of the upper estimate of the Hausdorff dimension is not easy to improve upon and we refer the reader to the original paper. However, the lower estimate by the Hölder continuity exponent of the initial condition is direct, as was observed by Handa (1993). We reproduce his argument below.

Using the Hopf-Cole formula, we can find the limit by the steepest descent argument (see, Lecture 3), that

$$u^0(x,t) := \lim_{\mu \to 0} u^\mu(x,t) = \frac{x - m(x,t)}{t}, \tag{7}$$

if the set $M(x,t)$ of points y at which the minimum

$$\min_y F(x,y,t) = \frac{x^2}{2t} + \min_y \left[\int_0^y (u(z,0) + z)dz - \frac{xy}{t} \right], \tag{8}$$

where

$$F(x,y,t) = \int_0^y u(z,0)dz + (x - y)^2/2t, \tag{9}$$

is attained consists of a unique point $m(x,t)$. Furthermore, $u^0(x,t)$ is discontinuous for those x where $M(x,t)$ has more than one point (see, Theorem 1). Note that in (8) there appears the Legendre transformation of $W_0(y)$ defined in (2). So, shock points can be interpreted as those x for which the graph of the convex hull $C(y)$ of $W_0(y)$ contains a straight segment with slope x/t.

Fix the value of t, say $t = 1$, and assume that the initial condition $u = u(z,0)$ is continuous, satisfies $\int_0^y u(z,0)\, dz = o(y^2)$ as $y \to \infty$, and vanishes for all $z \leq 0$. Hence,

$$W_0(y) = \frac{y^2}{2}, \qquad \text{for} \quad y \leq 0,$$

and we can define $F(y)$ via (3) for $y \geq 0$. Obviously $F(\cdot)$ is non-decreasing and the continuity of $u(\cdot, 0)$ implies that of $F(\cdot)$. Regarding $F(\cdot)$ as a Borel measure on $[0, \infty)$, introduce a closed subset $S(u)$ of $[0, \infty)$ defined by

$$S(u) = \operatorname{supp} F. \tag{10}$$

Denote by $\partial S(u)$ the topological boundary of $S(u)$. The above observation tells us that each point in $\partial S(u)$ contributes to a shock in the inviscid limit $u^0(x, 1)$. Such points are called *Lagrangian shock points.*

The above assumptions are satisfied almost surely if we take as $u(y,0), y \geq 0$ the Brownian motion, or more generally, the fractional Brownian motion $b_H(y)$ with scaling exponent $0 < H < 1$. So the Sinai's Theorem 1 can be formulated now as follows: *If $\{b(y); y \geq 0\}$ is the standard Brownian motion starting from 0 with time parameter $y \geq 0$. Then, with probability one, $S(b)$ has Lebesgue measure 0, and the Hausdorff dimension $\dim S(b) = 1/2$.* The result implies that $S(b)$ actually coincides with the set $\partial S(b)$ of all Lagrangian shock points. She, Aurell and Frisch's (1992) conjecture now reads that, with probability one,

$$\dim \partial S(b_H) = H.$$

Handa's result involves no assumption on randomness of the initial condition.

Proposition 1. *Suppose that the initial condition $u = u(y,0)$ satisfies*

$$|u(y,0) - u(z,0)| \leq C|y - z|^D, \qquad y, z \in I, \tag{11}$$

for some constants $0 < C < \infty$, $0 < D \leq 1$, and a bounded interval I. Then,

$$\dim (S(u) \cap I) \geq D, \tag{12}$$

whenever $F(I) > 0$, where $F(I)$ is the mass of the Stieltjes measure F on I.

PROOF. A simple (but crucial) observation is that

$$F(y) = \frac{dC}{dy} = \frac{dW_0}{dy} = u(y,0) + y$$

for $y \in \operatorname{supp} F = S(u)$. Combining this equality with (11), we have that

$$|F(y) - F(z)| \leq C_1|y - z|^D, \qquad y, z \in S(u) \cap I,$$

for some $0 < C_1 < \infty$. Simple calculations involving integration by parts yield that

$$\iint_{I \times I} \frac{dF(y)\,dF(z)}{|y - z|^{D-\epsilon}} \leq F(I)\left(C_1 + \frac{D}{\epsilon}\right) < \infty$$

for all $\epsilon > 0$. With the help of Frostman's lemma (see, e.g., Falconer (1990)) we obtain that

$$\dim (S(u) \cap I) \geq D - \epsilon,$$

provided that $F(I) > 0$.   ∎

It should be noted that this result is quite natural in view of the observation in She, Aurell and Frisch (1992) which was mentioned above. Proposition 1 connects the power-law with the Hausdorff dimension associated with shocks. We see this by considering the *fractional Brownian motion* $b_H(\cdot)$ $(0 < H < 1)$ which is a Gaussian process with mean 0 and covariance

$$\mathbf{E}\,|b_H(y) - b_H(z)|^2 = |y - z|^{2H}, \tag{13}$$

Furthermore, $b_H(\cdot)$ has the scaling law

$$b_H(cy) \overset{\text{law}}{=} c^H b_H(y), \qquad \text{for} \quad c > 0. \tag{14}$$

Using (13) together with the Gaussian property, we can verify by a standard argument that $b_H(\cdot)$ satisfies (11) with $D = H - \epsilon$, for arbitrary $\epsilon > 0$ and any bounded interval $I \subset [0, \infty)$. It is also easy to prove that, with probability one, $b_H(y) + y \to \infty$, as $y \to \infty$, and hence $W_0(y) \to \infty$ as $y \to \infty$. This implies the existence of a, possibly random, bounded interval I such that $F(I) > 0$, so that, by Proposition 1, $\dim S(b_H) \geq H - \epsilon$, a.s. Letting $\epsilon \to 0$ leads to the estimate $\dim S(b_H) \geq H$. Finally, note that, for all $0 < H < 1$, $S(b_H)$ has the Lebesgue measure 0. Thus, we can conclude that

$$\dim \partial S(b_H) \geq H, \qquad \text{a.s.},$$

and a proof of the lower bound estimate for the She, Aurell and Frisch's conjecture is complete. The upper estimate remains a challange.

Remark. Avellaneda and E (1995) proved a related result in the case of the white noise initial data. Namely, they demonstrated that $F(s) \propto s^{1/2}$, for $s \ll 1$, where $F(s)$ is the cumulative probability distribution of shock strengths. Also, recently, Molchan (1998) considered the fractional Brownian motion initial potential data $U(x) = b_H(x), u(x, o) = U_x(x)$, and proved that, in the zero viscosity limit, the cumulative distribution $F(x)$ of the shock strengths has the following asymptotic behavior as $x \to 0$ or $x \to \infty$: $0 < c_1 < F(x) \cdot x^{H-1} < c_2 < \infty, x < \epsilon$, and $-\infty < c_- < \ln(1 - F(x)) \cdot x^{2H-4} < c_+ < 0, x > e^{-1}$. This confirmed a conjecture of M.Vergassola, B. Dubrulle, U. Frisch and A. Noullez (1994) which was supported by computer simulations. A similar result has been independently obtained by Ryan (1998) using large deviations techniques.

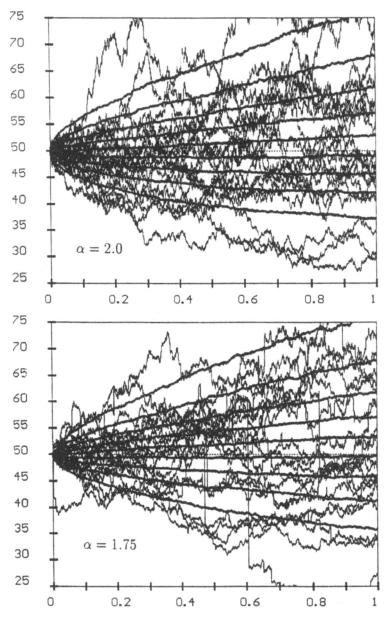

Fig. 5.4.2. A comparison of typical sample paths of the Brownian motion (with a drift), i.e. a 2-stable motion with typical smaple paths of the Lévy 1.75-stable motion (with a drift). For $\alpha < 2$ the sample paths are discontinuous with fewer jumps of larger size as α decreases. The thick lines represent deciles obtained from the ensemble of experiments.

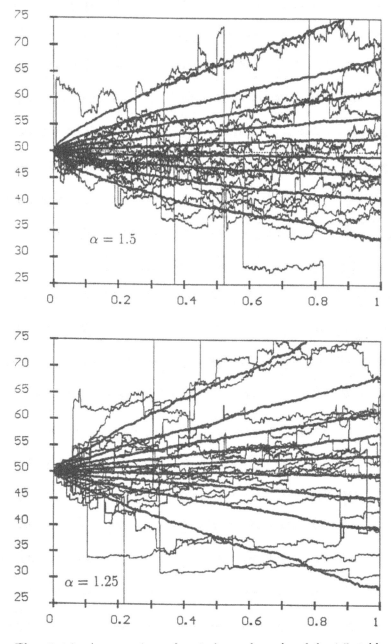

Fig. 5.4.3. A comparison of typical sample paths of the 1.5-stable motion with typical sample paths of the Lévy 1.25-stable motion (with a drift). For $\alpha < 2$ the sample paths are discontinuous with fewer jumps of larger size as α decreases. The thick lines represent deciles obtained from the ensemble of experiments.

Self-similar Lévy α-stable initial data. In this subsection we take a look at the Sinai theorem in the context of heavy–tailed self-similar data of the Lévy α–stable type with the index of self-similarity equal to $1/\alpha$. We concentrate on the interval $1 < \alpha \leq 2$, where the Hausdorff dimension of regular points can be established, but also take a brief look at the asymmetric α–stable intial data with $0 < \alpha < 1$. In absence of analytic results, we rely on computer experimentation and statistical estimation techniques, as She, Aurell and Frisch (1992) did for the Gaussian case. Our experiments suggest that it is possible to extend the Sinai's Theorem to α–stable initial data. The material below comes from Janicki and Woyczynski (1997).

The *Lévy α–stable motion* is placed somewhere inbetween the Brownian motion and the Poisson processes in the vast class of infinitely divisible processes, for which the structure of stochastic integrands and construction of the stochastic integrals are well understood (see, e.g., Kwapień and Woyczynski (1992)). For further details concerning theoretical properties of α–stable random variables and processes we refer to Samorodnitsky and Taqqu (1994). The numerical and statistical methods of their simulation are discussed in Janicki and Weron (1994).

The most common and convenient way to introduce α–stable random variables X is via their *characteristic functions* $\phi(\theta) = \mathbf{E}\exp(i\theta X)$, which depend on four parameters: α – the index of stability, β – the skewness parameter, σ – the scale parameter and μ – the shift. These functions are given by

$$\log \phi(\theta) = \begin{cases} -\sigma^\alpha |\theta|^\alpha \left\{1 - i\beta \operatorname{sgn}(\theta)\tan(\alpha\pi/2)\right\} + i\mu\theta, & \text{if} \quad \alpha \neq 1, \\ -\sigma|\theta| + i\mu\theta, & \text{if} \quad \alpha = 1, \end{cases} \tag{15}$$

where $\alpha \in (0,2]$, $\beta \in [-1,1]$, $\sigma \in \mathbf{R}_+$, $\mu \in \mathbf{R}$.

The fact that a random variable X has an α-stable distribution determined by (15) will be denoted $X \sim S_\alpha(\sigma,\beta,\mu)$. Note that $S_2(\sigma,0,\mu)$ and $S_1(\sigma,0,\mu)$ are, respectively, the Gaussian distribution $N(\mu,2\sigma^2)$ and the Cauchy distribution. The simplest case is that of a standard *symmetric α-stable distribution* $S_\alpha(1,0,0)$, $\alpha \in (0,2]$. Its characteristic function $\phi(\theta) = \exp[-|\theta|^\alpha]$. Working with α-stable distributions is complicated by the fact that, except for a few values of the parameters α, β, σ and μ, explicit expressions for their density functions are not known.

Recall that α-stable Lévy motion $\{L_\alpha(t) : t \geq 0\}$ is defined by the following properties:

    (i) $L_\alpha(0) = 0$ *a.s.*;

    (ii)   The process $\{L_\alpha(t) : t \geq 0\}$ has independent increments;

(iii) $L_\alpha(t) - L_\alpha(s) \sim S_\alpha((t-s)^{1/\alpha}, \beta, 0)$, $0 \le s < t < \infty$.

We are interested in investigation of statistical solutions of the Burgers equation (1) with a trajectory of an α–stable Lévy motion (with $\alpha \in (1, 2)$) as initial data, i.e. we put

$$u_0(a) = \begin{cases} L_\alpha(a), & \text{for } a \ge 0, \\ 0, & \text{for } a < 0. \end{cases} \tag{16}$$

This initial condition assures self-similarity of solutions of the Burgers equation. Indeed, as we will see later on, since, in distribution,

$$L_\alpha(Ct) \sim C^{1/\alpha} L_\alpha(t), \quad \text{for } C = \text{const} > 0, \text{ and } t > 0, \tag{17}$$

we also have

$$u(x, t) \sim t^{\frac{1}{\alpha-1}} u\left(x t^{\frac{\alpha}{\alpha-1}}, 1\right), \tag{18}$$

for $\alpha \in (1, 2]$.

Now, let us recall the *Lagrangian shock points* and *Legendre transform* framework for the inviscid limit in Burgers turbulence. In this situation, it is more convenient to deal with the limit of the velocity potential function $\psi = \psi(t, x; \nu) = -\int_{-\infty}^{x} u(t, z)\, dz$. Taking $\nu \to 0$ in the Hopf-Cole formula and using the steepest descent argument we get

$$\psi(t, x) = \max\{ \psi_0(a) - (x-a)^2/(2t) : a \in \mathbf{R}\}, \tag{19}$$

where $\psi_0 = \psi_0(a) = -\int_{-\infty}^{a} u_0(b)\, db$ is the initial potential.

Notice that with the use of the *Lagrangian potential*

$$\phi(t, a) =: -\frac{a^2}{2} + t\, \psi_0(a), \tag{20}$$

and its *Legendre transform*

$$H_\phi(t, x) =: \sup\{\phi(t, a) + xa : a \in \mathbf{R}\}, \tag{21}$$

the formula (19) can be written in the form

$$\psi(t, x) = \frac{H_\phi(t, x) - x^2/2}{t}. \tag{22}$$

Making use of the crucial fact that

$$H_\phi \equiv H_{\bar\phi},$$

where $\bar\phi = \bar\phi(t, a)$ denotes the convex hull (envelope) of ϕ with respect to space variable a, and defining the *Lagrangian map*

$$\mathcal{L}_t\, a =: -\frac{\partial}{\partial a} \bar\phi(t, a), \tag{23}$$

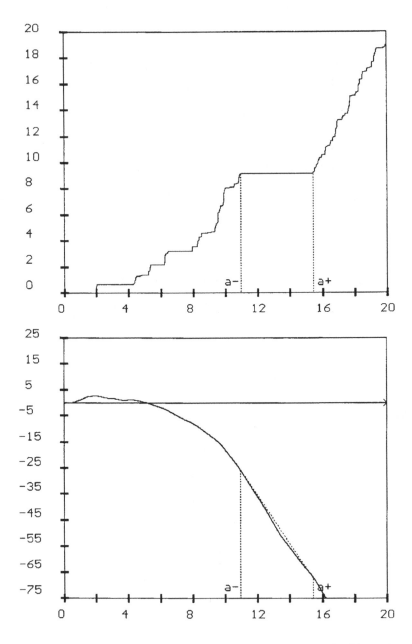

Fig. 5.4.4. The "Eulerian-Lagrangian machinery" at work. Plots of the Lagrangian map $\mathcal{L}_1 a = x(a)$ and the Lagrangian potential $\phi = \phi(a)$ as functions of a. They correspond to the solution of the inviscid Burgers equation at $t = 1$, with a trajectory of the α–stable Lévy motion ($\alpha = 1.75$) as the initial velocity (only the largest shock interval is indicated).

we get the solution to the inviscid (i.e. zero viscosity limit) Burgers' equation in the following form

$$u(t, x) = u_0(\mathcal{L}_t^{-1} x), \tag{24}$$

where $\mathcal{L}_t^{-1}$ is the inverse of $\mathcal{L}_t$.

Observe that from the Hopf-Cole formula and (17) we also get the following self-similarity properties (in distribution) of the velocity potential $\psi(x, t)$

$$\psi_0(Ca) \sim C^{\frac{\alpha+1}{\alpha}} \psi_0(a) \tag{25}$$

and from (19) we derive

$$\psi(x, t) \sim t^{\frac{\alpha+1}{\alpha-1}} \psi\left(x t^{\frac{\alpha}{\alpha-1}}, 1\right). \tag{26}$$

Let $a = a(t, x)$ denote any point where the maximum in (19) is attained. Function $a(t, x)$ as a function of the space variable x for a fixed time t is the principal object of study in this paper. In what follows $a = a(t, x)$ will be called the *inverse Lagrangian function* and $x = x(t, a)$ – the (usual) *Lagrangian function*. Notice that

$$a(t, x') - a(t, x'') \geq 0, \quad \text{for} \quad x' > x'', \tag{27}$$

which expresses the "sticky" property of shock fronts in Burgers' turbulence; they may not pass through each other, while they may coalesce on collision. For some values of x, called *Eulerian shock points*, there exists a whole interval $[a^-, a^+]$, with $a^- = a(t, x-)$ and $a^+ = a(t, x+)$, called the *Lagrangian shock–interval*, where the maximum is achieved. For such values of x the Eulerian velocity $u(x, t)$ is discontinuous and has a jump (*shock amplitude*) of the size

$$u^+ - u^- = -\frac{a^+ - a^-}{t}, \tag{28}$$

that is proportional to the length of the Lagrangian shock interval. The union of all Lagrangian shock–intervals $a \in [a^-, a^+]$ is called the set of *Lagrangian shock points*. The *set of Lagrangian regular points* is the complement of the union of the interiors (a^-, a^+) of all Lagrangian shock–intervals.

It follows from the mass conservation principle that all particles, initially located in the interval $[a^-, a^+]$, have coalesced by the time t into the single Eulerian point x. In such a situation, we extend the Lagrangian map by imposing

$$\mathcal{L}_t a = x, \quad \text{for} \quad a \in [a^-, a^+]. \tag{29}$$

In the numerical simulations presented below we construct directly the inverse Lagrangian map

$$\mathcal{L}_t^{-1} : x \mapsto a(t, x) \tag{30}$$

by searching for the points maximizing (19) with fixed $t = 1$, getting $a(x) = \mathcal{L}_1^{-1} x$. The Eulerian velocity is then obtained as

$$u(t, x) = u_0(a(t, x)) = \frac{x - a(t, x)}{t}. \tag{31}$$

All the above relations are illustrated in the series of figures obtained in Janicki and Woyczynski (1997) on the basis of a fixed realization of initial data (Fig. 5.4.5-8). The graphs for $\alpha = 2.0$, 1.75, and 1.5 show the familiar *devil's staircase* structures. A closer look indicates that for the α-stable Lévy motion $(1 < \alpha \leq 2)$ as the initial velocity, the total number of shocks per unit length is infinite and the Euler shocks are dense. The latter effect becomes more pronounced when α approaches 1, which is somewhat surprising and contradicts naive guesses. Due to self–similarity of the initial data ψ_0^τ and of the solution $u = u(t, x)$, we can restrict our attention to the solution at fixed time, say $t = 1$, and construct and study only functions

$$a = a(x) =: a(1, x), \quad \text{and} \quad x = x(a) =: x(1, a).$$

So, for a particle initially (at time $t = 0$) at position a, $x(a)$ denotes its position at time $t = 1$. Vice versa, if $a = a(x)$ is continuous at x, $a(x)$ denotes the initial position of a particle which at time $t = 1$ is located at x. If $a = a(x)$ is discontinuous at x, then the interval $[a(x-), a(x+)]$ describes initial positions of points a which form a "cluster" at x at time $t = 1$. The method based on construction of convex envelopes for given Lagrangian potential functions $\phi = \phi(a) = \phi(1, a)$ also leads to the same structure of shock fronts, even when ϕ is not differentiable (i.e., when the initial velocity field is discontinuous).

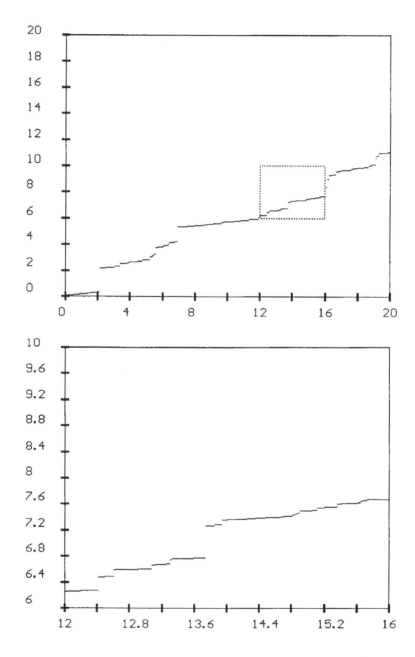

Fig. 5.4.5. *Top:* Inverse Lagrangian function $a = a(x)$ corresponding to the solution of the inviscid Burgers equation at $t = 1$, with a trajectory of the α-stable Lévy motion ($\alpha = 2.0$) as the initial velocity (a versus x plot). *Bottom:* Zooming–in on finer structures of the above graph.

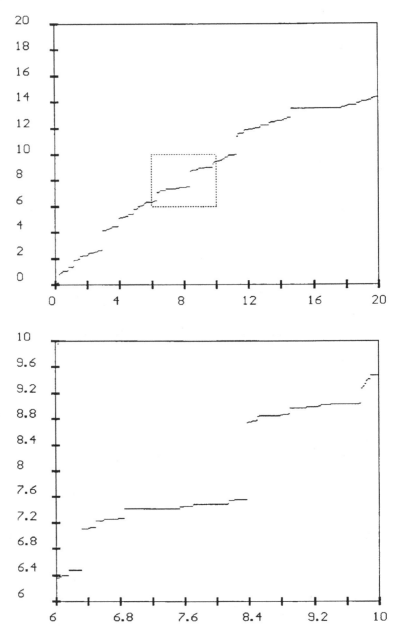

Fig. 5.4.6. *Top:* Inverse Lagrangian function $a = a(x)$ corresponding to the solution of the inviscid Burgers equation at $t = 1$, with a
trajectory of the α-stable Lévy motion ($\alpha = 1.5$) as the initial velocity (a versus x plot). *Bottom:* Zooming–in on finer structures of the
above graph.

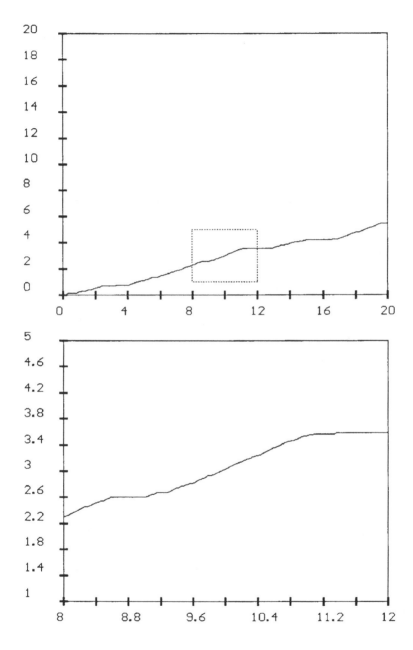

Fig. 5.4.7. *Top:* Inverse Lagrangian function $a = a(x)$ corresponding to the solution of the inviscid Burgers equation at $t = 1$, with a trajectory of the *totally skewed* α–stable Lévy motion ($\alpha = 0.75$, $\beta = -1$) as the initial velocity (a versus x plot). *Bottom:* Zooming–in on finer structures of the above graph.

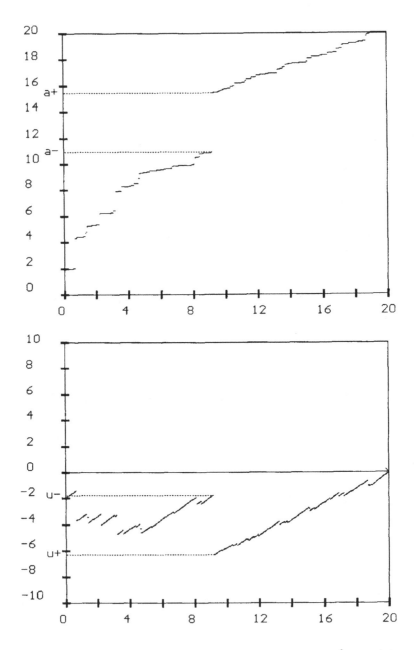

Fig. 5.4.8. Plots of the inverse Lagrangian function $\mathcal{L}_1^{-1}x = a(x)$ and the corresponding solution $u = u(x)$ as functions of x. Function u solves the inviscid Burgers equation at $t = 1$, with a trajectory of the α–stable Lévy motion ($\alpha = 1.75$) as the initial velocity (the largest jump of a and u, corresponding to the same value of x, is indicated).

These experiments give the following estimates for the Hausdorff dimension H of the set of regular Lagrangian points (i.e., those described by intervals (x^-, x^+)) for Lévy α-stable initial data:

$$H = 4/7 + \begin{cases} +.004 \\ -.020 \end{cases} \qquad \text{for} \qquad \alpha = 1.75;$$

$$H = 2/3 + \begin{cases} +.009 \\ -.033 \end{cases} \qquad \text{for} \qquad \alpha = 1.50;$$

$$H = 4/7 + \begin{cases} +.006 \\ -.024 \end{cases} \qquad \text{for} \qquad \alpha = 1.25.$$

They strongly indicate that the scaling properties of the data carry over to the fractal structure of regular points of the Lagrangian maps if $1 < \alpha < 2$. In particular, it suggests the following conjecture: *Let $1 < \alpha < 2$. The set of Lagrangian regular points of statistical solution of Burgers' equation (1) with the α-stable Lévy motion as initial data has Hausdorff dimension $1/\alpha$.* For completely asymmetric α-stable initial data with $0 < \alpha < 1$, the simulations indicate that the conjecture is no longer true. The conjecture was recently proved by Bertoin (1998).

5.5 Voronoi tessellation of shock fronts in $\mathbf{R}^d$

In the inviscid limit ($\nu \to 0$) in $\mathbf{R}^d$ the shock fronts of the velocity fields in unforced Burgers' turbulence form a Voronoi tessellation with growing cells seeded by the local maxima of the initial velocity potential

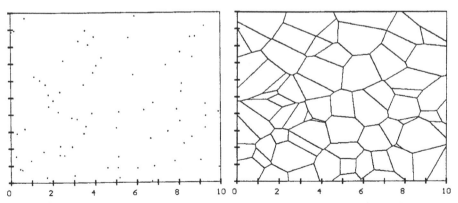

Fig. 5.5.1. Planar Poisson point process of local maxima of the initial velocity potential (left) providing seeds for the Voronoi tessellation of shock fronts (right) in unforced Burgers' turbulence in $\mathbf{R}^2$. The snapshot of the cellular structure was taken at $t = 0.1$.

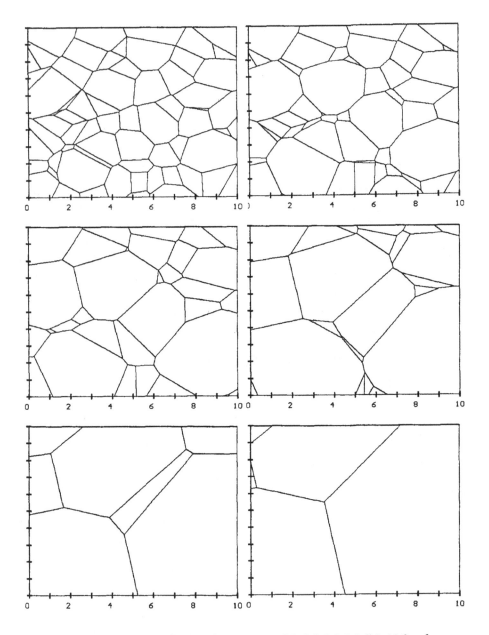

Fig. 5.5.2. Snapshots, taken at $t = 0.2, 0.5, 1.0, 2.0, 5.0, 10.0$, of evolution of the Voronoi cells created by shock fronts in an unforced planar Burgers' turbulence.

field. We shall discuss these structures in more detail in Lecture 6 in the context of general forced Burgers' turbulence. At this point (see, Fig. 5.5.1-2) we just provide snapshots of evolution of the Voronoi cells in $\mathbf{R}^2$, for the initial velocity potential of a shot noise type with cells' seeds created by a planar Poisson point process of local maxima of the initial velocity potential. The simulations were taken from Janicki, Surgailis and Woyczynski (1995).

Lecture 6
Forced Burgers Turbulence

6.1 Stationary regimes

We begin this lecture with a look at statistical properties of the multidimensional Burgers turbulence evolving in presence of a force field with random potential, which is *delta-correlated in time and smooth in space*, in the inviscid limit and at the physical level of rigorousness. Formal existence and uniqueness results for forced Burgers' flows with white noise type forcing are included in the last section of this lecture. The solution algorithm reduces to finding multistream fields describing the motion of an auxiliary gas of interacting particles in a force field. Consequently, the statistical description of the forced Burgers turbulence is obtained by finding the largest possible value of the least action for the auxiliary gas. The exponential growth of the number of streams is found to be a necessary condition for the existence of stationary regimes. The material of sections 6.1-6 is taken from Saichev and Woyczynski (1997a); its rigorous, mathematical treatment remains a challenge.

Consider the velocity field $v(x, t)$, $x \in \mathbf{R}^d, d \geq 1$, satisfying the multidimensional Burgers equation

$$\frac{\partial v}{\partial t} + (v \cdot \nabla)v = \mu \Delta v + f(x, t), \tag{1}$$

$$v(x, t = 0) = v_0(x),$$

where $\mu > 0$, and the initial velocity v_0 along with force f are known and random. We usually mean the potential velocity field

$$v(x, t) = \nabla S(x, t), \tag{2}$$

generated by potential S, which then satisfies the Hamilton-Jacobi type
equation

$$\frac{\partial S}{\partial t} + \frac{1}{2}(\nabla S)^2 = \mu \Delta S + U(\boldsymbol{x}, t), \tag{3}$$

where U is the potential of external forces, i.e.,

$$\boldsymbol{f}(\boldsymbol{x}, t) = \nabla U(\boldsymbol{x}, t). \tag{4}$$

The goal of the present section is to provide a quantitative study
of the statistically stationary regimes in Burgers' turbulence. To begin
with, let us review some conditions for existence of such equilibria.

Since dissipation leads to a decay of turbulence, to sustain it one
needs a supply of energy from outside. In the atmospheric hydrodynamic turbulence such an "engine" is powered by the solar energy,
which generates large-scale convective eddies. Their nonlinear descend-
ing cascade maintains in the dynamic equilibrium even smaller-scale,
turbulent rotational motions.

In the Burgers turbulence (1) the necessary input of energy is pro-
vided by the external random force field $\boldsymbol{f}(\boldsymbol{x}, t)$. Observe, however,
that not all force fields $\boldsymbol{f}(\boldsymbol{x}, t)$, even if they are stationary in time and
homogeneous in space, will lead to a stationary regime in Burgers' tur-
bulence. For that reason one would like to know conditions on forces
$\boldsymbol{f}(\boldsymbol{x}, t)$ which would guarantee the establishment of a stationary regime
as $t \to \infty$. A significant result in this direction has been obtained by
Sinai (1991) (see, also, Sinai (1996)) who gave a rigorous proof of the
fact that (in the 1-D case) there exists a broad class of random potentials $U(\boldsymbol{x}, t)$, periodic in space and delta-correlated in time, for which
the solution $\boldsymbol{v}(\boldsymbol{x}, t)$ of the Burgers equation converges (as $t \to \infty$) to a
solution $\boldsymbol{v}_\infty(\boldsymbol{x}, t)$ which is independent of the initial condition, stationary in time and periodic in space. So much for the *positive* results.

On the other hand, *negative* examples abound and, below, we display a case of random forces $\boldsymbol{f}(v, t)$ for which the stationary regime
is impossible in principle. We shall restrict ourselves here to the 1-D
Burgers equation

$$\frac{\partial v}{\partial t} + v \frac{\partial v}{\partial x} = \mu \frac{\partial^2 v}{\partial x^2} + f(x, t), \tag{5}$$

$$v(x, t = 0) = v_0(x),$$

for the velocity field $v(x, t)$, where $v_0(x)$ is a statistically homogeneous
stochastic process with zero mean and correlation function

$$\Gamma_0(z) = \langle v_0(x)v_0(x + z) \rangle, \tag{6}$$

and $f(x, t)$ is a Gaussian, delta-correlated in time and statistically homogeneous in space, random field with correlation function

$$\langle f(x, t) f(x + z, t + \tau) \rangle = \Gamma_f(z) \delta(\tau). \tag{7}$$

Above, and in what follows, the angled brackets denote the statistical averaging over the ensemble of realizations of the force and (if necessary) of the random initial data, which are assumed to be independent of each other.

The spatial correlation function

$$\Gamma(z; t) = \langle v(x, t) v(x + z, t) \rangle,$$

of the 1-D Burgers turbulence satisfies equation

$$\frac{\partial}{\partial t} \Gamma(z; t) + \frac{1}{2} \frac{\partial}{\partial z} \left[\Gamma_{12}(z; t) - \Gamma_{12}(-z; t) \right]$$

$$= 2\mu \frac{\partial^2}{\partial z^2} \Gamma(z; t) + \langle f(x, t) v(x + z, t) \rangle + \langle f(x + z, t) v(x, t) \rangle, \tag{1.8}$$

$$\Gamma(z; t = 0) = \Gamma_0(z),$$

where the third order moments

$$\Gamma_{12}(z; t) = \langle v(x, t) v^2(x + z, t) \rangle.$$

In what follows we will need the following *Furutsu-Novikov-Donsker formula*:

$$\langle f_i(x) R[f] \rangle = \int \langle f_i(x) f_k(x') \rangle \left\langle \frac{\delta R[f]}{\delta f(x') dx} \right\rangle dx',$$

where $\delta R[f]/\delta f$ is the variational derivative of a functional R. Introduced by Furutsu (1963) in the context of the statistical theory of electromagnetic waves in a fluctuating medium, and by Novikov (1964) in a study of randomly forced turbulence, the formula is a powerful tool in analysis of random processes and fields. Donsker (1964) obtained it independently while studying mathematical theory of path integrals. The formula explicitly calculates the correlation of arbitrary zero-mean Gaussian field $f(x) = (f_i(x))_i$ and its analytic functional $R[f]$, and is obtained by a straightforward formal comparison of the functional power series expansions of the left-hand side and the right-hand side.

We will illustrate its usefulness by applying it to evaluate the correlation $\langle f(x, t) v(x, t) \rangle$, where $f(x, t)$ is a Gaussian random field with zero mean and correlation function (7), and $v(x,t)$ is the solution of 1-D Burgers' equation (5). In this case, the Furutsu-Novikov-Donsker formula yields the following exact equality:

$$\langle f(x,t)v(x+z,t)\rangle = \int dy \int_0^t d\tau \langle f(x,t)f(y,\tau)\rangle \left\langle \frac{\delta v(x+z,t)}{\delta f(y,\tau)} \right\rangle. \quad (9)$$

Applying the variational derivative to the Burgers equation (5) gives

$$\frac{\partial}{\partial t}\left(\frac{\delta v(x,t)}{\delta f(y,\tau)}\right) + \frac{\partial}{\partial x}\left(v(x,t)\frac{\delta v(x,t)}{\delta f(y,\tau)}\right) = \mu\frac{\partial^2}{\partial x^2}\frac{\delta v(x,t)}{\delta f(y,\tau)} + \delta(x-y)\delta(t-\tau).$$

Now, taking into account the causality principle, one can replace the above linear equation for the sought variational derivative by the following Cauchy problem for the homogeneous equation

$$\frac{\partial}{\partial t}\left(\frac{\delta v(x,t)}{\delta f(y,\tau)}\right) + \frac{\partial}{\partial x}\left(v(x,t)\frac{\delta v(x,t)}{\delta f(y,\tau)}\right) = \mu\frac{\partial^2}{\partial x^2}\frac{\delta v(x,t)}{\delta f(y,\tau)},$$

$$\frac{\delta v(x,t=\tau)}{\delta f(y,\tau)} = \delta(x-y). \quad (10)$$

Substituting into (9) the correlation function (7), we obtain

$$\langle f(x,t)v(x+z,t)\rangle = \int dy\,\Gamma_f(y-x)\int_0^t \delta(t-\tau)\left\langle\frac{\delta v(x+z,t)}{\delta f(y,\tau)}\right\rangle d\tau,$$

or, using the probing property of the Dirac delta,

$$\langle f(x,t)v(x+z,t)\rangle = \frac{1}{2}\int dy\,\Gamma_f(y-x)\left\langle\frac{\delta v(x+z,t)}{\delta f(y,t)}\right\rangle.$$

So, finally, in view of equality (10),

$$\langle f(x,t)v(x+z,t)\rangle = \frac{1}{2}\Gamma_f(z).$$

In brief, the Furutsu-Novikov-Donsker formula gives for the cross-correlations in (8) the relation

$$\langle f(x,t)\,v(x+z,t)\rangle = \langle f(x+z,t)\,v(x,t)\rangle = \frac{1}{2}\Gamma_f(z).$$

As a result, equation (8) assumes the form

$$\frac{\partial}{\partial t}\Gamma(z;t) + \frac{1}{2}\frac{\partial}{\partial z}\left[\Gamma_{12}(z;t) - \Gamma_{12}(-z;t)\right] = 2\mu\frac{\partial^2}{\partial z^2}\Gamma(z;t) + \Gamma_f(z). \quad (11)$$

Introducing the spatial spectral density

$$G(\kappa;t) = \frac{1}{2\pi}\int \Gamma(z;t)e^{i\kappa z}dz$$

of the Burgers turbulence v, and the spatial spectral density

$$G_f(k) = \frac{1}{2\pi}\int \Gamma_f(z)e^{i\kappa z}dz,$$

of the force field f, we discover from (11) and with the help of a natural from the physical viewpoint assumption

$$\lim_{|z|\to\infty} \Gamma_{12}(z,t) = 0,$$

that, at $\kappa = 0$, the former satisfies equation

$$\frac{d}{dt}G(0;t) = G_f(0), \tag{12}$$

with the initial condition

$$G(0;t=0) = \frac{1}{2\pi}\int \Gamma_0(z)dz. \tag{13}$$

The solution of equation (12) is

$$G(0;t) = G(0,t=0) + G_f(0)t, \tag{14}$$

so that if $G_f(0) \neq 0$ then the spectral density of the Burgers turbulence grows linearly in time at $\kappa = 0$, which is clearly impossible in a stationary regime. Thus, we have arrived at the following result:

Ansatz 1. *A necessary condition for the existence of a stationary regime in forced Burgers' turbulence is that*

$$G_f(k=0) = \frac{1}{2\pi}\int \Gamma_f(z)\,dz = 0, \tag{15}$$

i.e. that the spectral density of the external force vanishes for $\kappa = 0$.

This condition and its multidimensional analog are fulfilled, in particular, if the *random force's potential $U(\boldsymbol{x},t)$ is statistically homogeneous in space*, and we will make this assumption in the remainder of Sections 1-6.

It also follows from (14) that the spectral density of the Burgers turbulence depends on $G(0,t=0)$. This means that if $G(0,t=0) \neq$

0 then the Burgers turbulence always "remembers" the initial field. Consequently:

Ansatz 2. *A necessary condition for the stationary regime in forced Burgers' turbulence to be ergodic (i.e. independent of the initial field) is that*

$$G(0; t = 0) = \frac{1}{2\pi} \int \Gamma_0(z) dz = 0. \tag{16}$$

Observe, that the necessary conditions (15-16) of the existence of an ergodic stationary regime are clearly satisfied for the class of forces and initial conditions studied by Sinai (1991).

Equation (11) also permits us to formulate the following, somewhat less obvious, result about statistical properties of stationary regimes in the forced Burgers turbulence. Its validity follows directly from (14-16).

Ansatz 3. *Assume that there exists an ergodic stationary regime of Burgers' turbulence and that the limits*

$$\Gamma^\infty(z) = \lim_{t \to \infty} \Gamma(z, t), \qquad \Gamma_{12}^\infty(z) = \lim_{t \to \infty} \Gamma_{12}(z, t)$$

exist. Then, its spectral density vanishes at $\kappa = 0$, i.e.,

$$G^\infty(\kappa = 0) = \frac{1}{2\pi} \int \Gamma^\infty(z)\, dz = 0. \tag{17}$$

Other propositions will answer the question whether a Gaussian stationary regime is feasible. To arrive at these results, observe that in the stationary regime, equation (11) takes the form

$$\frac{d}{dz}\Gamma_{12:odd}^\infty(z) = 2\mu\frac{d^2}{dz^2}\Gamma^\infty(z) + \Gamma_f(z), \tag{18}$$

where $\Gamma_{12:odd}^\infty(z)$ is the odd part of function $\Gamma_{12}^\infty(z)$. Multiplying the last equation by z^2, integrating it term-by-term over all z's, and taking into account equality (17), we get that

$$\int z\Gamma_{12:odd}^\infty(z)\, dz = -2\pi\frac{d^2}{d\kappa^2}G_f(\kappa)\,\Big|_{\kappa=0}, \tag{19}$$

where the spatial spectral density of the force $G_f(\kappa)$ was defined above. Since, for a Gaussian field, necessarily $\Gamma_{12:odd}^\infty(z) \equiv 0$, formula (19) implies the following proposition.

Ansatz 4. *For the existence of a Gaussian ergodic stationary regime in the forced Burgers' turbulence it is necessary that*

$$G_f(\kappa) = o(\kappa^2), \quad (\kappa \to 0).$$

Hence, from (18), we obtain another result.

Ansatz 5. *If a stationary regime in forced Burgers' turbulence is Gaussian then its spectral density satisfies condition*

$$G^\infty(\kappa) = \frac{1}{2\mu} \frac{G_f(\kappa)}{\kappa^2}. \tag{20}$$

Additional problems related to the energy dissipation mechanism in the inviscid limit ($\mu \to 0+$) and steady-state Burgers' turbulence are addressed in Section 6.6. Also, in the inviscid limit, we have another result which follows from (20).

Ansatz 6. *If an ergodic stationary regime exists for the inviscid forced Burgers' turbulence then it is non-Gaussian.*

6.2 Least action principle

The present section is devoted to a rather detailed discussion of solutions of the nonhomogeneous Burgers equation (6.1.1) with the potential force (6.1.4) in an arbitrary d-dimensional space ($\boldsymbol{x} \in \mathbf{R}^d$, $d \geq 1$).

For the sake of simplicity we will assume in this section that the potential $U(\boldsymbol{x}, t)$ is a sufficiently smooth function in both the space variable $\boldsymbol{x}$ and time variable t. Additionally, we will complement equation (6.1.1) by the zero initial condition

$$\boldsymbol{v}(\boldsymbol{x}, t = 0) = 0. \tag{1}$$

The nonzero initial conditions can be taken into account by a special choice of the external force's potential $U(\boldsymbol{x}, t)$.

By the Hopf-Cole transformation

$$\boldsymbol{v}(\boldsymbol{x}, t) = -2\mu \boldsymbol{\nabla} \ln \phi(\boldsymbol{x}, t),$$

equation (6.1.1) with the initial condition (1) is reduced to a linear Schrödinger-type diffusion equation

$$\frac{\partial \phi}{\partial t} = \mu \Delta \phi - \frac{1}{2\mu} U(\boldsymbol{x}, t)\phi, \tag{2a}$$

with the initial condition

$$\varphi(x, t = 0) = 1. \tag{2b}$$

Its solution can be written out in the form of the well-known Feynman-Kac formula

$$\phi(x, t) = \mathbf{E} \exp\left(-\frac{1}{2\mu} \int_0^t U\left(x - w(t) + w(\tau), \tau\right) d\tau\right), \tag{3}$$

where the averaging $\mathbf{E}$ is with respect to the ensemble of realizations of the vector-valued Wiener process $w(t) = (w_l(t))$ whose statistical properties are determined by conditions $w(0) = 0$, $\langle w_l(t) w_m(t) \rangle = 2\mu t \delta_{lm}$, $l, m = 1, 2, \ldots, d$, (see, e.g., Carmona, Lacroix (1990), for a formal derivation).

To make the further analysis more transparent, let us write (3) in the form of a path integral. For this purpose, consider a discretized form

$$U(x, t) = \varepsilon \sum_{p=0}^{\infty} U(x, p\varepsilon) \delta(t - p\varepsilon) \tag{4}$$

of the external force potential (6.1.4). Substituting it into (3) and assuming, for simplicity, that the time $t = (q + 1)\varepsilon - 0$, $q = 0, 1, 2, \ldots$, is also discrete, we obtain that

$$\phi(x, t) = \mathbf{E} \exp\left[-\frac{\varepsilon}{2\mu} \sum_{p=0}^{q} U\left(x - \sum_{r=p}^{q} \Omega_r, p\varepsilon\right)\right], \tag{5}$$

where

$$\Omega_r = w((r + 1)\varepsilon) - w(r\varepsilon), \qquad r = 0, 1, 2, \ldots,$$

are mutually independent Gaussian random vectors with the correlation tensor

$$\langle \Omega_{rl} \Omega_{rm} \rangle = 2\mu\varepsilon\delta_{lm}, \qquad l, m = 1, 2, \ldots, d.$$

Writing explicitly the average in (5) with respect to the Gaussian ensemble $\{\Omega_0, \Omega_2, \ldots, \Omega_q\}$ we get

$$\phi(x, t) =$$

$$\int \ldots \int \exp\left[-\frac{1}{2\mu}\left(\varepsilon \sum_{p=0}^{q} U\left(x - \sum_{r=p}^{q} z_r, p\varepsilon\right) + \sum_{p=0}^{q} \frac{z_p^2}{2\varepsilon}\right)\right] \mathcal{D}_{q+1}(z), \tag{6}$$

where each of the above integrals denotes integration over the d-dimensional space and

$$\mathcal{D}_{q+1}(z) = \left(\frac{1}{4\pi\mu\varepsilon}\right)^{d(q+1)/2} d^d z_0 \, d^d z_1 \ldots d^d z_q. \tag{7}$$

Remember that our final goal is to find not the auxiliary field $\phi(x, t)$ but the solution $v(x, t)$ of the nonhomogeneous Burgers equation (6.1.1), expressed through the former via the Hopf-Cole formula. In that solution, in addition to $\phi(x, t)$ itself, there also appears its gradient which we shall find by acting with the operator ∇ on the right-hand side of equality (6). Putting the derivatives under the integral signs, noticing that

$$\frac{\partial}{\partial x_l} \exp\left[-\frac{\varepsilon}{2\mu}\sum_{p=0}^{q} U\left(x - \sum_{r=p}^{q} z_r, p\varepsilon\right)\right]$$

$$= -\frac{\partial}{\partial z_{ql}} \exp\left[-\frac{\varepsilon}{2\mu}\sum_{p=0}^{q} U\left(x - \sum_{r=p}^{q} z_r, p\varepsilon\right)\right]$$

and integrating by parts the integral with respect to z_q, we obtain that

$$-2\mu\nabla\phi(x, t) = \tag{8}$$

$$\int \ldots \int \frac{z_q}{\varepsilon} \exp\left[-\frac{\varepsilon}{2\mu}\left(\sum_{p=0}^{q} U\left(x - \sum_{r=p}^{q} z_r, p\varepsilon\right) + \frac{1}{2}\left(\frac{z_p}{\varepsilon}\right)^2\right)\right] \mathcal{D}_{q+1}(z).$$

Let us change variables in integrals (6) and (8) from $\{z_p\}$ to

$$X_p = x - \sum_{r=p}^{q} z_r, \qquad p = 0, 1, \ldots, q, \qquad X_{q+1} = x,$$

so that $z_p = X_{p+1} - X_p$, $p = 0, 1, \ldots, q$, and, as a result, equalities (6) and (8) take the form

$$\phi(x, t) = \tag{9}$$

$$\int \ldots \int \exp\left[-\frac{\varepsilon}{2\mu}\sum_{p=0}^{q}\left(U(X_p, p\varepsilon) + \frac{1}{2}\left(\frac{X_{p+1} - X_p}{\varepsilon}\right)^2\right)\right] \mathcal{D}_{q+1}(X),$$

$$-2\mu\nabla\phi(x, t) = \int \ldots \int \frac{x - X_q}{\varepsilon} \times \tag{10}$$

$$\times \exp\left[-\frac{\varepsilon}{2\mu}\sum_{p=0}^{q}\left(U(X_p, p\varepsilon) + \frac{1}{2}\left(\frac{X_{p+1} - X_p}{\varepsilon}\right)^2\right)\right] \mathcal{D}_{q+1}(X).$$

Let us pass in the formulas (9) to the limit

$$\varepsilon \to 0, \qquad q = (t - \varepsilon)/\varepsilon \to \infty.$$

Remark, that $\boldsymbol{X}_p$ can be naturally regarded as values, for $\tau = p\varepsilon$, of a certain vector-valued process $\boldsymbol{X}(\tau)$: $\boldsymbol{X}_p = \boldsymbol{X}(p\varepsilon)$, so that the multiple integrals (9) can be interpreted as discretized functional integrals

$$\phi(\boldsymbol{x}, t) = \int \exp\left(-\frac{1}{2\mu} S[\boldsymbol{X}(\tau)]\right) \mathcal{D}[\boldsymbol{X}(\tau)], \tag{11}$$

$$-2\mu\nabla\phi(\boldsymbol{x}, t) = \int \frac{d\boldsymbol{X}(\tau)}{d\tau}\bigg|_{\tau=t} \exp\left(-\frac{1}{2\mu} S[\boldsymbol{X}(\tau)]\right) \mathcal{D}[\boldsymbol{X}(\tau)], \tag{12}$$

over all the sample paths $\boldsymbol{X}(\tau), \tau \in [0, t]$, satisfying the obvious condition

$$\boldsymbol{X}(\tau = t) = \boldsymbol{x}. \tag{13}$$

In (2.11), there appears the *action functional*

$$S[\boldsymbol{X}(\tau)] = \int_0^t \left[U(\boldsymbol{X}(\tau), \tau) + \frac{1}{2}\left(\frac{d\boldsymbol{X}}{d\tau}\right)^2\right] d\tau. \tag{14}$$

Substituting (11) in the Hopf-Cole formula, we obtain a solution of the nonhomogeneous Burgers equation (6.1.1), expressed through the functional integrals

$$\boldsymbol{v}(\boldsymbol{x}, t) = \frac{\int \frac{d\boldsymbol{X}(\tau)}{d\tau}\big|_{\tau=t} \exp\left(-\frac{1}{2\mu} S[\boldsymbol{X}(\tau)]\right) \mathcal{D}[\boldsymbol{X}(\tau)]}{\int \exp\left(-\frac{1}{2\mu} S[\boldsymbol{X}(\tau)]\right) \mathcal{D}[\boldsymbol{X}(\tau)]}. \tag{15}$$

For arbitrary $\mu > 0$, the above functional form of the nonhomogeneous Burgers equation's solution is poorly suited for analytic calculations. Nevertheless, for $\mu \to 0+$, expression (15) supplies a geometrically helpful Lagrangian picture of the corresponding generalized solution which is an analogue of the Feynman least-action principle in quantum electrodynamics.

Least-Action Principle for Forced Burgers' Turbulence. *In the inviscid limit,*

$$\boldsymbol{v}(\boldsymbol{x}, t) = \frac{d\boldsymbol{X}(\tau)}{d\tau}\bigg|_{\tau=t}, \tag{16}$$

where $\boldsymbol{X}(\tau)$ is the vector-valued process on which the action functional (14) takes the minimal absolute value.

Note, that analogous constructions of generalized solutions of first-order nonlinear partial differential equations can be found in the mathematical literature (see, e.g. Oleinik (1957), in the 1-D case, and Lions (1982), in the multidimensional case).

The extremals of functional (14) fulfill equations

$$\frac{d\boldsymbol{X}}{d\tau} = \boldsymbol{V}, \qquad \frac{d\boldsymbol{V}}{d\tau} = \boldsymbol{f}(\boldsymbol{X}, \tau), \tag{17}$$

together with boundary condition (13) combined with another obvious condition at $\tau = 0$:

$$\boldsymbol{V}(\tau = 0) = 0, \qquad \boldsymbol{X}(\tau = t) = \boldsymbol{x}. \tag{18}$$

Equations (17), along with equations

$$\frac{dS}{dt} = U(\boldsymbol{X}, \tau) + \frac{1}{2}\boldsymbol{V}^2, \tag{19}$$

$$S(\tau = 0) = 0,$$

for the action functional, form a system of characteristic equations corresponding to the following first-order pde's with respect to the field $S(\boldsymbol{x}, t)$ and its gradient $\boldsymbol{v}(\boldsymbol{x}, t) = \boldsymbol{\nabla} S(\boldsymbol{x}, t)$:

$$\frac{\partial S}{\partial t} + \frac{1}{2}(\boldsymbol{\nabla} S)^2 = U(\boldsymbol{x}, t), \tag{20}$$

$$\frac{\partial \boldsymbol{v}}{\partial t} + (\boldsymbol{v} \cdot \boldsymbol{\nabla})\boldsymbol{v} = \boldsymbol{f}(\boldsymbol{x}, t). \tag{21}$$

The latter have a clear-cut physical meaning as they describe the action and the velocity fields for a gas of noninteracting particles in the hydrodynamic limit.

If the external force $\boldsymbol{f}(\boldsymbol{x}, t)$ is a sufficiently smooth function of its arguments, then there exists a $t_1 > 0$, such that for $0 < t < t_1$ the solutions of equations (20) and (21) exist, are unique and continuous for any $\boldsymbol{x} \in \mathbf{R}^n$. At this initial stage, until the formation of discontinuities in the profile of generalized solution (16), it coincides with the solution of equation (21).

For $t > t_1$, the boundary-value problem (17-19) may, for some $\boldsymbol{x}$, have $N > 1$ solutions

$$\{\boldsymbol{X}_m(\tau), \boldsymbol{V}_m(\tau), S_m(\tau), m = 1, 2, \ldots, N\}. \tag{22}$$

Its values for $\tau = t$ and given m,

$$\boldsymbol{v}_m(\boldsymbol{x}, t) = \boldsymbol{V}_m(\tau = t), \qquad S_m(\boldsymbol{x}, t) = S_m(\tau = t),$$

can be conveniently thought of as values of a multistream solution of equations (20),(21) in the m-th stream. Let us enumerate the streams in the increasing order

$$S_1(\boldsymbol{x}, t) < S_2(\boldsymbol{x}, t) < \ldots < S_N(\boldsymbol{x}, t). \tag{23}$$

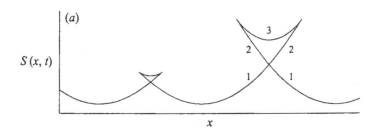

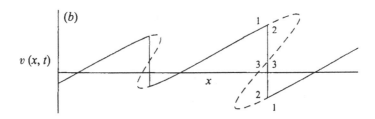

Fig. 6.2.1. A 1-D example of multistream fields of action $S(\boldsymbol{x}, t)$ (a), and of velocity $\boldsymbol{v}(\boldsymbol{x}, t)$ (b). The solid line in (b) indicates the stream that corresponds to the generalized solution of the inviscid Burgers equation. The numerals point out the stream-numbers.

Then the generalized solution (16), taking into account the appearance of discontinuities, can be written in the form

$$\boldsymbol{v}(\boldsymbol{x}, t) = \boldsymbol{v}_1(\boldsymbol{x}, t), \tag{24}$$

(see Fig. 6.2.1).

The above discussion can be summarized by the following statement which forms the basis of the discussion in Section 6.3-6:

Conclusion. *The physically significant inviscid limit solutions of the non-homogeneous Burgers equation are fully determined by multistream properties of the gas of noninteracting particles.*

6.3 Inviscid limit and multistream regimes

In this section we carry out a statistical analysis of solutions of the randomly forced Burgers equation. In what follows we shall assume that the potential field $U(x, t)$ of the random force $f(x, t)$ is a statistically homogeneous and isotropic in space, and delta-correlated in time, Gaussian random field with zero mean and correlation function

$$\langle U(x, t) U(x + y, t + \theta) \rangle = 2a(y)\delta(\theta). \tag{1}$$

Therefore, the random force $f(x, t)$ is also a statistically isotropic in space Gaussian field with correlation tensor

$$\left\langle f_l(x, t) f_m(x + y, t + \theta) \right\rangle = 2\delta(\theta) \left[b(y)\delta_{lm} + \frac{y_l y_m}{y} \frac{db(y)}{dy} \right], \tag{2}$$

where

$$b(y) = -\frac{1}{y} \frac{da(y)}{dy}, \quad l, m = 1, 2, \ldots, d.$$

Statistical description of the auxiliary multistreams. As we observed before, the statistical analysis of the inviscid Burgers turbulence reduces to the statistical analysis of the stochastic boundary-value problem (6.2.17-19). The presence of boundary conditions, even for the delta-correlated in time random force f, does not permit a direct use of the Markov processes apparatus in the analysis of statistics of solutions (6.2.17-19). To make those powerful tools available, one has to formulate initially an auxiliary Cauchy problem, the statistical properties thereof will determine the desired statistical properties of the boundary-value problem (6.2.17-19). As it will become clear from what follows, it is natural to take as such an auxiliary problem the Cauchy problem

$$\frac{dX}{dt} = V, \qquad \frac{dS}{dt} = U(X, t) + \frac{1}{2}V^2, \qquad \frac{dV}{dt} = f(X, t), \tag{3}$$

$$X(y, t = 0) = y, \qquad S(y, t = 0) = V(y, t = 0) = 0,$$

$$\frac{d\hat{J}}{dt} = \hat{K}, \qquad \frac{d\hat{K}}{dt} = \hat{g}(\hat{X}, t)\hat{J}, \tag{4}$$

$$\hat{J}(y, t = 0) = \hat{I}, \qquad \hat{K}(y, t = 0) = 0,$$

for the scalar field $S(\boldsymbol{y}, t)$, vector fields $\boldsymbol{X}(\boldsymbol{y}, t)$ and $\boldsymbol{V}(\boldsymbol{y}, t)$, and also tensor fields $\hat{J}(\boldsymbol{y}, t)$ and $\hat{K}(\boldsymbol{y}, t)$ with components

$$J_{lm}(\boldsymbol{y}, t) = \frac{\partial X_l}{\partial y_m}, \qquad K_{lm} = \frac{\partial V_l}{\partial y_m}.$$

The following notation has been used: $\hat{I}$ is the diagonal unit matrix and $\hat{g}$ is a random tensor with components

$$g_{lm}(\boldsymbol{x}, t) = \frac{\partial^2 U(\boldsymbol{x}, t)}{\partial x_l \partial x_m}. \tag{5}$$

The Cauchy problem (3) has a clear-cut intuitive physical interpretation. It describes the evolution of coordinates $\boldsymbol{X}$, action S, and velocity $\boldsymbol{V}$ of particles forced by $\boldsymbol{f}(\boldsymbol{x}, t)$. The notation clearly displays the dependence on the initial coordinates $\boldsymbol{y}$ of the particle. This dependence plays a fundamental role in further analysis. The Cauchy problems (3) and (4) together with arbitrarily distributed initial positions $\boldsymbol{y}$ can be naturally interpreted as a gas of noninteracting particles. The tensors $\hat{J}$ and $\hat{K}$ describe the deformation of an infinitesimal volume "frozen" in the gas. Recall that $\boldsymbol{y}$ are Lagrangian coordinates of this gas. Their connection with the Eulerian coordinates $\boldsymbol{x}$ is given by a vector equality

$$\boldsymbol{x} = \boldsymbol{X}(\boldsymbol{y}, t). \tag{6}$$

For given $\boldsymbol{x}$ and t it is an equation with respect to $\boldsymbol{y}$. Solving it, we obtain

$$\boldsymbol{y} = \boldsymbol{Y}(\boldsymbol{x}, t), \tag{7}$$

the Lagrangian coordinates of particles, which at time t arrive at a point with Eulerian coordinates $\boldsymbol{x}$. We should emphasize that in the general case, the gas of noninteracting particles has several, say $N(\boldsymbol{x}, t) \geq 1$, streams. It means that equation (6) may have several roots. In this case, equation (7) defines a multi-valued function assuming N values

$$\boldsymbol{Y}_1(\boldsymbol{x}, t), \quad \boldsymbol{Y}_2(\boldsymbol{x}, t), \quad \dots, \boldsymbol{Y}_N(\boldsymbol{x}, t). \tag{8}$$

Consider the joint probability density of the solutions of the auxiliary Cauchy problem (3-4):

$$\mathcal{P}(\boldsymbol{x}, s, \boldsymbol{v}, \hat{j}, \hat{\kappa}; \boldsymbol{y}, t) = \tag{9}$$

$$\left\langle \delta(\boldsymbol{X}(\boldsymbol{y}, t) - \boldsymbol{x}) \delta(S(\boldsymbol{y}, t) - s) \delta(\boldsymbol{V}(\boldsymbol{y}, t) - \boldsymbol{v}) \delta(\hat{J}(\boldsymbol{y}, t) - \hat{j}) \delta(\hat{K}(\boldsymbol{y}, t) - \hat{\kappa}) \right\rangle.$$

Let us transform the right-hand side of equality (9), using the well known identity

$$\delta(\boldsymbol{x} - \boldsymbol{X}(\boldsymbol{y}, t)) = \sum_{n=1}^{N(\boldsymbol{x},t)} \frac{\delta(\boldsymbol{Y}_n(\boldsymbol{x}, t) - \boldsymbol{y})}{|J(\boldsymbol{Y}_n, t)|}, \tag{10}$$

for the delta-function (see, e.g. Saichev, Woyczynski (1996)), where

$$J(\boldsymbol{y}, t) = \|\hat{J}(\boldsymbol{y}, t)\| = \left\| \frac{\partial X_l}{\partial y_m} \right\|, \tag{11}$$

is the Jacobian of the Eulerian-to-Lagrangian coordinate transformation. Substituting (10) into (9) and taking into account the probing property of the delta-function, we have

$$|j| \mathcal{P}(\boldsymbol{x}, s, \boldsymbol{v}, \hat{j}, \hat{l}; \boldsymbol{y}, t) = \tag{12}$$

$$\left\langle \sum_{n=1}^{N(\boldsymbol{x},t)} \delta(\boldsymbol{Y}_n(\boldsymbol{x}, t) - \boldsymbol{y}) \delta(s_n(\boldsymbol{x}, t) - s) \delta(\boldsymbol{v}_n(\boldsymbol{x}, t) - \boldsymbol{v}) \delta(\hat{j}_n(\boldsymbol{x}, t) - \hat{j}) \right.$$

$$\left. \times \delta(\hat{\kappa}_n(\boldsymbol{x}, t) - \hat{\kappa}) \right\rangle,$$

where

$$s_n(\boldsymbol{x}, t) = S(\boldsymbol{Y}_n, t), \quad \boldsymbol{v}_n(\boldsymbol{x}, t) = \boldsymbol{V}(\boldsymbol{Y}_n, t), \tag{13a}$$

$$\hat{j}_n(\boldsymbol{x}, t) = \hat{J}(\boldsymbol{Y}_n, t), \quad \hat{\kappa}_n(\boldsymbol{x}, t) = \hat{K}(\boldsymbol{Y}_n, t), \tag{13b}$$

are fields that describe state of the gas in the n-th of N streams which occur at point $\boldsymbol{x}$ at time t, and where j is the determinant of the matrix $\hat{j}$ ($j = \|\hat{j}\|$.) By the total probability formula, in view of (12),

$$|j| \mathcal{P}(\boldsymbol{x}, s, \boldsymbol{v}, \hat{j}, \hat{\kappa}; \boldsymbol{y}, t) = \sum_{N=1}^{\infty} P(N; \boldsymbol{x}, t) \sum_{n=1}^{N} W_n(\boldsymbol{y}, s, \boldsymbol{v}, \hat{j}, \hat{\kappa}; \boldsymbol{x}, t|N),$$

$$\tag{14}$$

where $P(N; \boldsymbol{x}, t)$ is the probability of the event that at a given point $\boldsymbol{x}$ at time t we have N streams present, and where $W_n(\boldsymbol{y}, s, \boldsymbol{v}, \hat{j}, \hat{\kappa}; \boldsymbol{x}, t|N)$ is the conditional joint probability density of random fields (8) and (13a,b) in the n-th stream, given that the total number of streams is N.

Approximations for the Burgers' turbulence statistics. In view of (6.2.22-24), the sought joint probability density of the least-action functional, corresponding Lagrangian coordinates $\boldsymbol{Y}(\boldsymbol{x}, t)$, the generalized solution $v(x, t)$ of the nonhomogeneous Burgers equation in the inviscid limit, and the auxiliary fields $\hat{j}, \hat{\kappa}$, are expressed in the following fashion through the components of sum (14):

$$W(y, s, v, \hat{j}, \hat{\kappa}; x, t) = \sum_{N=1}^{\infty} P(N; x, t) W_1(y, s, v, \hat{j}, \hat{\kappa}; x, t|N). \quad (15)$$

In the case of statistically homogeneous fields—in what follows we will restrict our attention to such fields—the probability density of the streams' number does not depend on x, and the probability density in (14-15) depends only on $x - y$. Hence, integrating equalities (14-15) over all $x, \hat{j}, \hat{\kappa}$, we arrive at the relations

$$\langle |J| \rangle_{sv} P(s, v; t) = \sum_{N=1}^{\infty} P(N; t) \sum_{n=1}^{N} W_n(s, v; t|N), \quad (16)$$

$$W(s, v; t) = \sum_{N=1}^{\infty} P(N; t) W_1(s, v; t|N), \quad (17)$$

more convenient for further analysis. Here $\langle ... \rangle_{sv}$ denotes the average under the condition that $S(y, t) = s$, $V(y, t) = v$ are given.

Unfortunately we cannot extract the partial sum (17), which is of interest to us, from the total sum (16). Such an operation is possible in principle, but to find (17) one has to have knowledge of all the joint probability densities for the Cauchy problem (3-4) under different initial conditions. These joint probability densities satisfy complex Kolmogorov equations whose solutions are not known. For that reason we will utilize a semi-qualitative method of finding probability densities of the forced Burgers turbulence.

Our main assumption is as follows: *there exists a number $\bar{S}(t)$— the largest value of the least-action—such that*

$$\int_{-\infty}^{\bar{S}(t)} W_1(s; t|N) ds \approx 1, \quad (18)$$

and

$$\int_{-\infty}^{\bar{S}(t)} W_n(s; t|N) ds \approx 0, \quad n = 2, 3, \ldots N,$$

where

$$W_n(s; t|N) = \int_{-\infty}^{\infty} W_n(s, v; t|N) d^d v, \quad n = 1, 2, \ldots, N.$$

If this assumption is satisfied, then the desired probability density

$$W(\boldsymbol{v};t) = \sum_{N=1}^{\infty} P(N;t) W_1(\boldsymbol{v};t|N)$$

of Burgers' turbulence can be approximated by integration of equality (16) over all the values of s in the interval $(-\infty, \bar{S}(t))$, that is

$$W(\boldsymbol{v};t) = \int_{-\infty}^{\bar{S}(t)} \langle |J| \rangle_{sv} \mathcal{P}(s, \boldsymbol{v};t)\, ds. \tag{19}$$

In addition, the value of $\bar{S}(t)$ can be determined from the normalization condition

$$1 = \int_{-\infty}^{\bar{S}(t)} \langle |J| \rangle_{s} \mathcal{P}(s;t)\, ds, \tag{20}$$

for probability density (19), where $\mathcal{P}(s;t)$ is the probability density of random action $S(\boldsymbol{y},t)$ satisfying the auxiliary Cauchy problem (3).

Closing this subsection we will make an additional assumption that the *random Jacobian field J (11) is statistically independent from the random fields $S(\boldsymbol{y},t)$ and $\boldsymbol{V}(\boldsymbol{y},t)$*. In such a case, the expressions (19) for the solutions of the nonhomogeneous Burgers equation and equation for the maximal value of absolute minima $\bar{S}(t)$ (20) take a particularly simple form

$$W(\boldsymbol{v};t) = \langle N(t) \rangle \int_{-\infty}^{\bar{S}(t)} \mathcal{P}(s, \boldsymbol{v};t)\, ds, \tag{21}$$

$$\langle N(t) \rangle \int_{-\infty}^{\bar{S}(t)} \mathcal{P}(s;t)\, ds = 1.$$

Note, that the last assumption is not really essential and has only a technical nature. If it is not satisfied then the following calculations do not change qualitatively, but they do get more complicated. In the test case considered in the next section we will verify that the statistical dependence between J and $S, \boldsymbol{V}$ does not significantly affect the final outcome. For that reason, in the remainder of this section we will always assume J to be statistically independent from the values of the vector $(S, \boldsymbol{V})$ and use expression (21) instead of a more correct, but much more complex formulas (19-20).

A model example. Let us test the conjecture underlying formulas (19-20) on the following simple model which, nevertheless, is relatively close to the problem we are considering. Let, for a given number of streams N, the values of actions of different streams $\{S_1, \ldots S_N\}$

form a family of statistically independent random variables with identical cumulative distribution functions

$$F(s) = P(S_n < s).$$

In each realization, as in (23), we will form an order statistic

$$S^1 \leq S^2 \leq \ldots \leq S^N,$$

and denote the cumulative distribution function of the n-th ordered variable S^n by

$$F_n^N(s) = P(S^n < s)).$$

It is well known that

$$F_n^N(s) = 1 - \sum_{l=0}^{n-1} \binom{N}{l} \left(\frac{z}{N}\right)^l \left(1 - \frac{z}{N}\right)^{N-l}, \qquad (22)$$

and in particular, that the cumulative distribution of the smallest $S_{min} = S^1$ is equal to

$$F_1^N(s) = P(S^1 < s) = 1 - \left(1 - \frac{z}{N}\right)^N, \qquad (23)$$

where $z = z(s) = NF(s)$. Besides, it is clear that

$$\sum_{n=1}^N F_n^N = NF = z. \qquad (24)$$

Within the framework of this example, the conditional normalizations (20) defining values of $\bar{S}$, reduce to the equality

$$NF = z = 1.$$

In addition, according to our assumption, conditions

$$F_1^N \Big|_{z=1} \approx 1, \quad \sum_{n=2}^N F_n^N \Big|_{z=1} \approx 0, \qquad (25)$$

analogous to (18) have to be fulfilled. Let us verify to what extent they are valid, substituting here corresponding expressions from (22-24). This gives

$$F_1^N(\bar{s}) = F_1^N \Big|_{z=1} = 1 - \left(1 - \frac{1}{N}\right)^N, \qquad (26)$$

$$R^N(\bar{S}) = \sum_{n=2}^N F_n^N \Big|_{z=1} = \left(1 - \frac{1}{N}\right)^N.$$

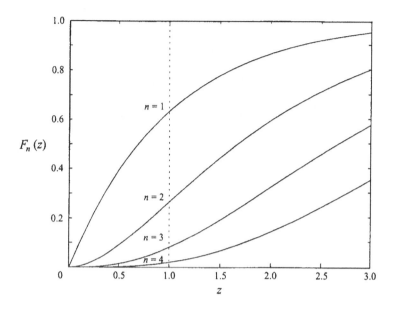

Fig. 6.3.2. Graphs of cumulative distributions of order statistics $F_n^\infty(z)$. The curves have a universal form and do not depend on the distributions of original random variables $S_1, S_2, \ldots$ One can see that, for $z = 1$, the value $F_1^\infty(1)$ is rather close to 1, while other cumulative distributions are still pretty small.

Now, for example, it follows that the values of $\{F_1^N(\bar{S})\}$ form a monotonically decreasing (as N increases) sequence, the limit thereof is

$$F_1^\infty(z = 1) = \lim_{N \to \infty} F_1^N(\bar{s}) = 1 - e^{-1}, \tag{27}$$

a quantity one can think of, in this context, as being "close" to 1. Correspondingly, $\{R^N(\bar{s})\}$ is a monotonically increasing sequence which, for $N \to \infty$, converges to a "small" number

$$R^\infty(z = 1) = \lim_{N \to \infty} R^N(\bar{s}) = e^{-1}. \tag{28}$$

In this fashion, in the model example under consideration, the relations (25) are satisfied with a "good" accuracy, expressed by the limit equalities (27-28).

Furthermore, notice that for arbitrary z and large $N \to \infty$, probability densities of the order statistics $\{S^1, S^2, \ldots\}$ are described (see Fig.

6.3.2) by the main asymptotics of expressions (22-24):

$$F_1^\infty(z) = 1 - e^{-z}, \quad F_n^\infty(z) = 1 - e^{-z} \sum_{l=0}^{n-1} \frac{z^l}{l!}, \quad (29)$$

$$R^\infty(z) = \sum_{m=2}^\infty F_n^\infty(z) = z - 1 + e^{-z}.$$

We should emphasize here that the original problem of finding statistical properties of solutions of the nonhomogeneous Burgers equation is related to the situation discussed in the above example at very large times, when the average number

$$\langle N(t) \rangle = \sum_{m=1}^\infty N P(N; t) = \langle |J| \rangle \quad (30)$$

of streams is much larger than 1. Indeed, for $\langle N \rangle \gg 1$ the law-of-large-numbers effects take over, the random number $N(t)$ of streams is not much different from the mean number

$$\sqrt{\langle N - \langle N \rangle \rangle^2} \ll \langle N \rangle,$$

and one can assume that, for $\langle N \rangle \gg 1$, the number of streams in each realization is the same and equal to $\langle N(t) \rangle$.

In addition, it is natural to assume that in the multistream regime $\langle N \rangle \gg 1$, the particles which arrive at a given time at point x, move along strongly dispersed paths, so that the forces acting on different particles $f(X_m(\tau), \tau); \tau \in [0, t]$, actually are statistically independent. Therefore, the values $\{S_1(x, t), S_2(x, t), \dots, S_{\langle N \rangle}(x, t)\}$ of their actions can be treated as independent parameters of the particles.

A test case: 2-D unforced Burgers' turbulence.   We shall illustrate the above general statistical approach in the relatively well understood case of the homogeneous Burgers turbulence. To be specific, we will restrict ourselves to the 2-D case $x \in \mathbf{R}^2$. Then, the potential

$$U(x, t) = S_0(x)\delta(t),$$

where $S_0(x)$ is the initial velocity field potential, that is

$$v_0(x) = \nabla S_0(x).$$

Taking this into account, the auxiliary Cauchy problem (3-4) takes the following form:

$$\frac{dX}{dt} = V, \quad \frac{dS}{dt} = \frac{1}{2}V^2, \quad \frac{dV}{dt} = 0,$$

$$X(y, t = 0) = y, \quad S(y, t = 0) = S_0(y), \quad V(y, t = 0) = v_0(y),$$

$$\frac{d\hat{J}}{dt} = \hat{K}, \qquad \frac{d\hat{K}}{dt} = 0, \tag{31}$$

$$\hat{J}(y, t = 0) = \hat{I}, \qquad \hat{K}(y, t = 0) = \hat{K}_0(y),$$

where $\hat{K}_0(x)$ is a tensor with components

$$K_{0lm}(x) = \frac{\partial^2 S_0(x)}{\partial x_l \partial x_m}.$$

Let $S_0(x)$ be a Gaussian, statistically isotropic field with zero mean and correlation function

$$\langle S_0(x) S_0(x + y) \rangle = \frac{\sigma_0^2}{\kappa^2} \exp\left(-\frac{1}{2}\kappa^2 y^2\right).$$

Then the fields $S_0(x)$ and $v_0(x)$ are statistically independent at the same spatial point, and the joint probability density of solutions S and V of the Cauchy problem (31) takes the form

$$\mathcal{P}(s, v; t) = w_v(v) w_s(s - v^2 t/2), \tag{32}$$

where $w_v(v)$ and $w_s(s)$ are respectively, the probability densities of fields $v_0(x), S_0(x)$ which, in the 2-D case, are

$$w_v(v) = \frac{1}{2\pi\sigma_0^2} \exp\left(-\frac{v^2}{2\sigma_0^2}\right), \tag{33}$$

$$w_s(s) = \frac{\kappa}{\sqrt{2\pi}\sigma_0} \exp\left(-\frac{s^2\kappa^2}{2\sigma_0^2}\right). \tag{34}$$

For convenience, let us introduce a dimensionless scalar field

$$u(x, t) = v^2(x, t)/2\sigma_0^2. \tag{35}$$

It follows from (21) and (32-34) that its probability density is given by the formula

$$W(u; t) = \frac{1}{2}\langle N(t) \rangle e^{-u} \text{erfc}\,(u\tau - \rho), \tag{36}$$

where the quantity ρ is determined from the normalization condition

$$\langle N(t) \rangle \int_0^\infty e^{-u} \text{erfc}\,(u\tau - \rho)du = 2,$$

which is not difficult to transform into the following, more convenient for our analysis, form:

$$\langle N(t) \rangle \left[\mathrm{erfc}\,(-\rho) - \exp\left(-\rho^2 + \left(\rho - \frac{1}{2\tau} \right)^2 \right) \mathrm{erfc}\left(\frac{1}{2\tau} - \rho \right) \right] = 2.$$
(37)

In (36-37), we have introduced the following dimensionless variables

$$\rho = \kappa \bar{S}/\sqrt{2}\sigma_0, \qquad \tau = \kappa \sigma_0 t/\sqrt{2}, \tag{38}$$

and the notation

$$\mathrm{erfc}\,(z) = 1 - \mathrm{erf}\,(z), \qquad \mathrm{erf}\,(z) = \frac{2}{\sqrt{\pi}} \int_0^z e^{-y^2} dy, \tag{39}$$

was used for the special error function.

Expressions (36-37) contain the mean value $\langle N(t) \rangle$ of the streams' number, which will be calculated below. For now, assuming that $\langle N(t) \rangle$ is known, observe that it is not very difficult to solve equation (37) numerically with respect to $\rho(\tau)$, and define the probability density (36) and corresponding moment functions for any τ. Here, we will restrict ourselves to the derivation of the asymptotic formulas for the late stage when multiple discontinuities coalesce ($\tau \gg 1$, $\langle N(t) \rangle \gg 1$) in the Burgers turbulence. At that stage, equation (37) can be replaced, with help of the asymptotic formula

$$\mathrm{erfc}\,(z) \sim \frac{1}{\sqrt{\pi}z} e^{-z^2}, \qquad z \to \infty, \tag{40}$$

by the asymptotic relation

$$\rho^2 e^{\rho^2} = \frac{\langle N(t) \rangle}{4\tau\sqrt{\pi}}. \tag{41}$$

If the right-hand side of this equality is much larger than 1, then we get the following asymptotic formula

$$|\rho| \sim \sqrt{\ln\left(\frac{\langle N(t) \rangle}{4\tau\sqrt{\pi}} \right)}, \qquad \rho < 0, \quad |\rho| \gg 1. \tag{42}$$

Let us substitute expression (42) into (36). Using (40), we arrive at the following result:

Ansatz 1. *For $\tau \gg 1$, $\langle N(t) \rangle \gg 1$, and $|\rho| \gg 1$, the dimensionless kinetic energy $u = v^2/2\sigma_0^2$ in unforced Burgers' turbulence has the probability density*

$$W(u; \tau) = 2|\rho|\tau \exp(-2|\rho|\tau u), \tag{43}$$

where ρ and τ are given by (38).

In particular, it follows that the average dimensionless kinetic energy $\langle u(x,t) \rangle$ in Burgers turbulence in the late stage of multiple shock coalescence, satisifes the asymptotic law

$$\langle u(\boldsymbol{x}, t) \rangle \sim 1/2|\rho|\tau. \tag{44}$$

In relations (42),(43-44), the principal role was played by the average number $\langle N(t) \rangle$ of streams in the gas of noninteracting particles. Let us calculate that number in the 2-D case under consideration. For that purpose recall that this average is connected by formula (30) with the statistical characteristics of the Jacobian $J(\boldsymbol{y}, t)$ (11):

$$\langle N(t) \rangle = \langle |J| \rangle.$$

It is known (see, e.g., Gurbatov, Malakhov, Saichev (1991)) that in the 2-D case the Jacobian is statistically equivalent with the following random quantity

$$J = (1 + 2\alpha)^2 - 2\beta,$$

where α, $-\infty < \alpha < \infty$, and $\beta \geq 0$ are statistically independent random quantities with probability densities

$$P(\alpha; \tau) = \frac{1}{\sqrt{2\pi}\tau} \exp\left(-\frac{\alpha^2}{2\tau^2}\right), \qquad Q(\beta; \tau) = \frac{1}{\sqrt{2\tau^2}} \exp\left(-\frac{\beta}{2\tau^2}\right).$$

The above two formulas permit us to obtain an exact expression for the probability density of the Jacobian:

$$P(j; \tau) = \frac{1}{8\sqrt{3}\tau^2} \exp\left(\frac{j}{4\tau^2} - \frac{1}{12\tau^2}\right) \times \tag{45}$$

$$\times \begin{cases} 2, & \text{if } j < 0; \\ 2 - \text{erf}\left(\sqrt{\frac{3j}{8\tau^2}} - \frac{1}{4\sqrt{3}\tau}\right) - \text{erf}\left(\sqrt{\frac{3j}{8\tau^2}} + \frac{1}{4\sqrt{3}\tau}\right), & \text{if } j > 0. \end{cases}$$

wherefrom, after simple calculations, we obtain that

$$\langle N(t) \rangle = 1 + \frac{8}{\sqrt{3}}\tau^2 \exp\left(-\frac{1}{12\tau^2}\right). \tag{46}$$

In particular, for $\tau \to \infty$, the average number of streams satisfies the following asymptotic power law:

$$\langle N(t) \rangle \sim \frac{8}{\sqrt{3}} \tau^2. \tag{47}$$

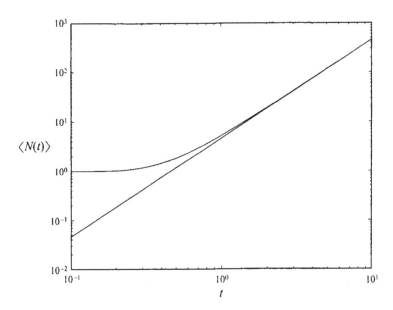

Fig. 6.3.3. Time evolution of the exact (top line; see (46)) and asymptotic (bottom line; see (47)) average number $\langle N(t) \rangle$ of streams in 2-D homogeneous Burgers turbulence. Initially, when the dimensionless time $\tau < 1$, the number of streams is close to 1, and the discontinuities of $v(x,t)$ are practically absent. In the late stages $\tau \gg 1$ of the multiple shock coalescence, the number of shocks is well described by the asymptotic formula (47).

Substituting it into (42), we find that

$$|\rho| \sim \sqrt{\ln \left(\tau / \sqrt{3\pi} \right)}, \qquad \tau \gg 1, \tag{48}$$

which gives the following result:

Ansatz 2. *The average kinetic energy (44) of the unforced 2-D inviscid Burgers turbulence decays, at sufficiently large times, as*

$$\langle u(x, t) \rangle \sim \left(2\tau \sqrt{\ln \left(\tau / \sqrt{3\pi} \right)} \right)^{-1},$$

where $\tau = \kappa \sigma_0 t / \sqrt{2}$.

Note, that the above result agrees well with the asymptotic expression for the average kinetic energy in Burgers' turbulence obtained for the 1-D problem by a different asymptotic approach in Tatsumi, Kida (1972) and Gurbatov, Malakhov, Saichev (1991). This also is an indirect evidence in support of our conjecture that S and V are actually statistically independent of the Jacobian J. Recall that this conjecture permitted us to replace the more precise expressions (19-20) by expressions (21) which are more convenient for calculations.

Remark 1. For $\tau \to \infty$, the probability density of the Jacobian (45) has the following *self-similar* property:

$$P(j; \tau) \sim \frac{1}{c\langle N\rangle} \mathcal{P}_\infty \left(\frac{j}{c\langle N\rangle} \right), \tag{49}$$

where

$$\mathcal{P}_\infty \left(\frac{z}{c} \right) = \frac{2}{3} \exp\left(\frac{z}{\sqrt{12}} \right) \times \begin{cases} 1, & \text{if } z < 0; \\ 1 - \operatorname{erf}\left(\sqrt{z\sqrt{3}} \right), & \text{if } z > 0, \end{cases}$$

and c is a normalizing constant, which in this case is

$$c = \frac{4}{\sqrt{15}} \left(3\sqrt{6} - 2\sqrt{5} \right).$$

The self-similarity (49) of the Jacobian probability density, which is clear in the homogeneous case, will be used later on in the multidimensional and forced case as an assumption under which we will find the rate of growth for the time evolution of $\langle N(t)\rangle$ and of the average kinetic energy. In the 1-D case, we will be able to use a more precise approach to study the convergence of Burgers' turbulence to a stationary regime.

6.4   Statistical characteristics

Statistics of noninteracting particles' action.   Let us apply the proposed above algorithm to the calculation of statistical properties of forced Burgers' turbulence. First, we shall study the probability density of solutions of the auxiliary Cauchy problem (6.3.3). It follows from (6.3.1-3) that action $S(\boldsymbol{y}, t)$ can be represented as a sum of two statistically independent summands

$$S(\boldsymbol{y}, t) = S_1(\boldsymbol{y}, t) + S_2(\boldsymbol{y}, t). \tag{1}$$

Moreover, the first summand is also independent of the random velocity $V(y, t)$ and has the probability density

$$P_1(s; t) = \frac{1}{2\sqrt{\pi a t}} \exp\left(-\frac{s^2}{4at}\right), \qquad a = a(0). \qquad (2)$$

Furthermore, the joint probability density of the second summand in (1) and the velocity field $V(y, t)$ satisfies the following Kolmogorov equation

$$\frac{\partial P_2}{\partial t} + \frac{1}{2}v^2\frac{\partial P_2}{\partial s} = b\Delta_v P_2, \qquad b = b(0), \qquad (3)$$

$$P_2(s, v; t = 0) = \delta(s)\delta(v).$$

Respectively, the joint probability density of the full action $S(y, t)$ and the velocity $V(y, t)$ is equal to

$$P(s, v; t) = P_1(s; t) \otimes P_2(s, v; t), \qquad (4)$$

where the symbol $\otimes$ means the convolution operation, here with respect to variable s.

Let us pass from (3) to an equation for the function

$$\theta(\mu, \nu; t) = \int_0^\infty ds \int_{-\infty}^\infty \cdots \int_{-\infty}^\infty P_2(s, v; t) \exp[-\mu s + i(\nu \cdot v)]d^d v. \qquad (5)$$

That equation has the form

$$\frac{\partial \theta}{\partial t} = \frac{\mu}{2}\Delta_\nu \theta - b\nu^2\theta, \qquad \theta(\mu, \nu; t = 0) = 1. \qquad (6)$$

We shall look for a solution of this Cauchy problem in the form

$$\theta(\mu, \nu; t) = \exp[q(\mu, t) - \frac{1}{2}p(\mu, t)\nu^2]. \qquad (7)$$

Substituting (7) into (6), we arrive at the following equation for q and p:

$$\frac{dq}{dt} + \frac{\mu d}{2}p = 0, \qquad q(\mu, 0) = 0,$$

$$\frac{dp}{dt} + \mu p^2 = 2b, \qquad p(\mu, 0) = 0,$$

the solutions thereof, under the initial conditions indicated above, are

$$p(\mu, t) = \frac{\tau}{\sqrt{\delta}}\tanh\sqrt{\delta}, \qquad q(\mu, t) = -\frac{d}{2}\ln(\cosh\sqrt{\delta}),$$

where new variables

$$\delta = 2\mu b t^2, \qquad \tau = 2bt, \tag{8}$$

have been introduced. Substituting the above expressions for p and q into (7), we obtain that

$$\theta(\mu, \boldsymbol{\nu}; t) = \left(\frac{1}{\cosh\sqrt{\delta}}\right)^{d/2} \exp\left(-\frac{\tau}{2}\nu^2\frac{\tanh\sqrt{\delta}}{\sqrt{\delta}}\right). \tag{9}$$

In particular, for $\nu = 0$, we have the expression

$$\theta_2(\mu; t) = \left(\frac{1}{\cosh\sqrt{\delta}}\right)^{d/2} \tag{10}$$

for the Laplace transform

$$\theta_2(\mu; t) = \int_0^\infty e^{-\mu s} \mathcal{P}_2(s; t)\, ds \tag{11}$$

of the probability density of the second action component S_2.

Finally, calculating the inverse Fourier transform with respect to $\boldsymbol{\nu}$, we pass from (9) to the following expression

$$\Phi(\mu, \boldsymbol{v}; t) = \left(\frac{\sqrt{\delta}}{2\pi\tau \sinh\sqrt{\delta}}\right)^{d/2} \exp\left(-\frac{v^2\sqrt{\delta}}{2\tau \tanh\sqrt{\delta}}\right) \tag{12}$$

for the Laplace transform

$$\Phi(\mu, \boldsymbol{v}; t) = \int_0^\infty e^{-\mu s} \mathcal{P}_2(s, \boldsymbol{v}; t)\, ds \tag{13}$$

of the probability density $\mathcal{P}_2(s, \boldsymbol{v}; t)$ with respect to variable s.

Introduce an auxiliary dimensionless random variable

$$G_2 = S_2(\boldsymbol{y}, t)/2bt^2. \tag{14}$$

It follows from (10) that probability density $\tilde{\mathcal{P}}_2(g)$ is independent of time and has the Laplace transform

$$\tilde{\theta}_2(\delta) = \int_0^\infty \tilde{\mathcal{P}}_2(g) e^{-\delta g} dg = \frac{1}{\cosh\sqrt{\delta}}. \tag{15}$$

Here, as in the previous section, we have taken $d = 2$. Using the inverse Laplace transform of (15) we get

$$\tilde{\mathcal{P}}_2(g) = \langle \delta(g - G_2)\rangle = \sum_{k=0}^\infty (-1)^k \frac{2k+1}{\sqrt{\pi g g}} \exp\left(-\frac{(2k+1)^2}{4g}\right). \tag{16}$$

Probability density of the full, normed with respect to (14), action is equal to the convolution

$$\tilde{P}(g;\tau) = \tilde{P}_2(g) \otimes \tilde{P}_1(g;\tau) \tag{17}$$

of the probability density (16), and the Gaussian probability density

$$\tilde{P}_1(g;\tau) = \frac{1}{\sqrt{2\pi\epsilon^2}} \exp\left(-\frac{g^2}{2\epsilon^2}\right), \tag{18}$$

obtained from (2) by passing to dimensionless variables τ and $g = s/2bt^2$. In (18), the dimensionless parameter

$$\epsilon = 2\sqrt{ab/\tau^3}. \tag{19}$$

For sufficiently large times, when $\epsilon \ll 1$, the probability density $\mathcal{P}_1$ (18) plays the role of a delta-function in convolution (17), and we can use an approximate formula

$$\tilde{P}(g) \approx \tilde{P}_2(g). \tag{20}$$

Asymptotics of the largest value of least-action. The above discussion of statistical properties of action of noninteracting particles will help us to find the largest value of least action $\bar{S}$ which, in turn, will determine statistical properties of the Burgers turbulence in the inviscid limit. Let us introduce, similar to (14), dimensionless value

$$\rho = \bar{S}/2bt, \tag{21}$$

For very large times, when $\epsilon \ll 1$ (see (2)) and additionally $\langle N(t) \rangle \gg 1$, it is sufficient to know the behavior of function (16) for small $g \ll 1$. For such g, the sum (16) is approximately equal to its first summand. As a result, we arrive at the asymptotic formula

$$\tilde{P}(g) \sim \frac{1}{\sqrt{\pi g}g} \exp\left(-\frac{1}{4g}\right), \qquad \epsilon \ll 1, \ g \ll 1. \tag{22}$$

Similarly, it is not difficult to show that in the space of arbitrary dimension d, the probability density of action is described by an asymptotic expression

$$\tilde{P}(g) \sim \sqrt{\frac{2^d}{\pi g}} \frac{d}{4g} \exp\left(-\frac{d^2}{16g}\right), \qquad \epsilon \ll 1, \ g \ll 1. \tag{23}$$

Consequently, equation for ρ

$$\langle N(t)\rangle \int_0^\rho \tilde{P}(g)\, dg = 1 \tag{24}$$

assumes the form

$$\langle N(t)\rangle \sqrt{2^d}\, \mathrm{erfc}\left(-d/4\sqrt{\rho}\right) = 1. \tag{25}$$

Utilizing the asymptotic formula (6.3.40), we can reduce (25) to the transcendental equation

$$\langle N(t)\rangle \frac{4}{d}\sqrt{\frac{2^d \rho}{\pi}}\, \exp\left(-\frac{d^2}{16\rho}\right) = 1, \tag{26}$$

the asymptotic solution thereof can be written in the form

$$\rho = d^2 \Big/ 16\ln\left(\langle N(t)\rangle \sqrt{\frac{2^{d+1}}{\pi}}\right). \tag{27}$$

Average energy of Burgers' turbulence. Now, we can return to an analysis of the desired statistical characteristics of the forced Burgers turbulence. First of all, let us take a look at the behavior of the average kinetic energy

$$\langle u(\boldsymbol{x}, t)\rangle = \frac{1}{2}\langle v^2(\boldsymbol{x}, t)\rangle.$$

Multiply (12) by $\langle N(t)\rangle \boldsymbol{v}^2/2$ and then integrate it over all the values of $\boldsymbol{v}$. As a result, we obtain the following auxiliary function

$$T(\delta, \tau) = \langle N(t)\rangle \tau d \left(\frac{1}{\cosh\sqrt{\delta}}\right)^{d/2} \frac{\tanh\sqrt{\delta}}{\sqrt{\delta}}. \tag{28}$$

To calculate the average kinetic energy, it is necessary to find the inverse Laplace transform of that function with respect to δ, and then to integrate the obtained expression with respect to g, over the interval $(0, \rho)$. To implement these steps note that the behavior of the desired original function for small values of g, which are of interest to us, is determined by the behavior of its Laplace transform (28) for large values of δ. For that reason, we will pass in (28) to the corresponding asymptotic expression

$$T(\delta, \tau) \sim \langle N(t)\rangle \tau d \sqrt{\frac{2^d}{\delta}} e^{-(d/2)\sqrt{\delta}} = -4\tau\langle N(t)\rangle \sqrt{2^d}\frac{d}{d\delta} e^{-(d/2)\sqrt{\delta}}, \quad \delta \gg 1.$$

Finding the inverse Laplace transform of this function, integrating it over g in the interval $(0, g)$, we arrive at the following asymptotic formula for a kinetic energy of the Burgers turbulence:

$$\langle u(\boldsymbol{x}, t)\rangle \sim \tau d\langle N\rangle \sqrt{\frac{2^d}{\pi}} \int_0^\rho \frac{d\mathring{g}}{\sqrt{g}} \exp\left(-\frac{d^2}{16g}\right).$$

Replacing the integral by its main asymptotics for $\rho \ll 1$, we have

$$\langle u(\boldsymbol{x}, t)\rangle \sim \tau\rho\langle N(t)\rangle \frac{16}{d} \sqrt{\frac{2^d\rho}{\pi}} \exp\left(-\frac{d^2}{16\rho}\right).$$

Comparing this expression with equation (26) we finally obtain the following result:

Ansatz 1. *Let $\epsilon = 2\sqrt{ab/\tau^3}$, $\tau = 2bt$, and $\langle N(t)\rangle$ be the average number of streams of the auxiliary gas of noninteracting particles. Then, for $\epsilon \ll 1$, $\langle N\rangle \gg 1$, the average dimensionless kinetic energy in forced Burgers' turbulence has the following asymptotic behavior:*

$$\langle u(\boldsymbol{x}, t)\rangle \approx 4\tau\rho, \tag{29}$$

where ρ (27) is the largest possible value of least-action.

The above conclusion and formula (27) give us an opportunity to formulate a necessary condition for existence of a stationary regime in forced Burgers' turbulence:

Ansatz 2. *A necessary condition for the existence of a stationary regime in forced Burgers' turbulence is the exponential growth*

$$\langle N(t)\rangle \sim Ce^{\gamma\tau} \tag{30}$$

of the average stream-number in the auxiliary gas of noninteracting particles. The exponent γ determines the limit average energy via the formula

$$u_\infty = \lim_{t\to\infty} \langle u(\boldsymbol{x}, t)\rangle = d^2/4\gamma. \tag{31}$$

In the multidimensional case, the verification of the exponential growth law (30) requires the knowledge of the joint $2d$-dimensional probability density $\mathcal{P}(\hat{\jmath}, \hat{k}; t)$ for components of tensors $\hat{J}$ and $\hat{K}$. This is a formidable problem, both analytically and numerically. Nevertheless, linearity of the corresponding stochastic equations for $\hat{J}$ and $\hat{K}$ enables us to reach some conclusions about the behavior of the forced Burgers turbulence for $t \to \infty$.

First of all, notice that in the 2-D case it is rather easy to derive the following exact equation for the second moment $\langle J^2 \rangle$ of the Jacobian:

$$\frac{d^6 \langle J^2 \rangle}{d\theta^6} - 14 \frac{d^3 \langle J^2 \rangle}{d\theta^3} - 2\theta \langle J^2 \rangle = 0, \tag{32}$$

where $\theta = c^{1/3} t$ is the dimensionless time, and c is the third coefficient in the power series expansion

$$a(y) = a - \frac{b}{2} y^2 + \frac{c}{8} y^4 - \cdots$$

of function $a(y)$ from (6.3.1). A suitable solution of equation (32) has the form

$$\langle J^2 \rangle = \left(1 - \sqrt{3/23}\right) \left[\exp(\beta_1 \theta) + 2 \exp(-\beta_1 \theta/2) \cos(\sqrt{3}\beta_1 \theta/2)\right]$$
$$+ \left(1 + \sqrt{3/23}\right) \left[\exp(-\beta_2 \theta) + 2 \exp(\beta_2 \theta/2) \cos(\sqrt{3}\beta_2 \theta/2)\right],$$

where $\beta_{1,2} = \sqrt{\sqrt{69} \pm 7}$, and it grows monotonically with θ. As $\theta \to \infty$, we have the exponential asymptotics

$$\langle J^2 \rangle \sim (1 - \sqrt{3/23}) e^{\beta_1 \theta}$$

(see Fig. 6.4.1).

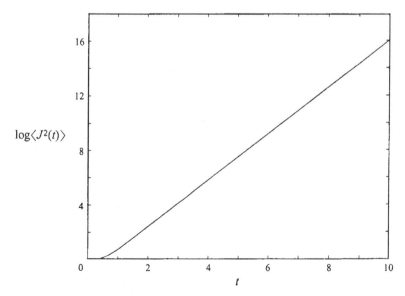

Fig. 6.4.1. The logarithmic plot of the time evolution of Jacobian's second moment. The exponential asymptotics is clearly visible.

In addition, it is also clear that

$$\langle N(t)\rangle = \langle|J|\rangle < \sqrt{\langle J^2\rangle} \sim \exp(\beta_1\theta/2). \tag{33}$$

It means that the *average energy of the forced Burgers turbulence is bounded from below and satisfies the following asymptotic inequality:*

$$\langle u(\boldsymbol{x},t)\rangle \geq d^2 b/\beta_1 c^{1/3}. \tag{34}$$

Remark 1. If the self-similarity property (6.3.49) is taken as a working hypothesis (it has been established in the previous section for the unforced Burgers turbulence), then the exponential law (30) follows with

$$\gamma = \beta_1 c^{1/3}/4b,$$

and the kinetic energy converges to the stationary value

$$u_\infty = d^2 b/\beta_1 c^{1/3}$$

which coincides with the right-hand side of bound (34). The 1-D case, where the crucial exponential law (30) can be derived by more precise methods, will be discussed in the next section.

6.5 Stream-number statistics for a 1-D gas of non-interacting particles

In this section we discuss statistical properties of the Jacobian (6.3.11) and find an asymptotic rate of growth of the average number of streams $\langle N(t)\rangle$ (6.3.29). We will restrict our attention to the 1-D case. Then, equations for the Jacobian (3.4) have a particularly simple form

$$\frac{dJ}{dt} = K, \qquad \frac{dK}{dt} = g(X,t)J. \tag{1}$$

In the delta-correlated approximation used in this lecture, the random field $g(x,t)$ can be replaced by a statistically equivalent Gaussian process $g(t)$ with zero mean and correlation function

$$\langle g(t)g(t+\theta)\rangle = 2c\delta(\theta). \tag{2}$$

We need to solve equations (1) with initial conditions

$$J(t=0) = 1, \qquad K(t=0) = 0. \tag{3}$$

Let us introduce an ordered sequence

$$0 < t_1 < t_2 < \ldots < t_m < \ldots \tag{4}$$

of times $\{t_m\}$ which are roots of the equation

$$J(t) = 0. \tag{5}$$

Take one of these times t_m as the initial time. Then, the sought solution of equation (1) for $t > t_m$ can be written in the form

$$J(t) = \tilde{K}(t_m)\tilde{J}(t|t_m), \qquad K(t) = \tilde{K}(t_m)\tilde{K}(t|t_m), \tag{6}$$

where $\tilde{J}(t|t_m)$ and $\tilde{K}(t|t_m)$ are solutions of equation (1) with the initial conditions

$$\tilde{J}(t = t_m|t_m) = 0, \qquad \tilde{K}(t = t_m|t_m) = 1. \tag{7}$$

Expressing, in turn, $\tilde{K}(t_m)$ by $\tilde{K}(t_{m-1})$ and so on, we arrive at the equality

$$\tilde{K}(t_m) = \prod_{p=1}^{m} K_p, \tag{8}$$

where

$$K_1 = K(t_1), \qquad K_p = \tilde{K}(t_p|t_{p-1}), \quad p > 1.$$

Additionally, observe that—according to (6)—the product of random variables (8) defines the value $J(t)$ of the solution of the initial value problem (1-3) at time $t > t_m$:

$$J(t) = \tilde{J}(t|t_m) \prod_{p=1}^{m} K_p. \tag{9}$$

We emphasize that, for a given value of m, all the factors in the products (8-9) are statistically mutually independent, since they are functionals of the white noise $g(t)$ on the nonoverlapping time intervals (t_{p-1}, t_p). It is not difficult to show that even a more general statement is true: elements of the sequence of random quantities $\{K_p, \tau_p\}$, where

$$\tau_p = t_p - t_{p-1},$$

with different indices p and p' are statistically independent, and the joint probability density with identical indices

$$w(\kappa, \tau) = \langle \delta(K_p - \kappa)\delta(\tau_p - \tau)\rangle, \qquad p > 1,$$

does not depend on the index p.

Recall that, in the final count, we are interested in the average stream-number $\langle N(t) \rangle$ (29)

$$\langle N(t) \rangle = \langle |J(t)| \rangle. \tag{10}$$

For sufficiently large times, when $\langle N(t) \rangle \gg 1$, using the law of large numbers one can assume that

$$m = t/\langle \tau_1 \rangle, \tag{11}$$

where $\langle \tau_1 \rangle$ is the mean length of the time interval between adjacent zeros of the process $J(t)$. In this fashion, taking into account (9), we obtain the following

Ansatz 1. *In the forced 1-D Burgers turbulence, the average stream-number*

$$\langle N(t) \rangle \sim C e^{\nu t}, \qquad t \gg \langle \tau_1 \rangle,$$

where the exponent

$$\nu = \frac{1}{\langle \tau_1 \rangle} \ln \left(\langle K \rangle \right), \tag{12}$$

and $\langle K \rangle$ is the statistical average of any of the random factors in the product (8) for $p > 1$.

Hence, the calculation of the exponent ν reduces to finding the averages $\langle \tau_1 \rangle$ and $\langle K \rangle$. These averages can be computed numerically. For that purpose we shall introduce a new dimensionless time

$$\theta = c^{1/3} t,$$

and transform (1) into dimensionless equations

$$\frac{dR}{d\theta} = K \qquad \frac{dK}{d\theta} = \alpha(\theta) R, \tag{13}$$

where $\alpha(\theta)$ is a Gaussian, delta-correlated process with correlation function

$$\langle \alpha(\theta) \alpha(\theta + \eta) \rangle = 2\delta(\eta).$$

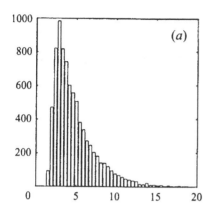 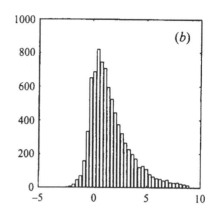

Fig. 6.5.1. The histograms of data $\{\theta_1^m\}$ (a) and $\{\log K^m\}$ (b) for $M \approx 8,000$. The logarithmic scale was needed in case (b) bacause of the huge variance of the data

The suggested scheme of numerical calculations of $\langle \tau_1 \rangle$ and $\langle K \rangle$ requires repeated numerical solutions of equations (13) with initial conditions $R(0) = 0$, $K(0) = 1$, for a large number $M \gg 1$ of statistically independent realizations of $\alpha(\theta)$. Stopping the calculations at the first moment $\theta = \theta_1 > 0$ when $R_1(\theta_1) = 0$, we obtain two data arrays $\{\theta_1^m\}$ and $\{K^m\}$, $m = 1, 2, \ldots, M$, $K^m = K^m(\theta_1^m)$, the means thereof give us approximate values of statistical averages of θ_1 and K. Notice, that $\langle \theta_1 \rangle$ is related to the above mentioned average $\langle \tau \rangle$ via an obvious equality $\langle \tau_1 \rangle = \langle \theta_1 \rangle c^{-1/3}$.

The histograms on Fig. 6.5.1 illustrate the results of $M \approx 8,000$ such numerical calculations. In particular, they provide the following estimates: $\langle \theta_1 \rangle \approx 4.83$, $\langle K \rangle \approx 81.26$, and as a result $\nu \approx 0.91c^{1/3}$.

6.6 Mechanism of energy dissipation in the inviscid 1-D Burgers turbulence

In this appendix we discuss, in the relatively simple 1-D case, the mechanism of energy dissipation in the inviscid Burgers turbulence and the corresponding problems of steady-state regimes maintained in presence of external forces.

First, let us consider the homogeneous Burgers equation

$$\frac{\partial v}{\partial t} + v \frac{\partial v}{\partial x} = \mu \frac{\partial^2 v}{\partial x^2},$$

$$v(x, t = 0) = v_0(x),$$

where $v_0(x)$ is a stationary and homogeneous random field. Then, obviously, the solution $v(x, t)$ of this equation is also a statistically homogeneous function of x. This means that the average energy

$$\langle u(x, t) \rangle = \frac{1}{2} \left\langle v^2(x, t) \right\rangle$$

obeys the equation

$$\frac{d \langle u \rangle}{dt} = -\bar{\epsilon}, \tag{1}$$

where the energy dissipation rate is defined by the formula

$$\bar{\epsilon} = \mu \left\langle g^2(x, t) \right\rangle, \tag{2}$$

where

$$g(x, t) = \frac{\partial v(x, t)}{\partial x}$$

is the Burgers turbulence velocity gradient. It is clear from (2) that, in the inviscid limit $\mu \to 0_+$, dissipation occurs only in the infinitesimal vicinities of Burgers' velocity shock fronts, where the velocity gradient has big jumps of size $\sim 1/\mu$. These large peaks balance the influence of the vanishing coefficient μ at the right hand-side of (2).

To recover the detailed mechanism of energy dissipation in the inviscid limit let us recall (see, e.g., Gurbatov, Malakhov, Saichev (1993)) the universal shape of the Burgers equation's solution in a small vicinity of the shock front of size a, moving with velocity V, and situated at the point $x^* = x - Vt + C$:

$$v_s(x - x^*, a) = V - \frac{a}{2} \tanh\left(\frac{a(x - x^*)}{4\mu} \right).$$

The corresponding velocity field gradient has, in the vicinity of this shock, the form

$$g_s(x - x^*, a) = -\frac{a^2}{8\mu} \cdot \frac{1}{\cosh^2\left(\frac{a(x-x^*)}{4\mu} \right)}. \tag{3}$$

It is physically natural to assume that, for sufficiently small viscosity μ, the gradient is of the same shape in the case of forced Burgers' velocity field. So, neglecting contribution to the dissipation rate of the gradient field realizations inbetween shocks, we can write these realizations in the form of a series of nonoverlapping peaks:

$$g(x,t) = \sum_k g_s(x - x_k, a_k),\qquad(4)$$

where x_k and a_k are coordinates and amplitudes of successive shocks. Substituting (4) into (2), and taking into account (3), we get:

$$\bar{\epsilon} = \left\langle \frac{\vartheta a^4}{64\mu} \int\limits_{-\infty}^{\infty} \frac{dx}{\cosh^4\left(ax/4\mu\right)} \right\rangle,$$

where $\vartheta(a,t)$ denotes the average spatial frequency of shocks with amplitude a at the time t, and angle brackets denote the statistical averaging over random shock amplitudes a_k. Evaluating the integral we get

$$\bar{\epsilon} = \frac{\langle\vartheta a^3\rangle}{12}.$$

In case of the forced 1-D Burgers equation (6.1.5) and delta-correlated Gaussian forces (6.1.7), in view of (6.1.9), the average energy obeys an equation similar to (1):

$$\frac{d\langle u\rangle}{dt} = -\bar{\epsilon} + \frac{1}{2}\Gamma_f,$$

where $\Gamma_f = \Gamma_f(z = 0)$. At the initial stage, when shocks are virtually absent ($\vartheta \approx 0$), we get

$$\frac{d\langle u\rangle}{dt} = \frac{1}{2}\Gamma_f,$$

and the energy of turbulence is increasing linearly:

$$<u> \approx \frac{t}{2}\Gamma_f.$$

Then the growth rate of $\bar{\epsilon}$ is reduced due the appearance of shock fronts in Burgers' velocity field realizations. Eventually, for the steady-state regime of forced Burgers' turbulence, the frequency of shocks, their amplitudes and the statistical properties of external forces are tied by the equality:

$$\langle\vartheta a^3\rangle = 6\Gamma_f.$$

6.7   Variational methods

In the next two sections we use rigorous variational methods to establish evolution of quasi-Voronoi (curved boundaries) tessellation structure of shock fronts for solutions of the inviscid nonhomogeneous Burgers equation in $\mathbf{R}^d$ in presence of random forcing due to a degenerate potential. The mean rate of growth of the quasi-Voronoi cells is calculated and a scaled limit random tessellation structure is found. Time evolution of the probability that a cell contains a ball of a given radius is also determined. The results are taken from Molchanov, Surgailis and Woyczynski (1997).

Consider the Cauchy problem for the Burgers equation with forcing

$$\frac{\partial v}{\partial t} + (v, \nabla)v = \frac{1}{2}\mu\Delta v - \nabla\Phi, \tag{1}$$

$$v(0, x) \equiv -\nabla S_0(x),$$

for the velocity field $v = v(t, x)$, $(t, x) \in [0, \infty) \times \mathbf{R}^d$, where $\Phi = \Phi(x)$ and $S_0(x)$ are given potential fields. The Hopf-Cole substitution

$$v(t, x) = -\mu\nabla \log u(t, x), \tag{2}$$

reduces (1) to the Cauchy problem for a linear *parabolic equation of the Schrödinger type*

$$\frac{\partial u}{\partial t} = \frac{1}{2}\mu\Delta u + \frac{1}{\mu}\Phi u, \tag{3}$$

$$u(0, x) \equiv e^{S_0(x)/\mu}.$$

Its solution is given by the usual *Feynman-Kac formula*

$$u(t, x) = E^W\left[\exp\left(\frac{1}{\mu}\int_0^t \Phi(x + \sqrt{\mu}W_s)ds + \frac{1}{\mu}S_0(x + \sqrt{\mu}W_t)\right)\right], \tag{4}$$

where $E^W[\ldots]$ is the expectation over trajectories of the standard Wiener process $W_s, s \geq 0, W_0 = 0$ in $\mathbf{R}^d$ (see, e.g., Freidlin, Wentzell (1984), Carmona and Lacroix (1990)). Substituting (4) into (2), one obtains

$$v(t, x) = -\left(E^W\left[\exp\left(\frac{1}{\mu}\int_0^t \Phi(x + \sqrt{\mu}W_s)ds + \frac{1}{\mu}S_0(x + \sqrt{\mu}W_t)\right)\right]\right)^{-1} \tag{5}$$

$$\times E^W\left[\int_0^t (\nabla\Phi(x + \sqrt{\mu}W_s)\,ds + \nabla S_0(x + \sqrt{\mu}W_t)) \times\right.$$

$$\times \exp\left(\frac{1}{\mu}\int_0^t \Phi(x + \sqrt{\mu}W_s)ds + \frac{1}{\mu}S_0(x + \sqrt{\mu}W_t)\right)\Big].$$

Zero viscosity limit. For $\mu = 0$, equation (1) reduces to the *Hamilton-Jacobi equation*

$$\frac{\partial S}{\partial t} - \frac{1}{2}(\nabla S, \nabla S) = \Phi, \tag{6}$$

$$S(0, x) = S_0(x),$$

for the velocity potential $S(t, x)$, $(t, x) \in [0, \infty) \times \mathbf{R}^d$ satisfying

$$\boldsymbol{v}(t, x) = -\nabla S(t, x), \qquad \boldsymbol{v}(0, x) = -\nabla S_0(x). \tag{7}$$

The solution of (6) is given by

$$S(t, x) = \sup_{\gamma \in \Gamma_{x,t}} S(t, x; \gamma), \tag{8}$$

where the action functional

$$S(t, x; \gamma) = \int_0^t \left(\Phi(\gamma(s)) - \frac{1}{2}|\dot{\gamma}(s)|^2\right)ds + S_0(\gamma(t)) \tag{9}$$

is the difference of the potential and kinetic energy, and the supremum is taken in the class $\Gamma_{x,t}$ of all paths

$$\gamma : [0, t] \longrightarrow \mathbf{R}^d, \qquad \gamma(0) = x,$$

which are absolutely continuous and satisfying the condition

$$\int_0^t |\dot{\gamma}(s)|^2 ds < \infty. \tag{10}$$

In particular, for $\Phi \equiv 0$, the extremal (Lagrangian) paths are linear:

$$\gamma(s) = x + \frac{(y - x)s}{t}$$

yielding the well known "geometric" solution

$$S(t, x) = \sup_y \left(S_0(y) - \frac{1}{2t}|x - y|^2\right). \tag{11}$$

However, the physical inviscid (limit) solution of equation (1) is defined as a limit of the Hopf-Cole solution (5) for $\mu \to 0$ (see, e.g.,

Vergassola, Dubrulle, Frisch and Nullez (1994)). Clearly, finding the limit is related to the variational problem of maximizing the integral in the exponent of formula (5), and it is this problem that we will concentrate on in what follows. In the case of the homogeneous (unforced) Burgers equation with random initial condition, an analogous, but much simpler extremal problem was discussed in Lecture 5.

Consider the variational problem (8-9) and assume that $\Phi(.), S_0(.) \in C^1(\mathbf{R}^d)$ and that the extremal path $\gamma^* \in \Gamma_{x,t}$ in (8) exists and is unique. Then, as is well known (see Courant, Hilbert (1953)) the extremal path γ^* satisfies the Euler equation

$$\ddot{\gamma}^*(s) = -\nabla\Phi(\gamma^*(s)), \qquad s \in [0, t], \tag{12}$$

and the boundary conditions

$$\gamma^*(0) = x, \qquad \dot{\gamma}^*(t) = \nabla S_0(\gamma^*(t)). \tag{13}$$

Note that, for $S_0 \equiv 0$, the extremal path *stops* at the end of time $s = t$, or perhaps earlier.

Theorem 1. *Let* $\Phi(.), S_0(.) \in C^1(\mathbf{R}^d)$,

$$\Phi(x) \le C + C_1|x|^2, \tag{14}$$

$$S_0(x) \le C + C_2|x|^2, \tag{15}$$

and

$$|\nabla\Phi(x)| \le Ce^{C|x|^2}, \tag{16}$$

where $C, C_1, C_2 < \infty$ *are constants with* $C_1 < (4t^2)^{-1}, C_2 < (4t)^{-1}$. *Furthermore, suppose that the variational problem (8) has a unique solution* $\gamma^* \in \Gamma_{x,t}$. *Then, the zero viscosity limit solution*

$$v(t, x) = \lim_{\mu \to 0} v(t, x; \mu) \tag{17}$$

of the Burgers equation (1) exists, and is given by

$$v(t, x) = -\int_0^t \nabla\Phi(\gamma^*(s))\, ds - \nabla S_0(\gamma^*(t)) = -\dot{\gamma}^*(0). \tag{18}$$

PROOF. Set

$$T_\mu(t, x) = \mu^{-1}\left(\int_0^t \Phi(x + \sqrt{\mu}W_s)\, ds + S_0(x + \sqrt{\mu}W_t)\right)$$

where $W_s, s \geq 0$, $W_0 = 0$, is the Wiener process (see (4)). Let $\Psi(x), \Psi_0(x), x \in \mathbf{R}^d$, be continuous, possibly vector-valued, functions such that

$$|\Psi(x)| + |\Psi_0(x)| \leq Ce^{C|x|^2},$$

for some constant $C < \infty$. Then

$$\lim_{\mu \to 0} \frac{E^W[(\int_0^t \Psi(x + \sqrt{\mu}W_s)\, ds + \Psi_0(x + \sqrt{\mu}W_t))e^{T_\mu(t,x)}]}{E^W[e^{T_\mu(t,x)}]}$$

$$= \int_0^t \Psi(\gamma^*(s))\, ds + \Psi_0(\gamma^*(t)). \tag{19}$$

Relation (19) can be proved as in Schilder (1966), Theorem A, where only the one-dimensional case $d = 1$ was considered; see also Freidlin, Wentzell (1984). From (14) and the Feynman-Kac formula (5), with $\Psi(x) = \nabla\Phi(x)$, $\Psi_0(x) = \nabla S_0(x)$, one immediately obtains the existence of the limit (17) and the first equality of (18). Furthermore, from (12) and the boundary conditions (13), it follows that

$$\int_0^t \nabla\Phi(\gamma^*(s))\, ds + \nabla S_0(\gamma^*(t)) = -\int_0^t \ddot{\gamma}^*(s)\, ds + \nabla S_0(\gamma^*(t)) = \dot{\gamma}^*(0),$$

which completes the proof of Theorem 1.   ∎

Remark 1. If, in addition to the conditions of Theorem 1, one assumes that the variational problem (8) has a unique solution for every x from an open neighborhood U then $S(t, \, . \,) \in C^1(U)$ and

$$-\nabla S(t, x) = \boldsymbol{v}(t, x),$$

where $\boldsymbol{v}(t, x)$ is given by (18). Also, it is worthwhile to note that the results of this section can be properly interpreted within the framework of viscosity solutions for general Hamilton-Jacobi equations (see, e.g., Lions (1982), Chapter 11) which, however, is unnecessary in our relatively simple situation of the Burgers equation.

The case of point potential.   In the present subsection our aim is to obtain an explicit description, including the structure of the shock-fronts (discontinuities), of the zero-viscosity solution (18) in the case of the degenerate "discrete" potential

$$\Phi(x) = \sum_{j \in I} h_j \mathbf{1}(x = x_j), \qquad x \in \mathbf{R}^d, \tag{20}$$

which is a superposition of zero-volume "sticks" of height $h_j > 0$ located at points x_j. The index set I is assumed to be countable and the set set $\{x_j\}_{j \in I} \subset \mathbf{R}^d$, is assumed locally finite. To simplify the problem, we consider the case of zero initial velocity, or $S_0(x) = 0$, although a discrete potential $S_0(x)$ of a similar form can easily be included (see comments at the end of next section). The explicit form of our solutions also permits us to study the evolution of their discontinuities (shock fronts).

Obviously, Theorem 1 can not be applied directly, nor can equation (18), since (20) is not even continuous. Thus, a natural approach is to approximate $\Phi(x)$ appearing in formula (20), by smooth potentials $\Phi_n(x)$ converging to $\Phi(x)$ in a certain sense, and then to define the inviscid solution $v(t, x)$ as the limit of corresponding solutions $v_n(t, x)$, i.e.

$$v(t, x) = \lim_{n \to \infty} v_n(t, x), \tag{21}$$

where

$$v_n(t, x) = - \int_0^t \nabla \Phi_n(\gamma_n^*(s))\, ds = -\dot{\gamma}_n^*(0), \tag{22}$$

and where $\gamma_n^* \in \Gamma_{x,t}$ is the solution of the variational problem (8), with $\Phi(.)$ replaced by $\Phi_n(.)$, and $S_0(x) \equiv 0$.

Theorem 2. *For $\Phi(x)$ from (20) and satisfying condition (14), the maximal action functional*

$$S(t, x) \equiv \sup_{\gamma \in \Gamma_{x,t}} \int_0^t \left(\sum_j h_j \mathbf{1}(\gamma_s = x) - \frac{1}{2} |\dot{\gamma}(s)|^2 \right) ds$$

$$= \sup_j \left(t h_j - \sqrt{2 h_j} |x - x_j| \right) \vee 0 \tag{23}$$

is the upper envelope (i.e., supremum) of cones

$$c_j(t, x) = \left(t h_j - \sqrt{2 h_j} |x - x_j| \right) \vee 0 \tag{24}$$

of height $t h_j$ and centered at x_j.

PROOF. Initially, consider the case when the set $\{x_j\}_{j \in I}$ consists of a single point x_1, i.e.

$$\Phi(x) = h_1 \mathbf{1}(x = x_1). \tag{25}$$

We want to show that

$$S(t, x) = \begin{cases} t h_1 - \sqrt{2 h_1} |x - x_1|, & \text{if } t h_1 > \sqrt{2 h_1} |x - x_1|; \\ 0, & \text{otherwise.} \end{cases} \tag{26}$$

It is easy to check that the right-hand side of (26) equals the action along the linear motion from x to x_1 with constant speed $|v| = \sqrt{2h_1}$, until reaching x_1, and then staying at x_1 for the rest of time, or the action for the trivial trajectory $\gamma \equiv x$, depending on which of the two cases take place.

To prove (26), assume that $\gamma \in \Gamma_{x,t}$ does not visit x_1. Then,

$$S(t, x; \gamma) = h_1 \int_0^t \mathbf{1}(\gamma(s) = x_1)\, ds - \frac{1}{2} \int_0^t |\dot{\gamma}(s)|^2 ds < 0,$$

unless $\gamma(s) \equiv x$. Hence, the optimal trajectory $\gamma^* \in \Gamma_{x,t}$, whenever it exists, either stays at x all the time, or visits x_1. In the latter case, γ^* obviously remains at x_1 after first hitting it. In other words, if $\gamma^*(s) \not\equiv x$, then set $\tau_1 = \inf\{s : \gamma^*(s) = x_1\} \leq t$, and

$$\gamma^*(s) = x_1, \qquad \tau_1 \leq s \leq t. \tag{27}$$

By the Cauchy-Schwartz inequality, for any $\gamma \in \Gamma_{x,t}$,

$$\int_0^{\tau_1} |\dot{\gamma}(s)|^2 ds \geq \tau_1^{-1} \left| \int_0^{\tau_1} \dot{\gamma}(s) ds \right|^2$$

$$= \tau_1^{-1} |\gamma(\tau_1) - \gamma(0)|^2 = \tau_1^{-1} |x_1 - x|^2 = \tau_1 |v|^2, \tag{28}$$

where

$$v = \frac{x_1 - x}{\tau_1}. \tag{29}$$

Hence, the trajectory that minimizes the left-hand side of (28), has to move from x to x_1 with constant velocity (29). To find τ_1, let us maximize the corresponding action

$$S(t, x; \gamma) = h_1(\tau - \tau_1) - \frac{1}{2}|v|^2 \tau_1$$

$$= h_1(\tau - \tau_1) - \frac{1}{2}\tau_1^{-1}|x - x_1|^2 \tag{30}$$

over such rectilinear paths. This yields

$$\tau_1 = \frac{|x - x_1|}{\sqrt{2h_1}}, \tag{31}$$

or

$$|v| = \sqrt{2h_1}. \tag{32}$$

This proves the special case (26).

Now, consider the general case of potential $\Phi(x)$ defined in (20). It is clear from the above discussion that $S(t, x)$ is not smaller than the right-hand side of (23). Moreover, the latter is finite and the supremum, unless zero, is achieved for some cone $c_{j*}(t, x)$, which follows from condition (14) and the fact that $\{x_j\}$ is locally finite. Indeed, $c_j(t, x) = 0$ unless $\Phi(x_j) = h_j > 2|x - x_j|^2/t^2$ or, according to (14), unless $C + |x_j|^2/(4t^2) > 2|x - x_j|^2/t^2$. The last inequality implies $|x_j| < C_4$ for some $C_4 = C_4(t, x, C) < \infty$ and any t, x, C fixed, i.e., $c_j(t, x) = 0$ for all but finitely many points x_j in view of the assumption of local finiteness of $\{x_j\}_{j \in I}$. Hence, we can assume, without loss of generality that the set $\{x_j\}_{j \in I} \equiv \{x_j\}$ is finite. Indeed, adding new points to the set $\{x_j\}$ can only increase the left-hand side of (23), while the right-hand side remains the same $(= c_{j*}(t, x))$.

For any subset $\{y_j\} \subset \{x_j\}$, and any $0 \leq \sigma \leq t$, introduce the class $\Gamma_{x,t}(\{y_j\}, \sigma) \subset \Gamma_{x,t}$ of all paths which visit *all* points of $\{y_j\}$ and which stay at those points total time σ. In other words,

$$\Gamma_{x,t}(\{y_j\}) = \Big\{\gamma \in \Gamma_{x,t} : \exists \tau_j \in [0, t] \text{ s.t. } \gamma(\tau_j) = y_j \ \forall j$$

$$\text{and } \sum_j \int_0^t \mathbf{1}[\gamma(s) = y_j]\, ds = \sigma\Big\}. \tag{33}$$

Let $y_{j*} \in \{y_j\}$ be the point of the maximal peak in this set (we suppose, for simplicity, that it is unique), i.e.,

$$h_{j*} = \max\{h_j : y_j \in \{y_j\}\}. \tag{34}$$

Then, if $\gamma^* \in \Gamma_{x,t}$ is the optimal trajectory and $\gamma^* \in \Gamma_{x,t}(\{y_j\}, \sigma)$ for some subset $\{y_j\} \subset \{x_j\}$, then γ^* has to stay at y_{j*} after it first hits it, as otherwise the action will decrease:

$$h_{j*}(t - \tau_{j*}) > \int_{\tau_{j*}}^t \Big(\sum_j h_j \mathbf{1}[\gamma(s) = y_j] - \frac{1}{2}|\dot{\gamma}(s)|^2\Big) ds,$$

for any

$$\gamma \in \Gamma_{x,t}, \quad \gamma(\tau_{j*}) = y_{j*}, \quad \gamma(s) \neq y_{j*}, \ s \in [\tau_{j*}, t].$$

Let

$$\Gamma_{x,t}^*(\{y_j\}, \sigma) \subset \Gamma_{x,t}(\{y_j\}, \sigma)$$

be the set of trajectories $\gamma(s)$ having the property that γ visits y_{j*} as its last point, and then stays at it until time t.

By the above argument, one can find $\{y_j\} \subset \{x_j\}$ and $0 \leq \sigma \leq t$ such that

$$S(t, x) = \sup_{\gamma \in \Gamma^*_{x,t}(\{y_j\}, \sigma)} S(t, x; \gamma). \tag{35}$$

Let $\{y_j\} = \{y_1, \ldots, y_n\}$ so that $y_{j^*} = y_n$. Note that, for any $\gamma \in \Gamma^*_{x,t}(\{y_j\}, \sigma)$,

$$S(t, x; \gamma) \leq S(t, x; \tilde{\gamma}), \tag{36}$$

where $\tilde{\gamma} \in \Gamma^*_{x,t}(\{y_j\}, \sigma)$ visits the points $y_1, \ldots, y_n$ in the same order as γ, but stays the *whole* time σ at $y_{j^*} = y_n$. Note, that $\tilde{\gamma}$ can be easily constructed by pasting together the parts of γ between the visits, and putting $\tilde{\gamma}(s) = y_n$ for $s \in [t - \sigma, t]$. This, however, leads to the situation discussed at the beginning of the proof, where the potential had a single peak at y_n: instead of going along $\tilde{\gamma}$, it makes more sense to go straight to y_n with speed $|v| = \sqrt{2h_n}$, and stay there afterwards. The corresponding action is then given by (26), with h_1, x_1 replaced by h_n, y_n, respectively, and y_n chosen among all x_j's so that the action is minimal. This proves (23), and Theorem 2. ∎

Theorem 3. *Assume that the following four conditions hold true:*
(i) Approximating potentials $\Phi_n(.) \in C^1(\mathbf{R}^d)$, $n \geq 1$, and they satisfy conditions (14) and (16) of Theorem 1;
(ii) There exist unique solutions γ^*_n, and $\gamma^* \in \Gamma_{x,t}$ of the variational problem (8) corresponding to potentials $\Phi_n(.)$ and $\Phi(.)$; respectively.
(iii) Potentials $\Phi_n(x) \to \Phi(x)$ decrease monotonically for each $x \in \mathbf{R}^d$;
(iv) Gradients $\nabla \Phi_n(x) \to 0$ uniformly on each compact set in $\mathbf{R}^d \setminus \{x_j\}_{j \in I}$.
Then, for a given $t > 0$ and $x \in \mathbf{R}^d$, $x \notin \{x_j\}_{j \in I}$, the limit relation (21) is valid, and the limit solution

$$v(t, x) = \begin{cases} \sqrt{2h_{j^*}} \frac{x_{j^*} - x}{|x_{j^*} - x|}, & \text{if } h_{j^*} > \frac{2|x_{j^*} - x|^2}{t^2}; \\ 0, & \text{otherwise}, \end{cases} \tag{37}$$

where (x_{j^*}, h_{j^*}) is the point which maximizes the corresponding action (23), i.e.,

$$(th_{j^*} - \sqrt{2h_{j^*}}|x - x_{j^*}|) = \sup_j (th_j - \sqrt{2h_j}|x - x_j|). \tag{38}$$

PROOF. Clearly, the right-hand side of (24) coincides with $\dot{\gamma}^*(0)$; see the above proof of Theorem 2. Therefore, by (18), the convergence (21) is equivalent to

$$\dot{\gamma}^*(0) = \lim_{n \to \infty} \dot{\gamma}^*_n(0) \tag{39}$$

Let us show that the sequence $\{\gamma_n^*(.)\}$ is relatively compact in $C([0,t];\mathbf{R}^d)$, which follows from the condition

$$\sup_n \int_0^t |\dot\gamma_n^*(s)|^2 ds < \infty; \tag{40}$$

see Freidlin, Wentzell (1984), p. 78. Write $S_n(t,x), S_n(t,x;\gamma)$ for the action functionals (8-9), with $\Phi(.)$ replaced by $\Phi_n(.)$, and $S_0(.) \equiv 0$. Since $S_n(t,x) = S_n(t,x;\gamma_n^*) \geq S_n(t,x;\gamma(.) \equiv x) = \Phi_n(x) \geq \Phi(x) \geq 0$,

$$\frac{1}{2}\int_0^t |\dot\gamma_n^*(s)|^2 ds = -S_n(t,x) + \int_0^t \Phi_n(\gamma_n^*(s))\,ds \leq \int_0^t \Phi_n(\gamma_n^*(s))\,ds$$

$$\leq \int_0^t (C + C_1|\gamma_n^*(s)|^2)\,ds \qquad \text{(see (14), Theorem 3(}i\text{))}$$

$$= \int_0^t (C + C_1|x + \int_0^s \dot\gamma_n^*(u)\,du|^2)\,ds$$

$$\leq Ct + 2C_1|x|^2 t + 2C_1 \int_0^t \left|\int_0^s \dot\gamma_n^*du\right|^2 ds$$

$$\leq C_3 + 2C_1 \int_0^t s\int_0^s |\dot\gamma_n^*|^2 du\,ds \leq C_3 + C_1 t^2 \int_0^t |\dot\gamma_n^*(u)|^2 du, \tag{41}$$

where $C_3 = C_3(t,x) < \infty$ and C_1 are independent of n; see Theorem 3, (i). As $C_1 < (4t^2)^{-1}$, this proves (40). Consequently, without loss of generality, we can assume that there exists $\gamma_\infty^* \in \Gamma_{x,t}$ such that

$$\gamma_n^* \longrightarrow \gamma_\infty^* \qquad \text{in } C([0,t];\mathbf{R}^d). \tag{42}$$

We claim that

$$\gamma_\infty^* = \gamma^*, \tag{43}$$

which follows from the fact that

$$S(t,x;\gamma_\infty^*) = S(t,x), \tag{44}$$

and the assumption (ii) in Theorem 3, to the effect that the least action is achieved at a unique point. In turn, (44) follows from the inequalities

$$\limsup_{n\to\infty} S_n(t,x;\gamma_n^*) \leq S(t,x;\gamma_\infty^*) \tag{45}$$

and

$$S(t,x) \leq S_n(t,x;\gamma_n^*), \tag{46}$$

which will be proved below.

Let us prove inequality (45) first. By (42),

$$\int_0^t |\dot\gamma_\infty^*(s)|^2 ds \le \liminf_{n\to\infty} \int_0^t |\dot\gamma_n^*(s)|^2 ds, \qquad (47)$$

see Freidlin, Wentzell (1989), Lemma 2.1(a). Next, as $\Phi_n(x)$ are continuous and monotonically decrease to $\Phi(x)$, it is easy to show that for any $K < \infty$, any $\epsilon > 0$ and $\delta > 0$, one can find an n_0 such that for all $n > n_0$,

$$\Phi_n(x) \le \begin{cases} \epsilon, & \text{if } |x - x_j| \ge \delta, \ |x| \le K; \\ h_j + \epsilon, & \text{if } |x - x_j| \le \delta, \ |x| \le K. \end{cases} \qquad (48)$$

Therefore, and with (42) in mind,

$$\int_0^t \Phi_n(\gamma_n^*(s))\, ds \le \epsilon t + \sum_j (h_j + \epsilon) \int_0^t \mathbf{1}[|\gamma_n^*(s) - x_j| < \delta]\, ds$$

$$\le \epsilon t + \sum_j (h_j + \epsilon) \int_0^t \mathbf{1}[|\gamma_\infty^*(s) - x_j| < 2\delta]\, ds, \qquad (49)$$

provided $n > n_0$ is chosen sufficiently large. As $\epsilon > 0$ and $\delta > 0$ are arbitrary, from (47) and (49) we infer (45).

On the other hand, (46) follows easily from the inequality $\Phi_n(x) \ge \Phi(x)$, which holds true for all $x \in \mathbf{R}^d$, and which implies, of course, that

$$S_n(t, x; \gamma) \ge S(t, x; \gamma)$$

for any $\gamma \in \Gamma_{x,t}$, and consequently

$$S_n(t, x) = S_n(t, x; \gamma_n^*) \ge S(t, x; \gamma)$$

for any $\gamma \in \Gamma_{x,t}$, including $\gamma = \gamma^*$, which yields (46).

It remains to prove (39). By (42-43),

$$\gamma_n^* \longrightarrow \gamma^* \qquad \text{in } C([0, t]; \mathbf{R}^d), \qquad (50)$$

where $\gamma^*(s)$ is a rectilinear motion from x, with a constant velocity $v = \dot\gamma^*(0)$, at least for some time interval $0 \le s \le \tau \le t$. Write

$$\left(\gamma_n^*(s) - \gamma_n^*(0)\right) - \left(\gamma^*(s) - \gamma^*(0)\right) = \int_0^s \left(\dot\gamma_n^*(u) - v\right) du \qquad (51)$$

$$= \int_0^s \left(\dot\gamma_n^*(u) - \dot\gamma_n^*(0)\right) du + s((\dot\gamma_n^*(0) - v)).$$

For fixed $t \geq s > 0$, the left-hand side of (51) tends to 0 as $n \to \infty$ according to (50). Hence, it remains to show that, for some $s > 0$ (s can be arbitrarily small),

$$\lim_{n \to \infty} \int_0^s \left(\dot{\gamma}_n^*(u) - \dot{\gamma}_n^*(0) \right) du = 0, \tag{52}$$

or that

$$\sup_{0 \leq u \leq s} |\dot{\gamma}_n^*(u) - \dot{\gamma}_n^*(0)| \longrightarrow 0 \qquad (n \to \infty). \tag{53}$$

But

$$|\dot{\gamma}_n^*(u) - \dot{\gamma}_n^*(0)| = \left| \int_0^u \ddot{\gamma}_n^*(r)\, dr \right| = \left| \int_0^u \nabla \Phi_n(\gamma_n^*(r))\, dr \right| \leq u \cdot \sup_{|x-y| \leq \delta} |\nabla \Phi_n(y)|, \tag{54}$$

where $\delta > 0$ is chosen so that

$$\sup_{0 \leq r \leq s} |\gamma_n^*(r) - x| \leq \delta. \tag{55}$$

Let $x \notin \{x_j\}$. Then, using (50) for a given $\delta < (1/2)$ dist $(x, \{x_j\})$, one can find $s > 0$ and $n_0 > 0$ such that (55) holds for all $n > n_0$. According to assumption (iv) of Theorem 3,

$$\sup_{|x-y| \leq \delta} |\nabla \Phi_n(y)| \longrightarrow 0 \qquad (n \to \infty),$$

which proves (53) in view of (54). Thus, the proof of Theorem 3 is complete.   ∎

6.8   Quasi-Voronoi tessellation of shock fronts

Note, that limit velocity (6.7.24) equals

$$\boldsymbol{v}(t, x) = -\nabla S(t, x), \tag{1}$$

with $S(t, x)$ given by (23). In particular, the discontinuities of (1) correspond to intersections of the cones $c_j(t, x)$ with other cones or with the zero level $\Phi(x) = 0$. If all heights $h_1 = h_2 = \ldots$ are equal and the (bases of) cones cover all space $\mathbf{R}^d$, the set of discontinuities is independent of t and coincides with the classical straight-edged *Voronoi tessellation* of $\mathbf{R}^d$ with centers x_j. Different heights $h_i \neq h_j$, lead to a more complicated *quasi-Voronoi tessellation* with curved boundaries (see Fig. 6.8.1, for the evolution of level curves for the action cones from Theorem 6.7.2). For general information on the subject of Voronoi tessellations we refer to the monograph by Okabe, Boots, Sugihara (1992) and the lecture notes by Møller (1994).

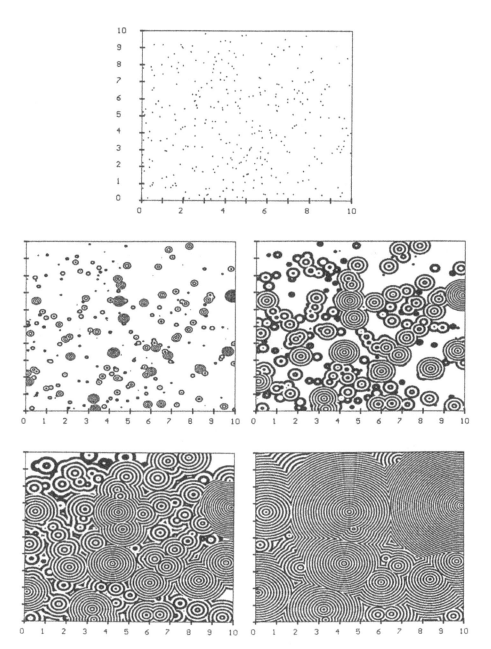

Fig. 6.8.1. Temporal evolution of shock fronts in multi-dimensional forced Burgers turbulence. *Top:* A Poisson ensemble. *Middle and bottom:* Shock fronts form a quasi-Voronoi tessellation. The boundaries between black and white areas are level curves for action cones, i.e., function $S_\infty(t, x)$ from Theorem 6.7.2, at $t = 0.2, 0.5, 1.0, 3.0$. From Janicki, Surgailis, Woyczynski (1995).

In its general features, this picture agrees with the situation observed in the unforced Burgers turbulence at large Reynolds numbers and discussed in Lecture 5 (see also, e.g., Kraichnan (1959, 1968), Gurbatov, Malakhov, Saichev (1991), Albeverio, Molchanov, Surgailis (1994), Molchanov, Surgailis, Woyczynski (1995), etc.) with the important difference that in the case of forced turbulence, the velocity *does not* decay in time—it remains constant as long as x belongs to the same Voronoi cell. At the moment in the time-evolution of the system when the cell is 'engulfed' by a larger one, the velocity increases in absolute value by an amount proportional to the square root of the height of the cone of the larger cell at time $t = 1$.

Random Poisson potential: evolution of the cell structure. Below, we consider the case of a random Poisson point potential $\Phi(x)$ (see (6.7.20)), i.e., we assume that $\{(x_j, h_j)\}$ is a (marked) Poisson process in $\mathbf{R}^{d+1}$, with intensity measure $\lambda\, dx\, dF(h)$, with F being a probability distribution function on $\mathbf{R}_+ = (0, \infty)$, and $\lambda > 0$ being a parameter. In other words, $\{x_j\}$ is a homogeneous Poisson process in $\mathbf{R}^d$ with intensity λ, while $h_j > 0$ are independent and identically distributed according to F. We assume that F satisfies condition

$$\int_{\mathbf{R}^d} \left(1 - F(|x|^2)\right) dx < \infty, \tag{2}$$

which guarantees the growth condition (6.7.14), or finiteness of the action functional $S(t, x)$, for almost all realizations $\{(x_j, h_j)\}$. Indeed, for given constants $K, K_1 < \infty$, consider the set

$$A_{K,K_1} := \{\{(x_j, h_j)\} : h_j < K + K_1|x_j|^2 \text{ for all } j\}.$$

Then, by the well-known properties of the Poisson process,

$$P(A_{K,K_1}) = E \prod_j F(K + K_1|x_j|^2)$$

$$= E \exp\left[\sum_j \log F(K + K_1|x_j|^2)\right]$$

$$= \exp\left[-\lambda K_1^{-d/2} \int_{\mathbf{R}^d} (1 - F(K + |x|^2))\, dx\right], , \tag{3}$$

which, in view of (2), implies that $P(A_{K,K_1}) \to 1$ as $K \to \infty$, so that, consequently,

$$P(\lim_{K \to \infty} A_{K,K_1}) = \lim_{K \to \infty} P(A_{K,K_1}) = 1.$$

Note, that in the final expression of (3) the negative of the term inside the exponential represents the total intensity of "bad" events $\{\{(x_j, h_j)\} : h_j \geq K + K_1|x_j|^2\}$.

With each point (x_j, h_j) we associate a, possibly empty, set

$$C_j(t) = \{x \in \mathbf{R}^d : c_j(t, x) = S(t, x)\}, \tag{4}$$

where $c_j(t, x)$ is the corresponding cone (24) with the vertex at (x_j, h_j). Any connected component $C \subset C_j(t)$ will be called a *cell*. We shall distinguish between the *first order cells*, containing the base x_j of the vertex as an interior point, and the rest, which we call *second order cells*. Apparently, only first order cells are important in the formation of the cellular structure in the presence of forcing due to a degenerate potential (20).

Let $\{(x_j^*(t), h_j^*(t))\}$ be the point process of vertices of *first order cones*, i.e., the cones above first order cells. In other words, for any $f \in C_0(\mathbf{R}^d \times \mathbf{R})$,

$$\sum_j f(x_j^*(t), h_j^*(t)) = \sum_j f(x_j, h_j) \times$$

$$\times \mathbf{1}(th_j > th_k - \sqrt{2h_k}|x_j - x_k|, \text{ for all } k \neq j). \tag{5}$$

The introduced process is a subprocess of the original Poisson process $\{(x_j, h_j)\}$; it is strictly stationary but *not* Poisson (unless $t = 0$). By the ergodicity of the Poisson process, for any fixed $t > 0$, the process $\{(x_j^*(t), h_j^*(t))\}$ is also ergodic (the σ-algebra of its shift-invariant sets is contained in the corresponding σ-algebra of the Poisson process, see also Surgailis (1981), Remark 3.2), with intensity

$$\lambda(t) = E\, N((0, 1]^d; t) = \lim_{R \to \infty} (2R)^{-d} N((-R, R]^d; t), \tag{6}$$

where $N(A; t) = \#\{j : x_j^*(t) \in A\}$ is the number of points in $A \subset \mathbf{R}^d$.

To analytically evaluate $\lambda(t)$, we shall introduce random variables $g_j = \sqrt{h_j}$, with the distribution function

$$P(g_j \leq u) = G(u) = F(u^2). \tag{7}$$

Using the relation $\lambda(t) = P(N(dx; t) = 1)/dx$, we obtain

$$\lambda(t) = P(x_j \in dx, tg_j^2 > tg_k^2 - \sqrt{2}g_k|x_j - x_k|, \text{ for all } k \neq j)/dx$$

$$= \lambda P(g_k < |x_k - x|/\sqrt{2}t + \sqrt{|x_k - x|^2/2t^2 + g_j^2}, \text{ for all } k \neq j \mid x_j \in dx)$$

$$= \lambda \int_0^\infty E \prod_k G(|x_k - x|/\sqrt{2}t + \sqrt{|x_k - x|^2/2t^2 + u^2}) \, dG(u),$$

where we can put $x = 0$. Hence, as in (3),

$$\lambda(t) = \lambda \int_0^\infty \exp\left[-\lambda(\sqrt{2}t)^d \int_{\mathbf{R}^d} (1 - G)(|x| + \sqrt{|x|^2 + u^2}) \, dx\right] dG(u). \quad (8)$$

The last formula implies that $\lambda(t) \to 0$ as $t \to \infty$; moreover, the decay rate is determined by tail behavior of the probability distribution function $G(u)$ as $u \to \infty$ (the probabilities of "high peaks"). The decay rate of $\lambda(t)$ can be rigorously obtained under the assumption that $G(.)$ is asymptotically *max-stable* (see below, also Albeverio, Molchanov, Surgailis (1994), Section 5). The inverse $1/\lambda^{1/d}(t)$ gives the order of the typical distance between the first-order cones, or the *linear scale of cells* of the quasi-Voronoi tessellation. (Indeed, we do know that there are on the average $\lambda(t) \cdot \text{Leb}(A)$ first-order cones in a large box $A \subset \mathbf{R}^d$ so, assuming that they are positioned more or less regularly, the typical distance between them should be of the order $1/\lambda^{1/d}(t)$. It would be interesting to give a rigorous interpretation of the above heuristic argument.)

Below, we assume that the tail $1 - G(u)$ of the distribution function G is continuous, strictly monotone and strictly positive for all sufficiently large u. Its inverse $(1 - G)^{-1}(.)$ is well-defined, continuous and strictly monotone on $(0, \delta)$, for some $\delta \in (0, 1)$, and $(1 - G)^{-1}(0+) = +\infty$. Put

$$H_{1,T}(u) := T(1 - G)(A(T) + uB(T)), \quad (9)$$

$$H_{2,T}(u) := T(1 - G)\left(\sqrt{A^2(T) + u^2 B^2(T)} + uB(T)\right), \quad (10)$$

$T \geq 1$, where $A(T), B(T) > 0$ are normalizing constants to be specified below. Also, recall, that a real-valued function $L(t), t > 0$, is said to be *regularly varying with exponent* $\theta \in \mathbf{R}$ if, for any $a > 0$, the ratio $L(at)/L(t) \to a^\theta$ as $t \to \infty$ (see, e.g., Bingham, Goldie, Teugels (1987)).

Theorem 1. *Assume that there exist* $A(T) = (1 - G)^{-1}(1/T)$ *and a regularly varying at infinity function* $B(T) > 0$, $T \geq 1$, *with exponent* $\theta \in [0, 1/d)$, *such that for any* $u \in \mathbf{R}$ *there exist limits*

$$\lim_{T \to \infty} H_{1,T}(u) \equiv H(u) \in [0, +\infty], \quad (11)$$

and

$$\lim_{T \to \infty} \int_{\mathbf{R}^d} H_{2,T}(|x|) \, dx \equiv h(\theta) \in (0, \infty). \quad (12)$$

Then, $\lambda(t)$, *defined in (8), regularly varies as* $t \to \infty$ *with exponent* $-d/(1 - \theta d)$; *i.e., there exists a slowly varying function* $L(t)$ *such that*

$$\lambda(t) = L(t)t^{-d/(1-\theta d)}. \tag{13}$$

Remark 1. Condition (11) implies that G is asymptotically max-stable, see e.g. Leadbetter, Lindgren, Rootzen (1983), Bingham, Goldie, Teugels (1987). Namely, for any $u \in \mathbf{R}$,

$$\lim_{n \to \infty} G^n(A(n) + uB(n)) = e^{-H(u)}.$$

The limit function has one of the three well-known parametric forms (type I, II or III extreme value distributions). Under the assumptions of Theorem 1, there are only two possibilities: either

$$H(u) = e^{-cu}, \tag{14}$$

$u \in \mathbf{R}$, with $c > 0$ (type I distribution), or

$$H(u) = (1 + cu)^{-\gamma}, \tag{15}$$

if $u > -1/c$, $H(u) = +\infty$ if $u \le -1/c$ (type II distribution), where $c, \gamma > 0$ are parameters ($\gamma > d$, according to condition (12)). In the latter case, $1 - G(u)$ is necessarily regularly varying with exponent $-\gamma$ (see, e.g., Leadbetter, Lindgren, Rootzen (1983)). In particular,

$$1 - G(u) \sim c_1 u^{-\gamma} \qquad (u \to \infty, c_1 > 0)$$

satisfies conditions of Theorem 1 with

$$A(T) = B(T) = (c_1 T)^{1/\gamma}$$

$$H(u) = (1 + u)^{-\gamma}, \qquad u > -1,$$

$$h(\theta) = h(1/\gamma) = \int_{\mathbf{R}^d} (|x| + \sqrt{|x|^2 + 1})^{-\gamma} dx,$$

yielding

$$\lambda(t) \sim c_2 t^{-d\gamma/(\gamma-d)}, \tag{16}$$

with

$$c_2 = \lambda^{-d/(\gamma-d)} \Gamma\left(\frac{2\gamma - d}{\gamma - d}\right) 2^{-d\gamma/2(\gamma-d)} h(1/\gamma)^{-\gamma/(\gamma-d)} c_1^{-d/(\gamma-d)},$$

see the proof of Theorem 1 in Surgailis, Molchanov and Woyczynski (1997).

The class of probability distributions attracted to a type I distribution (14) contains many familiar distributions such as normal, exponential, fractional (stretched) exponential (Weibull), etc. For example,

$$1 - G(u) \sim \exp[-c_3 u^\alpha], \qquad (u \to \infty, \alpha, c_3 > 0)$$

satisfies Theorem 1 with

$$\theta = 0, \qquad A(T) = c_3^{-1/\alpha} (\log T)^{1/\alpha}, \qquad B(T) = \alpha^{-1} c_3^{-1/\alpha} (\log T)^{1/\alpha - 1},$$

$$H(u) = e^{-u}, \qquad h(0) = \int_{\mathbf{R}^d} e^{-|x|} dx = 2\pi^{d/2} \Gamma(d)/\Gamma(d/2),$$

so that

$$\lambda(t) \sim c_4 t^{-d} (\log t)^{d(\alpha-1)/\alpha}, \tag{17}$$

with

$$c_4 = d^{d(\alpha-1)/\alpha} \alpha^d 2^{-1-d/2} \pi^{-d/2} c_3^{d/\alpha} \Gamma(d/2)/\Gamma(d).$$

By strengthening slightly the assumptions of Theorem 1 one can show that the process $\{(x_j^*(t), g_j^*(t))\}$ itself converges in distribution, after an appropriate scaling, to a limit process, giving rise to a limit quasi-Voronoi tessellation.

Let $\mathcal{N}(X)$ be the set of all locally finite point measures on an open set $X \subset \mathbf{R}^n$, $n \geq 1$, with the topology of vague convergence of measures (see, e.g. Kallenberg (1986)). Write $\Rightarrow$ for the convergence in distribution of random elements in $\mathcal{N}(X)$ (the weak convergence of point processes).

Theorem 2. *Assume, in addition to conditions imposed in Theorem 1, that there exists the limit*

$$\lim_{T \to \infty} \frac{A(T)}{B(T)} \equiv R \in [0, +\infty]. \tag{18}$$

Then, one can find normalizing constants $a_T, b_T \to \infty$ $(T \to \infty)$, $b_T^d \sim \text{const}/\lambda(T)$, such that the rescaled process

$$\{(x_j^*(tT)/b_T, (g_j^*(tT) - a_T)T/b_T)\}, \qquad t > 0, \tag{19}$$

converges, as $T \to \infty$, in the sense of weak convergence of finite-dimensional distributions, to an $\mathcal{N}(\mathbf{R}^d \times \mathbf{R})$-valued process

$$\{(x_{j,\infty}^*(t), g_{j,\infty}^*(t))\}, \qquad t > 0. \tag{20}$$

For each $t > 0$, the limit process (20) can be identified with the set of vertices of the cone envelope

$$S_\infty(t,x) = \begin{cases} \sup_j(2tg_{j,\infty} - \sqrt{2}|x - x_{j,\infty}|), & \text{if } R = \infty; \\ \sup_j(t(g_{j,\infty} + R)^2 - \sqrt{2}(g_{j,\infty} + R)|x - x_{j,\infty}|), & \text{if } R < \infty, \end{cases}$$
(21)

where $\{(x_{j,\infty}, g_{j,\infty})\}$ is a Poisson process on $\mathbf{R}^d \times (H_-, \infty)$, $H_- =: \inf\{u : H(u) < +\infty\}$, having intensity measure $-\lambda\, dx\, dH(u)$.

Of course, the mean density $\lambda(t)$ (6) of cells is the simplest statistical parameter of the quasi-Voronoi tessellation $\{C_j(t)\}$, with many others (e.g. the distributions of the volume, surface area, length of edges, etc., of a typical cell, and the corresponding averages) are of interest. However, exact analytic formulas are often difficult to obtain even for the classical Voronoi tessellation, usually being replaced by Monte-Carlo simulations, (see, e.g. Møller (1994), van de Weygaert (1991), and Janicki, Surgailis, Woyczynski (1995), where the correlation dimension and geometric thermodynamic temperature were estimated for the associated passive tracer flows). Sahni, Sathyaprakash, Shandarin (1994) obtained numerical histograms of the distribution of the void (cell) diameters for different times, in the adhesion model of the large scale structure of the Universe without forcing.

The distribution of the cell diameter in our model of forced Burgers' turbulence can be characterized in terms of what we call the *Palm cell function* which, by definition is the conditional probability

$$p(r,t) = P\Big(\{|y - x_j| \le r\} \subset C_j(t) \mid x_j = x \in \{x_j^*(t)\}\Big) \qquad (22)$$

that a cell $C_j(t)$ contains a ball of radius $r > 0$ centered at x_j, under the condition that the point $x_j = x$ is fixed. By stationarity, (22) does not depend on x which we can always assume to be 0. To evaluate $p(r,t)$ analytically, note that the condition

$$\{|y - x_j| \le r\} \subset C_j(t)$$

is equivalent to the condition

$$tg_k^2 - \sqrt{2}(|x_k - x| - r)g_k < tg_j^2 - \sqrt{2}rg_j$$

for all $k \ne j$. Then, as in (7-8), we obtain

$$p(r,t) = \frac{\lambda(r,t)}{\lambda(0,t)}, \qquad (23)$$

where $\lambda(0,t) = \lambda(t)$ and

$$\lambda(r,t) = P\Big(2tg_k < \sqrt{2}(|x_k - x| - r)$$

$$+\sqrt{2(|x_k - x| - r)^2 + 4tg_j(tg_j - \sqrt{2}r)}, \text{ for all } k \neq j, x_j \in dx\Big)/dx$$

$$= \lambda \int_{\sqrt{2r}/t}^{\infty} \exp\Big[-\lambda(\sqrt{2}t)^d \int_{\mathbf{R}^d}(1 - G)\Big(|x| - r/\sqrt{2}t$$

$$+\sqrt{(|x| - r/\sqrt{2}t)^2 + u(u - \sqrt{2}r/t)}\Big)dx\Big]dG(u). \tag{24}$$

Theorem 3. *Under the conditions and notation of Theorem 2, for any $r,t > 0$, there exists the limit*

$$\lim_{T\to\infty} p(rb_T, tT) = p_\infty(r,t), \tag{25}$$

which coincides with the Palm cell function for the scaling limit quasi-Voronoi tessellation function generated by $S_\infty(t,x)$ (21). In particular, in the case $R = +\infty$ and $H(u) = e^{-u}$, one has

$$p_\infty(r,t) = e^{-\sqrt{2}r/t}. \tag{26}$$

The proofs of Theorems 1-3 rely on the variational methods of Section 6.7 and can be found in Surgailis, Molchanov and Woyczynski (1997).

Nonhomogeneous initial data. The "geometric" solution of the variational problem (3) in Theorem 6.7.2 can be extended to a nonzero "discrete" initial potential $S_0(x)$ of the form

$$S_0(x) = \sum_j \xi_j 1[x = y_j], \tag{27}$$

where $\xi_j > 0$ and $y_j \in \mathbf{R}^d$ are isolated points. Namely, under the growth conditions (6.7.14-15), $S(t,x)$ coincides with the upper envelope

$$S(t,x) = \sup_j c_j(t,x) \vee p_j(t,x) \tag{28}$$

of cones $c_j(t,x)$ (6.7.23) and paraboloids

$$p_j(t,x) = (\xi_j - |x - y_j|^2/2t) \vee 0. \tag{29}$$

Then, the corresponding inviscid solution $v(t, x) = -\nabla S(t, x)$ can be approached by a smooth approximation $\Phi_n(.)$, $\xi_n(.)$, as in Theorem 6.7.3.

In the unforced case $\Phi(.) \equiv 0$, (28) yields the well-known formula

$$v(t, x) = \begin{cases} (x - y_{j*})/t, & \text{if } \xi_{j*} > |x - y_{j*}|^2/2t; \\ 0, & \text{otherwise,} \end{cases} \tag{30}$$

where (y_{j*}, ξ_{j*}) satisfies

$$\xi_{j*} - \frac{|x - y_{j*}|^2}{2t} = \sup_j \left(\xi_j - \frac{|x - y_j|^2}{2t} \right), \tag{31}$$

see Lecture 5 and, e.g., Albeverio, Molchanov, Surgailis (1994), Molchanov, Surgailis, Woyczynski (1995). The corresponding quasi-Voronoi tessellation consists of (connected) cells

$$D_j(t) = \{x \in \mathbf{R}^d; S(t, x) = p_j(t, x)\}.$$

Statistical properties of the point process $\{(y_j^*(t), \xi_j^*(t))\}$ of apexes of paraboloids $p_{j*}(t, x)$, centers of our quasi-Voronoi cells $D_j(t)$, were discussed in in Lecture 5 (see, also Albeverio, Molchanov, Surgailis (1994)), under the Poisson hypothesis of the initial process $\{(y_j, \xi_j)\}$, and similar conditions on the p.d.f. $Q(u) := P(\xi_j \leq u)$. Let

$$\nu(t) = P(y_j^*(t) \in dx)/dx$$

be the corresponding density; $\nu = \nu(0) = P(y_j \in dx)/dx$. Then, as in (8),

$$\nu(t) = \nu \int_0^\infty \exp\left[-\nu(\sqrt{2t})^d \int_{R^d}(1 - Q)(|x|^2 + u)\, dx\right] dQ(u), \tag{32}$$

which suggests a much slower decay compared with $\lambda(t)$; roughly

$$\nu(t) \approx O(\sqrt{\lambda(t)}) \qquad (t \to \infty). \tag{33}$$

In other words, *typical cells in the unforced Burgers turbulence are much smaller, roughly the square root of the size of cells in the forced turbulence*, indicating that in the latter case the formation of the cell structure occurs much faster.

Relation (33) can be rigorously established under additional conditions on the p.d.f. $Q(u)$ and $F(u) = P(h_j \leq u)$. For example, for

$$1 - Q(u) \sim 1 - F(u) \sim c_1 u^{-\gamma/2} \qquad (\gamma > d, u \to \infty),$$

we get that

$$\lambda(t) \sim c_2 t^{-d\gamma/(\gamma-d)},$$

whereas

$$\nu(t) \sim c_3 t^{-d\gamma/2(\gamma-d)},$$

where the constants $c_2, c_3 > 0$ can be explicitly found; see Remark 1. For exponentially decaying tails

$$1 - Q(u) \sim 1 - F(u) \sim \exp[-c_4 u^{\alpha/2}] \qquad (c_4, \alpha > 0, u \to \infty),$$

one obtains that

$$\lambda(t) \sim c_5 t^{-d}(\log t)^{d(\alpha-1)/\alpha},$$

whereas

$$\nu(t) \sim c_6 t^{-d/2}(\log t)^{d(\alpha-1)/\alpha},$$

which again confirms the hypothesis (33), up to a slowly varying factor.

The material presented in the last two section is only the first attempt to rigorously discuss the formation and evolution of the cellular structure in forced Burgers turbulence; our model (6.7.20) being rather a " caricature" of a more realistic potential (e.g., Gaussian) for which the problem remains unsolved. However, a discussion of such potentials may require more advanced techniques, in particular, the methods of localization theory and spectral analysis for Schrödinger operators, see, e.g., Molchanov (1994), and Molchanov, Surgailis and Woyczynski (1995a).

6.9   White noise forcing: existence and Feynman-Kac formula issues

In this section we discuss existence problems for the Burgers equation forced by spatially correlated temporal white noise (compare Sections 6.1-6) and related issues of rigorous formulation of the Feynman-Kac formalism. The material is taken from Handa (1996). Related developments can be found in Bertini, Cancrini and Jona-Lasinio (1994), and Holden, Øksendal, Ubøe and Zhang (1996).

Consider a multi-dimensional forced Burgers' equation

$$\frac{\partial \boldsymbol{v}}{\partial t} = \mu \Delta \boldsymbol{v} - \lambda(\boldsymbol{v}, \nabla)\boldsymbol{v} - \nabla \dot{w}, \qquad (1)$$

where μ is a positive constant, λ is a parameter, and $\dot{w} = \dot{w}_t(x)$ $(t > 0, x \in \mathbf{R}^d)$ is a random noise to be specified later. If the solution random field $v(t, x)$ of (1) is of the form $v = -\nabla h$ for some scalar field $h = h(t, x)$, then h satisfies the following *nonlinear Langevin equation*, called KPZ equation (see Lecture 1),

$$\frac{\partial h}{\partial t} = \mu \Delta h + \frac{\lambda}{2} |\nabla h|^2 + \dot{w} \tag{2}$$

In turn, the function $u(t, x)$ defined by

$$u(t, x) = \exp\left(\frac{\lambda}{2\mu} h(t, x)\right) \tag{3}$$

solves the linear equation with multiplicative noise term

$$\frac{\partial u}{\partial t} \mu \Delta u + \frac{\lambda}{2\mu} u\dot{w}. \tag{4}$$

This equation is often called *directed polymers in random media equation*. Indeed, regarding t as one of the space variables, we can consider a continuous path ω in

$$W_{0,0}^{t,x} := \{\omega \in C([0, t], \mathbf{R}^d); \omega(0) = 0, \omega(t) = x\}$$

as a directed polymer connecting the origin $(0, \mathbf{0})$ with (t, x) in $d + 1$-dimensions. Statistical mechanics of the ensembles of such polymers in the random potential $-\lambda \dot{w}_t(x)$ can be given in terms of the Boltzmann weight of the form

$$u(t, x) = \int_{W_{0,0}^{t,x}} \mathcal{D}\omega \exp\left[-\frac{1}{2\mu} \int_0^t ds \left\{\frac{1}{2} \left|\frac{d}{ds} \omega(s)\right|^2 - \lambda \dot{w}_s(\omega(s))\right\}\right]$$

where $\mathcal{D}\omega$ denotes the (formal) Feynman measure on $W_{0,0}^{t,x}$, and $u(t, x)$ gives the formal solution to the equation (4). Note that the constant μ here plays the role of the temperature up to some positive constant.

The goal of this section is to make the above observation rigorous and then to obtain a solution of the nonlinear equation (1). We expect that starting from the linear equation (4), the solution v of Burgers equation (1) could be constructed via the transformations

$$v = -\nabla h, \qquad h = \frac{2\mu}{\lambda} \log u, \tag{5}$$

the latter being the inverse of (3). To make this idea rigorous, two points have to be carefully considered.

The first one is that the nonlinear terms in (1) and in (2) make sense only when v or equivalently ∇h is in the usual function space. This excludes the space-time white noise, so we require the random noise to have some regularity in the space variables. Relevant issues in the context of fractal diffusions will be considered in Lecture 8.

The second difficulty is due to the fact that the change of variables formula (*Itô's formula*) for diffusion processes involves terms coming from the quadratic variations. In other words, the nonlinear transformation (5) does not result in the equation (1) as long as we understand the multiplication of the noise in (4) in the Itô's sense. It turns out, as in the finite-dimensional case, that the *Stratonovich integrals* are better adapted for our purpose, so the noise term in (4) is understood in the Stratonovich sense.

In view of the reasons mentioned above, we will introduce the notion of a spatially correlated noise, which we denote by $\eta_t(x)$, to distinguish it from the space-time white noise. Thus we will consider the *stochastic partial differential equation*

$$du(t,x) = \frac{1}{2}\Delta u(t,x)dt + u(t,x) \circ d\eta_t(x) \qquad (6)$$

instead of (4), where $\circ d\eta_t(x)$ is the Stratonovich differential. For the sake of notational simplicity we take $\mu = \frac{1}{2}, \lambda = 1$ in the rest of this section. The spatially correlated noise $\dot{\eta}_t(x)$ and precise meaning of (6) will be defined below. The notation is as follows: Given $r \geq 0$ and $m \in \mathbf{Z}_+$, let H_r^m be the completion of the pre-Hilbert space $C_0^\infty(\mathbf{R}^d)$ endowed with the scalar product $(f,g)_{r,m}$ and the norm $\|f\|_{r,m} = (f,f)_{r,m}^{1/2}$:

$$(f,g)_{r,m} = \sum_{|\alpha| \leq m} \int_{\mathbf{R}^d} D^\alpha f(x) D^\alpha g(x) e^{-r|x|} dx$$

where $\alpha = (\alpha_1, \cdots, \alpha_d) \in (\mathbf{Z}_+)^d, |\alpha| = \alpha_1 + \cdots + \alpha_d$ and

$$D^\alpha = (\partial^{\alpha_1}/\partial x_1^{\alpha_1}) \cdots (\partial^{\alpha_d}/\partial x_d^{\alpha_d}).$$

Stochastic partial differential equations with multiplicative spatially correlated noise. By *spatially correlated noise* we mean a Gaussian random field $\dot{\eta}_t(x)$ on $(0, \infty) \times \mathbf{R}^d$ specified by the characteristic functional of $\dot{\eta}_t(x)$ of the form

$$E\left[\exp(\sqrt{-1}\langle \Phi, \dot{\eta}\rangle)\right] = \exp\left[-\frac{1}{2}\int_0^\infty dt(Q\Phi(t,\cdot), \Phi(t,\cdot))\right] \qquad (7)$$

for test functions $\Phi(t, x)$, where Q is a positive self-adjoint operator on $L^2(\mathbf{R}^d)$ of trace class and $(\cdot, \cdot)$ denotes the scalar product in $L^2(\mathbf{R}^d)$. The following realization of $\dot{\eta}_t(x)$ or $\eta_t(x)$ will be utilized. Let λ_n be the eigenvalues of the square root $Q^{1/2}$ and $\{\psi_n\}$ be an orthonormal basis in $L^2(\mathbf{R}^d)$ such that

$$Q^{1/2}. = \sum_n \lambda_n(\psi_n, \cdot)\psi_n.$$

We fix independent standard $\{\mathcal{F}_t\}$-Brownian motions $\{W_t^n\}_{n=1}^\infty$ with $W_0^n = 0$ defined on a probability space with the right continuous filtration $\{\mathcal{F}_t\}$. Then

$$\eta_t(x) := \sum_n \lambda_n \psi_n(x) W_t^n \qquad (8)$$

satisfies (7). More precisely, defining $\langle \Phi, \dot{\eta} \rangle$ by the stochastic integrals

$$\langle \Phi, \dot{\eta} \rangle = \sum_n \lambda_n \int_0^\infty (\Phi(t, \cdot), \psi_n) dW_t^n, \qquad (9)$$

we have the identity (7). For each $m \in \mathbf{Z}_+$, we denote by $\mathcal{C}_b^m$ the class of above η's corresponding to the operators Q with the properties

$$\psi_n \in C^m(\mathbf{R}^d), \quad n = 1, 2, \cdots \qquad (10)$$

and

$$\sup_x a_\alpha(x) < \infty, \quad \text{for } \alpha \in (\mathbf{Z}_+)^d \text{ with } |\alpha| \leq m \qquad (11)$$

where $a_\alpha(x) = \sum_n \lambda_n^2 |D^\alpha \psi_n(x)|^2$. In the case when $\alpha = (0, \cdots, 0)$, $a_\alpha(x)$ is simply denoted by $a(x)$.

Assume that $\eta \in \mathcal{C}_b^m$ is given as in (8). To rigorously define the solution of the stochastic partial differential equation (6) transform we will formally rewrite it as the integral equation

$$u(t, x) = \int q(t, x, y) u_0(y) dy + \int_0^t \int q(t - s, x, y) u(s, y) \circ d\eta_s(y) dy \qquad (12)$$

where $u_0(\cdot)$ is the initial condition, and $q(t, x, y)$ is the heat kernel

$$q(t, x, y) = \left(\frac{1}{\sqrt{2\pi t}}\right)^d \exp\left(-\frac{|x - y|^2}{2t}\right).$$

The second term in the right-hand side of (6) is the *Stratonovich integral* with respect to η, and it can be rewritten in terms of the usual Itô integral as

$$\int_0^t \int q(t - s, x, y) u(s, y) \left\{\frac{1}{2} a(y) ds dy + d\eta_s(y) dy\right\}.$$

The meaning of the stochastic integral above is obvious in view of Funaki (1991). Namely, in general, by definition

$$\int_0^t \int f(s,y)d\eta_s(y)dy := \sum_n \lambda_n \int_0^\infty (f(s,\cdot),\psi_n)dW_s^n, \qquad (13)$$

provided that the right-hand side is well defined. To summarize, we say that an $\{\mathcal{F}_t\}$ -adapted $u(t,x)$ is a *mild solution* of the stochastic partial differential (6), with the initial condition $u_0(\cdot)$, if, with probability 1,

$$u(t,x) = \int q(t,x,y)u_0(y)dy$$

$$+ \int_0^t \int q(t-s,x,y)u(s,y)\left\{\frac{1}{2}a(y)dsdy + d\eta_s(y)dy\right\} \qquad (14)$$

Since this equation is linear, the following result can be shown by the standard arguments using (10-11), see, e.g., Funaki (1991).

Theorem 1. *Let $m \in \mathbf{Z}_+$ and $r \geq 0$. Suppose that $\eta \in C_b^m$. Then for all $u_0(\cdot) \in H_r^m$, there exists a mild solution $u(t,x)$ of (6) with the initial condition $u_0(\cdot)$, such that $u(t,\cdot) \in H_r^m$, $t > 0$ a.s. In H_r^0 this solution is pathwise unique.*

SKETCH OF THE PROOF. Given $u_0(\cdot) \in H_r^m$, define

$$U_{0,t}(u_0;x) = \int q(t,x,y)u_0(y)dy.$$

For an H_r^m-valued $\{\mathcal{F}_t\}$-adapted process $\{u(t,\cdot); \ t \geq 0\}$, set

$$U_{1,t}(u;x) = \frac{1}{2}\int_0^t ds \int q(t-s,x,y)u(s,y)a(y)dy,$$

$$U_{2,t}(u;x) = \int_0^t \int q(t-s,x,y)u(s,y)d\eta_s(y)dy.$$

Noting that the equation (14) is equivalent to

$$u(t,x) = U_{0,t}(u_0;x) + U_{1,t}(u;x) + U_{2,t}(u;x),$$

we perform the successive approximations to get a solution of (14). This is done by showing

$$\|U_{0,t}(u_0;\cdot)\|_{r,l} \ \leq \ \text{const } \|u_0\|_{r,l}$$

$$\|U_{1,t}(u;\cdot)\|_{r,l}^2 \ \leq \ \text{const } \int_0^t \|u(s,\cdot)\|_{r,l}^2 ds$$

$$E\left[\|U_{2,t}(u;\cdot)\|_{r,l}^2\right] \ \leq \ \text{const } \int_0^t E\left[\|u(s,\cdot)\|_{r,l}^2\right] ds$$

for all $0 \leq t \leq T$ and $0 \leq l \leq m$, where the constants depend on T, r and m. The pathwise uniqueness of the solution in H_r^0 is also proved by these inequalities (with $l = 0$) with the help of Gronwall's lemma. ∎

Feynman-Kac formula. Let the spatially correlated noise $\eta \in C_b^m$ be defined by (8). Our Feynman-Kac type formula will require introduction of the d- dimensional Brownian motion $\mathbf{B}_t = (B_t^1, \cdots, B_t^d)$ which is independent of $\{W_t^n\}_{n=1}^{\infty}$. Assuming that $\mathbf{B}_t$ is defined on the d-dimensional Wiener space $(\mathcal{W}, P^{\mathcal{W}})$, where $\mathcal{W} = \{\omega \in C([0, \infty), \mathbf{R}^d); \omega(0) = \mathbf{0}\}$ and $P^{\mathcal{W}}$ is the Wiener measure on $\mathcal{W}$, we can regard $\{W_t^n\}_{n=1}^{\infty}$ and $\mathbf{B}_t$ as independent stochastic processes defined on the product measure space $(\Omega \times \mathcal{W}, P \times P^{\mathcal{W}}) =: (\tilde{\Omega}, \tilde{P})$. Feynman-Kac type representations for solutions of stochastic partial differential equations of certain type can be found in, e.g., Kunita (1990), Chapter 6. But our case is not entirely covered by those results.

Theorem 2. *Suppose that u_0 is in H_r^m for some $m \in \mathbf{Z}_+$ and $r \geq 0$. Let $u(t, x)$ be the mild solution of (6) with the initial condition u_0 such that $u(t, \cdot) \in H_r^m$. Define random variables $\Gamma(t, x), t > 0, x \in \mathbf{R}^d$ on $(\tilde{\Omega}, \tilde{P})$ by*

$$\Gamma(t, x) = \sum_n \lambda_n \int_0^t \psi_n(\mathbf{B}_{t-s} + x) dW_s^n. \tag{15}$$

Then

$$u(t, x) = E^{P^{\mathcal{W}}} [u_0(\mathbf{B}_t + x) \exp \Gamma(t, x)] \tag{16}$$

Before proving Theorem 2, we will need

Lemma 1. *Let $\Gamma(t, x)$ be as in Theorem 2.*

(i) Given $\{\mathbf{B}_t; \ t \geq 0\}$, $\Gamma(t, x)$ is a Gaussian random variable such that $E^P[\Gamma(t, x)] = 0$, and

$$E^P[\Gamma(t, x)^2] = \int_0^t a(\mathbf{B}_{t-s} + x) ds \leq \sup_x a(x) \cdot t. \tag{17}$$

(ii) Denote by $v(t, x)$ the right-hand side of (16). Then

$$E\left[\|v(t, \cdot)\|_{r,0}^2\right] \leq \text{const } \|u_0\|_{r,0}^2, \tag{18}$$

where the constant depends on t and r.

PROOF. The assertion (i) is easily seen from the definition (15) of $\Gamma(t, x)$. By the Schwarz inequality and the Gaussian property in (i),

$$
\begin{aligned}
E\left[\|v(t, \cdot)\|_{r,0}^2\right] &\leq \int E^{P \times P^{\mathcal{W}}}\left[|u_0(\mathbf{B}_t + x)|^2 \exp(2\Gamma(t, x))\right] e^{-r|x|} dx \\
&\leq \text{const.} \int \int |u_0(y)|^2 q(t, x, y) e^{-r|x|} dx dy \\
&\leq \text{const.} \|u_0\|_{r,0}^2
\end{aligned}
$$

∎

PROOF OF THEOREM 2.   We continue to denote by $v(t, x)$ the right-hand side of (16). Note that since $v(t, x)$ is an H_r^0-valued $\{\mathcal{F}_t\}$-adapted process by Lemma 1 (ii), we only have to show that $v(t, x)$ satisfies the equation (14). A key to the proof of this is the following. For a fixed $\omega \in C([0, \infty), \mathbf{R}^d)$, set

$$
\Gamma_s(t, x) = \sum_n \lambda_n \int_0^s \psi_n(\omega_{t-u} + x) dW_u^n, \quad 0 \leq s \leq t.
$$

Then by the Itô formula

$$
\begin{aligned}
e^{\Gamma_t(t,x)} - 1 &= \sum_n \lambda_n \int_0^t \psi_n(\omega_{t-u} + x) e^{\Gamma_u(t,x)} dW_u^n \\
&\quad + \frac{1}{2} \int_0^t a(\omega_{t-u} + x) e^{\Gamma_u(t,x)} du, \qquad P\text{-a.s.}
\end{aligned}
$$

Multiplying both sides by $u_0(\omega_t + x)$, integrating over $\mathcal{W}$ by $P^{\mathcal{W}}(d\omega)$ and using the Markov property of the Brownian motion $\{\mathbf{B}_t; t \geq 0\}$, we obtain (14) with $u(t, x)$ replaced by $v(t, x)$. ∎

Theorem 3. *Let $\eta \in C_b^1$. Suppose that $u_0(\cdot) \in H_r^1$ is nonnegative and that there exists a Borel set A in $\mathbf{R}^d$ with positive Lebesgue measure such that*

$$
u_0(x) > 0 \quad \text{for all} \quad x \in A.
$$

Let $u(t, x)$ be the solution of (6) with the initial condition u_0 as in Theorem 1.

(i) Then, with probability 1, $u(t, x) > 0$ for all $t > 0$ and $x \in \mathbf{R}^d$.
(ii) If we set

$$
h(t, x) = \log u(t, x), \quad v(t, x) = -\nabla h(t, x), \tag{19}
$$

then

$$
\frac{\partial h}{\partial t} = \frac{1}{2}\Delta h + \frac{1}{2}|\nabla h|^2 + \dot{\eta}, \tag{20}
$$

and

$$\frac{\partial v}{\partial t} = \frac{1}{2}\Delta v - (v, \nabla)v - \nabla\dot{\eta},\tag{21}$$

in the sense of distribution, i.e., the nonlinear terms have the usual meaning and the equations are fulfilled when multiplied by test functions and then integrated.

PROOF. (i) By (16) and the Schwarz inequality,

$$u(t,x) \ge E^{P^W}\left[u_0(\mathbf{B}_t + x)^{\frac{1}{2}}\right]^2 / E^{P^W}\left[\exp(-\Gamma(t,x))\right].$$

Here, $E^{P^W}\left[u_0(\mathbf{B}_t + x)^{1/2}\right] \ge \int_A q(t,x,y)u_0(y)^{1/2} > 0$ by the assumption. On the other hand, observe that under P

$$E^{P^W}\left[\exp(-\Gamma(t,x))\right] \overset{\text{law}}{=} E^{P^W}\left[\exp\Gamma(t,x)\right].$$

The Feynman-Kac type representation (16) implies that the right-hand side in above equality is the mild solution of (6) with the initial condition $u_0(\cdot) \equiv 1$, and in particular is finite for all $t > 0$ and $x \in \mathbf{R}^d$, with probability 1.

(ii) Since $u(t,x)$ satisfies (14), we have the weak form of (6): for all $\phi \in C_0^\infty(\mathbf{R}^d)$

$$(u(t,\cdot),\phi) - (u_0,\phi)$$
$$= \frac{1}{2}\int_0^t (u(s,\cdot),(\Delta + a)\phi)ds + \int_0^t \langle u(s,\cdot)\phi, d\eta_s\rangle$$

where $(\cdot,\cdot)$ denotes the scalar product in $L^2(\mathbf{R}^d)$ and the last term is the stochastic integral

$$\int_0^t \langle u(s,\cdot)\phi, d\eta_s\rangle = \sum_n \lambda_n \int_0^t (u(s,\cdot)\phi,\psi_n)dW_s^n$$

In particular, if we set $u_\phi(t,x) = (u(t,\cdot) * \phi)(x)$, then for each $f \in C^2(\mathbf{R})$, by Itô's formula,

$$f(u_\phi(t,x)) - f((u_0 * \phi)(x))$$
$$= \frac{1}{2}\int_0^t f'(u_\phi(s,x))\Delta u_\phi(s,x)ds$$
$$+ \frac{1}{2}\int_0^t f'(u_\phi(s,x))((u(s,\cdot)a) * \phi)(x)ds$$
$$+ \int_0^t f'(u_\phi(s,x))\langle u(s,\cdot)\phi(x - \cdot), d\eta_s\rangle$$
$$+ \frac{1}{2}\int_0^t f''(u_\phi(s,x))\sum_n \lambda_n^2 |((u(s,\cdot)\psi_n) * \phi)(x)|^2 ds.$$

Take $\psi \in C_0^\infty(\mathbf{R}^d)$ to get

$$(f(u_\phi(t, \cdot)), \psi) - (f(u_0 * \phi), \psi)$$

$$= \frac{1}{2} \int_0^t (f'(u_\phi(s, \cdot))\Delta u_\phi(s, \cdot), \psi) \, ds$$

$$+ \frac{1}{2} \int_0^t (f'(u_\phi(s, \cdot))((u(s, \cdot)a) * \phi), \psi) \, ds$$

$$+ \sum_n \lambda_n \int_0^t (f'(u_\phi(s, \cdot))(u(s, \cdot)\psi_n) * \phi, \psi) \, dW_s^n$$

$$+ \frac{1}{2} \int_0^t \left(f''(u_\phi(s, \cdot)) \sum_n \lambda_n^2 |(u(s, \cdot)\psi_n) * \phi|^2, \psi \right) ds.$$

By the relation

$$\Delta f(u_\phi(s, \cdot))(x) = f''(u_\phi(s, x)) |\nabla u_\phi(s, x)|^2$$
$$+ f'(u_\phi(s, x))\Delta u_\phi(s, x)$$

the first term in the right-hand side of the previous identity is equal to

$$\frac{1}{2} \int_0^t (f(u_\phi(s, \cdot)), \Delta \psi) \, ds - \frac{1}{2} \int_0^t \left(f''(u_\phi(s, \cdot)) |\nabla u_\phi(s, \cdot)|^2, \psi \right) ds.$$

We now take $f(z) = \log z$, and consider $h(t, x) = \log u(t, x)$. Since $f'(z) = z^{-1}, f''(z) = -z^{-2}$, letting ϕ converge to the δ-function at $\mathbf{0}$ leads to

$$(h(t, \cdot), \psi) - (h(0, \cdot), \psi)$$

$$= \frac{1}{2} \int_0^t (h(s, \cdot), \Delta \psi) \, ds + \frac{1}{2} \int_0^t \left(u(s, \cdot)^{-2} |\nabla u(s, \cdot)|^2, \psi \right) ds$$

$$+ \frac{1}{2} \int_0^t \left(u(s, \cdot)^{-1} \times u(s, \cdot)a, \psi \right) ds$$

$$+ \sum_n \lambda_n \int_0^t \left(u(s, \cdot)^{-1} \times u(s, \cdot)\psi_n, \psi \right) dW_s^n$$

$$- \frac{1}{2} \int_0^t \left(u(s, \cdot)^{-2} \sum_n \lambda_n^2 |u(s, \cdot)\psi_n|^2, \psi \right) ds.$$

$$= \frac{1}{2} \int_0^t (h(s, \cdot), \Delta \psi) \, ds + \frac{1}{2} \int_0^t \left(|\nabla h(s, \cdot)|^2, \psi \right) ds + \int_0^t \langle \psi, d\eta_s \rangle$$

(Here we have used the assumption $u_0(\cdot) \in H_r^1$ to show the convergence of the nonlinear term.) This is the weak form of the required equation (20). Next, replace ψ by $-D^i \psi$. Then the above equation yields

$$\frac{\partial}{\partial t} D^i h = \frac{1}{2} \Delta D^i h + \sum_{j=1}^d D^j h \cdot D^i(D^j h) + D^i \dot\eta$$

in the sense of distribution. This proves that $v := -\nabla h$ satisfies the white noise forced Burger's equation (21). ∎

Lecture 7
Passive Tracer Transport in Burgers' and Related Flows

7.1 Burgers' turbulent diffusion, stochastic interpretation

In this lecture we will analyze evolution of passive tracer density and concentration fields advected in the velocity field governed by multidimensional Burgers and other related equations. We begin with a model description and study it at the physical level of rigorousness. This model field satisfies the mass conservation law and, in the zero viscosity limit, coincides with the generalized solution of the continuity equation. A numerical and analytical study of the evolution of such a model density field is much more convenient than the standard method of simulation of transport of passive tracer particles in the fluid. A rigorous mathematical study of the problem is an open problem.

In the 1-D case, a more general KdV-Burgers equation is suggested as a model which permits an analytical treatment of the density field in a strongly nonlinear model of compressible gas which takes into account dissipative and dispersive effects as well as pressure forces, the former not accounted for in the standard Burgers framework.

The dynamical and statistical properties of the density field are studied. In particular, utilizing the above model in the 2-D case, and most interesting for us situation of small viscosity, we can follow the creation and evolution of the cellular structures in the density field and the subsequent creation of the "quasi-particles" clusters of matter of enormous density. In addition, it is shown that in the zero viscosity limit the density field spectrum has a power tail $\propto k^{-n}$, with different exponents in different regimes. The material of the first five sections is taken from Saichev and Woyczynski (1996a ,b).

Consider a *density field* $\rho(x, t)$, governed by the *velocity field* $v(x, t)$ in $\mathbf{R}^d$ satisfying the multidimensional *Burgers equation*:

$$\frac{\partial v}{\partial t} + (v \cdot \nabla)v = \mu \Delta v, \tag{1}$$

$$v(\vec{x}, t = 0) = v_0(x).$$

We deliberately avoid writing at this point the usual continuity equation for the density field $\rho(x, t)$ since our model will subtly but significantly differ from it.

For a scalar field $S_0(y)$, and

$$\phi(y, x, t) := \frac{(y - x)^2}{2t} + S_0(y), \tag{2}$$

let us introduce the *spatial probability distribution function* (we will advisedly use the term "probability distribution function" instead of the "probability density function", to avoid confusion with the density fields that play the central role in this paper).

$$f_\mu(y; x, t) := \frac{\exp[-\frac{1}{2\mu}\phi(y, x, t)]}{\int \exp[-\frac{1}{2\mu}\phi(y, x, t)]dy}, \tag{3}$$

in y variable, which obviously satisfies the normalization condition over y:

$$\int f_\mu(y; x, t)dy \equiv 1.$$

The *spatial averaging* of an arbitrary function $g(y)$ by means of spatial distribution function $f_\mu(y; x, t)$ will be denoted by the double square brackets

$$[\![g(y)]\!] = [\![g]\!](x, t) = \int g(y)f_\mu(y; x, t)dy. \tag{4}$$

Whenever the initial field $v_0(x)$ is of potential type, multidimensional Burgers equation itself has an explicit solution. More precisely, in terms of the above spatial averaging, that fact is stated in the following theorem.

Theorem 1. *If*
$$v_0(x) = \nabla S_0(x),$$

then the Burgers equation (1) has a solution of the form

$$v(x, t) = \frac{x - [\![y]\!](x, t)}{t}, \tag{5}$$

where, in accordance with definition (4),

$$\llbracket y \rrbracket (x, t) = \int y f_\mu(y; x, t) dy. \tag{5a}$$

PROOF. Actually, formula (5) is a, convenient for our purposes, reformulation of the Hopf-Cole formula

$$v(x, t) = \frac{1}{t} \frac{\int (x - y) \exp[-\frac{1}{2\mu}\phi(y, x, t)] dy}{\int \exp[-\frac{1}{2\mu}\phi(y, x, t)] dy}, \tag{6}$$

for irrotational solutions of the Burgers equation, where ϕ is defined by (2). ■

Remark 1. It is important to notice that, for each x and $t > 0$, in the inviscid limit ($\mu \to 0+$), the spatial distribution function (2.3) weakly converges to the delta-function:

$$\lim_{\mu \to 0} f_\mu(y; x, t) = f_0(y; x, t) = \delta[y - y(x, t)], \tag{2.7}$$

where $y(x, t)$ is a y-coordinate of the absolute minimum of function $\phi(y; x, t)$ defined in (2), and the velocity field (5) is transformed into the multidimensional Hopf form of a generalized solution

$$v(x, t) = \frac{x - y(x, t)}{t} \tag{8}$$

of the Riemann equation

$$\frac{\partial v}{\partial t} + (v \cdot \nabla)v = 0. \tag{9}$$

The crucial observation, which permits a formulation of a rather simple approximate model of the density field ρ is that the vector function

$$\llbracket y \rrbracket (x, t) = x - v(x, t)t \tag{10}$$

(see (5a)), gives quasi-Lagrangian coordinates for particles in Burgers' turbulence. Indeed, the following result holds true.

Theorem 2. *The vector function $\llbracket y \rrbracket (x, t)$ satisfies equation*

$$\frac{\partial \llbracket y \rrbracket}{\partial t} + (v \cdot \nabla)\llbracket y \rrbracket = \mu \Delta \llbracket y \rrbracket, \qquad \llbracket y \rrbracket (x, t = 0) = x. \tag{11}$$

Moreover, the mapping

$$x \longmapsto y = [\![y]\!](x, t) \tag{12}$$

establishes a one-to-one correspondence between coordinate frames y *and* x.

PROOF. The validity of (11) is obtained by a direct verification. The second statement follows from the fact that the determinant $\|a_{ij}\|$ of the matrix

$$[a_{ij}] = [a_{ij}(x, t)] := \left[\frac{\partial [\![y_i]\!](x, t)}{\partial x_j} \right] = \left[\frac{1}{2\mu t} \big([\![y_i y_j]\!] - [\![y_i]\!][\![y_j]\!] \big) \right], \tag{13}$$

is positive for all $x \in \mathbf{R}^d$ since the last expression in (13) is the covariance matrix of random vector $y/\sqrt{2\mu t}$ with the d-dimensional probability distribution function $f_\mu(y, x, t)$ defined in (3) which is nondegenerate for $\mu > 0$. ∎

Remark 2. It follows from (2.13), that the matrix $[a_{ij}]$ is symmetric, i.e., $a_{ij} \equiv a_{ji}$. This implies that the vector field $[\![y]\!](x, t)$ is a potential field. Indeed, $[\![y]\!](x, t) = \nabla \psi(x, t)$, with

$$\psi(x, t) = \frac{x^2}{2} - S(x, t)t,$$

where

$$S(x, t) = -2\mu \log U(x, t)$$

is the potential of velocity field v, i.e., $v = \nabla S$.

Now, we are in a position to define our *analytical model* of the density field ρ. Suppose, that the initial density field

$$\rho(x, t = 0) = \rho_0(x),$$

is known. Consequently, in the quasi-Lagrangian coordinates, evolution of the density field is described by the formula

$$\rho(x, t) = \rho_0\big([\![y]\!](x, t)\big) \|a_{ij}(x, t)\|. \tag{14}$$

Notice, that the model density field (14) satisfies the mass conservation law. Indeed, in view of Theorem 2, the mass integral

$$m = \int \rho_0([\![y]\!](x, t)) \|a_{ij}(x, t)\| dx$$

does not depend on time t; one can change variables of integration from x to $y = [\![y]\!]$ to get that

$$m = \int \rho_0(y) dy = \text{const.}$$

The main attraction of the above model is that it is able to *analytically* predict the cellular intermittent structure of the density field associated with the Burgers flow. This program is carried out in the next few sections. Also, observe that numerical computations related to the model density field (14), which just involve evaluations of integrals of type (4), are much simpler than standard procedures (Shandarin, Zeldovich (1989), and Weinberg, Gunn (1990)) which compute the density evolution for the large-scale structure of the Universe by numerically solving a huge number of nonlinear differential equations

$$\frac{dX_i}{dt} = v(X_i, t), i = 1, 2, \ldots, N,$$

where $v(x, t)$ is a known solution of the Burgers equation (not to mention the complexities of dealing with the original N-body problem; see Kofman et al. (1992)).

Remark 3. Notice, that in our approach, equation (11) was *not* introduced as a softening, via addition of an extra Laplacian term, of the usual equation

$$\frac{\partial y}{\partial t} + (v \cdot \nabla)y = 0, \qquad y(x, t = 0) = x, \tag{15}$$

for exact Lagrangian coordinates (with which (11) coincides in the case $\mu = 0$) but it appeared as a natural and unavoidable description of the quasi-Lagrangian vector field $[\![y]\!](x, t)$ given in (10). Physical arguments justifying the replacement of the Lagrangian coordinate $y(x, t)$ by the quasi-Lagrangian $[\![y]\!](x, t)$ are provided in the next section.

Stochastic interpretation of the model density field. There exists an elegant and physically convincing interpretation of the introduced above model density field which takes into account the Brownian motion of passive tracer particles resulting from their collisions with the molecules of the surrounding medium. In this section we provide a detailed analysis of the indicated, stochastic intepretation of the model density field (14).

In the theory of turbulent diffusion one often uses the following stochastic equation

$$\frac{d\boldsymbol{X}}{dt} = \boldsymbol{v}(\boldsymbol{X},t) + \boldsymbol{\xi}(t), \tag{16}$$

$$\boldsymbol{X}(\boldsymbol{y},t=0) = \boldsymbol{y},$$

to describe the evolution of the passive tracer (see, e.g., Csanady (1980)). Above, $\boldsymbol{X}(\boldsymbol{y},t)$ denoted the coordinates of a passive tracer particle, and $\boldsymbol{v}(\boldsymbol{x},t)$—the hydrodynamic velocity field of the continuous medium in which the particle is carried. In this section we will assume that the field $\boldsymbol{v}(\boldsymbol{x},t)$ is deterministic, twice continuously differentiable everywhere, with bounded first spatial derivatives:

$$\left|\frac{\partial v_i}{\partial x_j}\right| < M < \infty, \qquad \boldsymbol{x} \in \mathbf{R}^d.$$

The white noise stochastic vector process $\boldsymbol{\xi}(t)$ in (16) which reflects the influence of random collisions of passive tracer particles with molecules of the surrounding medium will be assumed to be zero-mean Gaussian, delta-correlated, with the correlation matrix

$$\langle \xi_i(t)\xi_j(t+\tau)\rangle = 2\mu\delta_{ij}\delta(\tau), \tag{17}$$

where $\langle . \rangle$ denotes the averaging over the statistical ensemble of the realization of the process $\boldsymbol{\xi}(t)$.

It should be emphasized that in the case of a continuous medium at rest $(\boldsymbol{v} \equiv 0)$ equation (16) implies the well-known property of the Brownian motion:

$$\langle \boldsymbol{X} \rangle = \boldsymbol{y}, \qquad \langle (\boldsymbol{X} - \langle \boldsymbol{X} \rangle)^2 \rangle = 6\mu t. \tag{17a}$$

Consider the probability distribution function

$$\mathcal{P}(\boldsymbol{x};t|\boldsymbol{y}) = \langle \delta(\boldsymbol{X}(\boldsymbol{y},t) - \boldsymbol{x}) \rangle \tag{18}$$

of the coordinates of the passive tracer particle. It is a well-known fact in the theory of Markov processes that the above distribution function satisfies the forward Kolmogorov equation

$$\frac{\partial \mathcal{P}}{\partial t} + \nabla(\boldsymbol{v}\mathcal{P}) = \mu\Delta\mathcal{P}, \tag{19}$$

$$\mathcal{P}(\boldsymbol{x};t=0|\boldsymbol{y}) = \delta(\boldsymbol{x}-\boldsymbol{y}).$$

It is also well known that if the initial $(t = 0)$ density field was deterministic and equal to $\rho_0(\boldsymbol{x})$, then the mean density $\rho(\boldsymbol{x}, t)$ of the passive tracer at time t is given by the formula

$$\rho(\boldsymbol{x}, t) = \int \rho_0(\boldsymbol{y}) \mathcal{P}(\boldsymbol{x}; t | \boldsymbol{y}) \, d\boldsymbol{y}. \tag{20}$$

Multiplying equation (19) by $\rho_0(\boldsymbol{y})$, and integrating it over all $\boldsymbol{y}$'s, we obtain equation

$$\frac{\partial \rho}{\partial t} + \nabla(\boldsymbol{v}\rho) = \mu \Delta \rho, \tag{21}$$

$$\rho(\boldsymbol{x}; t = 0 | \boldsymbol{y}) = \rho_0(\boldsymbol{x}),$$

for the density field averaged over the ensemble of realizations of process $\boldsymbol{\xi}(t)$.

Notice that the above equation differs from the classical hydrodynamic continuity equation by the "superfluous" diffusion term, which takes into account the Brownian motion of the particle due to random molecular collisions. It is useful to recall that the diffusive term in (21) takes into account the discrete, molecular structure of the medium, that is completely ignored in the hydrodynamic derivation of the continuity equation.

Let us additionally provide another, equivalent to (20), form of the density of the passive tracer, averaging over the statistical ensemble of random molecular diffusions. For that purpose observe that if the conditions imposed above on the velocity field $\boldsymbol{v}(\boldsymbol{x}, t)$ are satisfied then, for each separate realization of the process $\boldsymbol{\xi}(t)$, the equality

$$\boldsymbol{x} = \boldsymbol{X}(\boldsymbol{y}, t) \tag{22}$$

defines a continuously differentiable and one-to-one mapping of $\boldsymbol{y} \in \mathbf{R}^d$ onto $\boldsymbol{x} \in \mathbf{R}^d$. Consequently, there exists an inverse mapping

$$\boldsymbol{y} = \boldsymbol{y}(\boldsymbol{x}, t), \tag{23}$$

with the same properties, and the Jacobian

$$\|\tilde{a}_{ij}\|, \qquad \tilde{a}_{ij} = \frac{\partial y_i(\boldsymbol{x}, t)}{\partial x_j}, \tag{24}$$

is continuous, bounded everywhere and strictly positive:

$$0 < \|\tilde{a}_{ij}\| < \infty.$$

Under the above conditions, the following distribution-theoretic formula is valid:

$$\delta(\boldsymbol{X}(\boldsymbol{y},t) - \boldsymbol{x}) = \|\tilde{a}_{ij}\|\delta(\boldsymbol{y}(\boldsymbol{x},t) - \boldsymbol{y}). \tag{25}$$

Averaging this equality over the ensemble of realization of $\boldsymbol{\xi}(t)$ we arrive at another useful expression for the probability distribution (18):

$$\mathcal{P}(\boldsymbol{x};t|\boldsymbol{y}) = \langle\|\tilde{a}_{ij}\|\delta(\boldsymbol{y} - \boldsymbol{y}(\boldsymbol{x},t))\rangle. \tag{25a}$$

Substituting it into (20) and using the defining property of the Dirac delta-function, we arrive at the promised expression for the, averaged over the ensemble of $\boldsymbol{\xi}(t)$, density of the passive tracer:

$$\rho(\boldsymbol{x},t) = \langle\|\tilde{a}_{ij}\|\rho_0(\boldsymbol{y}(\boldsymbol{x},t))\rangle \tag{26}$$

which is an alternative to (20).

Let us replace this exact equation by an approximate equation using a, commonly encountered in physics, *mean field approach*. Its essence is the drastic replacement of the averages of functions of random arguments by functions of their averages. For example,

$$\left\langle f\left(\boldsymbol{y}(\boldsymbol{x},t), \frac{\partial y_i}{\partial x_j}\right)\right\rangle \Longrightarrow f\left(\langle\boldsymbol{y}\rangle, \frac{\partial\langle y_i\rangle}{\partial x_j}\right).$$

Applying the mean field approach to the right-hand side (26) we arrive at the equality

$$\rho(\boldsymbol{x},t) = \|a_{ij}\|\rho_0(\langle\boldsymbol{y}(\boldsymbol{x},t)\rangle), \tag{27}$$

where

$$\|a_{ij}\| = \|\tilde{a}_{ij}\| \mid_{\boldsymbol{y}=\langle\boldsymbol{y}\rangle} = \left\|\frac{\partial\langle y_i\rangle}{\partial x_j}\right\|.$$

Leaving aside the question of the degree of validity of the mean field approach, let us turn our attention to the fact that the equality (27) formally coincides, up to the replacement of $\langle y\rangle$ by $[\![y]\!]$, with equality (14). This suggests that the proposed stochastic interpretation of the model density field is as follows: *Model density field (14) coincides with the average (with respect to the Brownian motion) of the passive tracer density, computed by the mean field approach.* To convince ourselves about the validity of this statement it suffices to prove the following

Theorem 3. *The statistical average $\langle\boldsymbol{y}(\boldsymbol{x},t)\rangle$ of the random mapping (23) satisfies equation (11).*

PROOF. Consider an auxiliary function (Wronskian)

$$J(\boldsymbol{y}, t) = \left\| \frac{\partial X_i(\boldsymbol{y}, t)}{\partial y_j} \right\|, \tag{28}$$

satisfying equation

$$\frac{dJ}{dt} = J(\nabla \cdot \boldsymbol{v}), \qquad J(\boldsymbol{y}, t = 0) = 1, \tag{29}$$

where

$$(\nabla \cdot \boldsymbol{v}) = \frac{\partial v_i(\boldsymbol{X}(\boldsymbol{y}, t), t)}{\partial X_i}.$$

Clearly, the vector $\{\boldsymbol{X}(\boldsymbol{y}, t), J(\boldsymbol{y}, t)\}$ forms a $(d + 1)$-dimensional Markov process with joint probability distribution function

$$\mathcal{P}(\boldsymbol{x}, j; t | \boldsymbol{y}) = \langle \delta(\boldsymbol{X}(\boldsymbol{y}, t) - \boldsymbol{x}) \delta(J(\boldsymbol{y}, t) - j) \rangle, \tag{30}$$

satisfying the forward Kolmogorov equation

$$\frac{\partial \mathcal{P}}{\partial t} + \nabla(\boldsymbol{v}(\boldsymbol{x}, t) \mathcal{P}) + (\nabla \cdot \boldsymbol{v}(\boldsymbol{x}, t)) \frac{\partial}{\partial j}(j \mathcal{P}) = \mu \Delta \mathcal{P}, \tag{31}$$

$$\mathcal{P}(\boldsymbol{x}, j; t = 0 | \boldsymbol{y}) = \delta(\boldsymbol{x} - \boldsymbol{y}) \delta(j - 1).$$

Using the obvious identity

$$J(\boldsymbol{y}, t) \|\tilde{a}_{ij}\| \Big|_{\boldsymbol{x} = \boldsymbol{X}(\boldsymbol{y}, t)} \equiv 1,$$

we can rewrite the relation (25) in the form

$$\delta(\boldsymbol{y}(\boldsymbol{x}, t) - \boldsymbol{y}) = J(\boldsymbol{y}, t) \delta(\boldsymbol{X}(\boldsymbol{y}, t) - \boldsymbol{x}).$$

Therefore it follows that the probability distribution function

$$\mathcal{Q}(\boldsymbol{y}; t | \boldsymbol{x}) = \langle \delta(\boldsymbol{y}(\boldsymbol{x}, t) - \boldsymbol{y}) \rangle$$

of the vector stochastic process $\boldsymbol{y}(\boldsymbol{x}, t)$ is related to the probability distribution (30) by the equality

$$\mathcal{Q}(\boldsymbol{y}; t | \boldsymbol{x}) = \int j \mathcal{P}(\boldsymbol{x}, j; t | \boldsymbol{y}) \, dj.$$

Multiplying equation (31) by j and integrating it over j, we arrive at the conclusion that $\mathcal{Q}(\boldsymbol{y}; t | \boldsymbol{x})$ satisfies the following *backward Kolmogorov equation*:

$$\frac{\partial \mathcal{Q}}{\partial t} + (\boldsymbol{v}(\boldsymbol{x}, t) \cdot \nabla) \mathcal{Q} = \mu \Delta \mathcal{Q}, \tag{32}$$

$$Q(\boldsymbol{y}; t | \boldsymbol{x}) = \delta(\boldsymbol{y} - \boldsymbol{x}).$$

The above equation implies, in particular, that the statistical average

$$\langle \boldsymbol{y}(\boldsymbol{x}, t) \rangle = \int \boldsymbol{y} Q(\boldsymbol{y}; t | \boldsymbol{x}) \, d\boldsymbol{y}, \tag{33}$$

which is of principal interest to us, is a solution of the Cauchy problem

$$\frac{\partial \langle \boldsymbol{y} \rangle}{\partial t} + (\boldsymbol{v} \cdot \nabla) \langle \boldsymbol{y} \rangle = \mu \Delta \langle \boldsymbol{y} \rangle, \tag{34}$$

$$\langle \boldsymbol{y}(\boldsymbol{x}, t = 0) \rangle = \boldsymbol{x}.$$

This concludes the proof of Theorem 3.   ∎

Remark 4. The formulation of Theorem 3.1 (replacing $\langle \boldsymbol{y} \rangle$ by $[\![\boldsymbol{y}]\!]$) is similar to the formulation of the first part of Theorem 2. The essential difference is that the field $\boldsymbol{v}(\boldsymbol{x}, t)$ in Theorem 3 need not be a solution of the Burgers' equation (1).

Remark 5. In the case velocity field $\boldsymbol{v}(\boldsymbol{x}, t)$ satisfies the Burgers equation with the coefficient μ identical to the coefficient of molecular diffusion in (17), then the average field $\langle \boldsymbol{y}(\boldsymbol{x}, t) \rangle$ is expressed by the velocity field with the help of a simple formula (10)). This serendipitous coincidence provided us with analytical advantages of the suggested model density field.

Remark 6. We have derived equation (34) utilizing equality (33), where Q satisfied equation (32). In the case of the Burgers velocity field $\boldsymbol{v}(\boldsymbol{x}, t)$, where $[\![\boldsymbol{y}]\!](\boldsymbol{x}, t)$ (see (5a)) satisfies equation (11), it is natural to assume that f_μ, like Q, also satisfies equation (32). A direct substitution shows that, indeed, this is the case. Hence, the spatial probability distribution function f_μ, like the density field itself, has a clear-cut stochastic interpretation: $f_\mu(\boldsymbol{y}; \boldsymbol{x}, t)$ *is the probability distribution of the Lagrangian coordinates of passive tracer particle driven by the Burgers velocity field and subjected to Brownian motion.*

Remark 7. Another important quantity in the theory of turbulent diffusion is the *mean concentration*

$$C(\boldsymbol{x}, t) = \int C_0(\boldsymbol{y}) Q(\boldsymbol{y}; t | \boldsymbol{x}) \, d\boldsymbol{y},$$

of the passive tracer which satisfies equation

$$\frac{\partial C}{\partial t} + (\boldsymbol{v}(\boldsymbol{x}, t) \nabla) C = \mu \Delta C,$$

$$C(\boldsymbol{x}, t = 0) = C_0(\boldsymbol{x}).$$

Recall, that the density ρ is proportional to the number of passive tracer particles in the unit volume, whereas the concentration C is proportional to the *ratio* of the number of passive tracer particles in the unit volume to the number of particles of the surrounding fluid in the same volume. In view of the preceding remark, in the case of Burgers' velocity field $\boldsymbol{v}(\boldsymbol{x}, t)$, we also obtain an *exact* solution

$$C(\boldsymbol{x}, t) = \int C_0(\boldsymbol{y}) f_\mu(\boldsymbol{y}; \boldsymbol{x}, t)\, d\boldsymbol{y}$$

of the equation for the concentration field. We will return to these issues later on.

Remark 8. The 1-D case occupies a special place. Then the determinant $\|\tilde{a}\|$ is tied to $y(x, t)$ via a simple linear equation

$$\|\tilde{a}\| = \frac{\partial y(x, t)}{\partial x}, \tag{35}$$

which implies that in the 1-D case we have the following identity:

$$\langle \|\tilde{a}\| \rangle \equiv \|a\|.$$

If, in addition, the initial density is the same everywhere ($\rho_0 = \text{const}$), then the "real" density

$$\rho(x, t) = \rho_0 \int \mathcal{P}(x; t|y)\, dy$$

is *exactly* equal to the density calculated via the mean field approach:

$$\rho(x, t) = \rho_0 \|a\| = \rho_0 \frac{\partial \langle y \rangle (x, t)}{\partial x}. \tag{35a}$$

7.2 2-D cellular structures

In general, as we observed above, the model density field $\rho(\boldsymbol{x}, t)$ described in (7.1.14) may be found only by numerical calculations of integrals similar to (7.1.4). However, in the most interesting for astrophysical applications case of small viscosity μ (Shandarin and Zeldovich (1989), Weinberg and Gunn (1990), Kofman et al. (1992), Vergassola et al. (1994)), which corresponds to the strongly nonlinear regime of evolution of the velocity and density fields, it is possible to obtain a physically acceptable analytical approximations for $\rho(\boldsymbol{x}, t)$. To escape unwieldy formulas which arise in the 3-D case, we shall demonstrate this fact on the 2-D case. Hence, for the remainder of this section, $\boldsymbol{x} \in \mathbf{R}^2$.

Assume that the initial velocity field potential $S_0(\boldsymbol{x})$ has second partial derivatives which are bounded from below. This implies that there exists a $t_1 > 0$ such that if $0 < t < t_1$ then, for any given $\boldsymbol{x}$, function $\phi(\boldsymbol{y}, \boldsymbol{x}, t)$ defined in (7.1.2) has only one local minimum. Let's denote coordinates of this minimum by $\boldsymbol{y}(\boldsymbol{x}, t)$. As far as the distribution $f_\mu(\boldsymbol{y}; \boldsymbol{x}, t)$ is concerned, the existence of a unique minimum of $\phi(\boldsymbol{y}, \boldsymbol{x}, t)$ means that at any $\boldsymbol{x}$, the Gaussian approximation is asymptotically valid for small μ. It corresponds to an application for $f_\mu(\boldsymbol{y}; \boldsymbol{x}, t)$ of the steepest descent method for calculation of integrals similar to (7.1.4). The method justifies an approximation of $\phi(\boldsymbol{y}, \boldsymbol{x}, t)$ by the first three terms of its Taylor expansion:

$$\phi(\boldsymbol{y}, \boldsymbol{x}, t) \approx \phi\big(\boldsymbol{y}(\boldsymbol{x}, t), \boldsymbol{x}, t\big) + \vec{\kappa} \cdot \left(\boldsymbol{v}_0\big(\boldsymbol{y}(\boldsymbol{x}, t)\big) + \frac{\boldsymbol{y}(\boldsymbol{x}, t) - \boldsymbol{x}}{t}\right)$$

$$+ \frac{1}{2t} A_{ij}(\boldsymbol{x}, t) \kappa_i \kappa_j, \tag{1}$$

where

$$A_{ij}(\boldsymbol{x}, t) = \left(\delta_{ij} + t \frac{\partial^2 S_0(\boldsymbol{y})}{\partial y_i \partial y_j}\right)\bigg|_{\boldsymbol{y} = \boldsymbol{y}(\boldsymbol{x}, t)},$$

and $\boldsymbol{\kappa} = \boldsymbol{y} - \boldsymbol{y}(\boldsymbol{x}, t)$. Since $\boldsymbol{y}(\boldsymbol{x}, t)$ are coordinates of a minimum of a smooth function $\phi(\boldsymbol{y}, \boldsymbol{x}, t)$ over $\boldsymbol{y}$'s, we have that

$$\boldsymbol{v}_0\big(\boldsymbol{y}(\boldsymbol{x}, t)\big) + \frac{\boldsymbol{y}(\boldsymbol{x}, t) - \boldsymbol{x}}{t} = 0. \tag{2}$$

Consequently, in this approximation, the spatial distribution function (7.1.3) is described by the asymptotic formula

$$f_\mu(\boldsymbol{y}; \boldsymbol{x}, t) \approx \frac{1}{4\pi\mu t} \sqrt{j(\boldsymbol{x}, t)} \exp\left(-\frac{1}{4\mu t} A_{ij}(\boldsymbol{x}, t) \kappa_i \kappa_j\right). \tag{3}$$

Here, again $\boldsymbol{\kappa} = \boldsymbol{y} - \boldsymbol{y}(\boldsymbol{x}, t)$, where $\boldsymbol{y}(\boldsymbol{x}, t)$ is a root of equation (2), and

$$j(\boldsymbol{x}, t) = J(\boldsymbol{y}, t)\big|_{\boldsymbol{y} = \boldsymbol{y}(\boldsymbol{x}, t)} = \|A_{ij}\|\big|_{\boldsymbol{y} = \boldsymbol{y}(\boldsymbol{x}, t)}$$

$$= \left[\left(1 + t \frac{\partial^2 S_0(\boldsymbol{y})}{\partial y_1^2}\right)\left(1 + t \frac{\partial^2 S_0(\boldsymbol{y})}{\partial y_2^2}\right) - t^2 \left(\frac{\partial^2 S_0(\boldsymbol{y})}{\partial y_1 \partial y_2}\right)^2\right]\bigg|_{\boldsymbol{y} = \boldsymbol{y}(\boldsymbol{x}, t)}. \tag{4}$$

It is worth recalling that, as $\mu \to 0$, function f_μ from (3) weakly converges to the delta-function (7.1.7). Nevertheless, for the density field itself, we have from (7.1.13-14)

$$\rho(\boldsymbol{x}, t) = \left(\frac{1}{2\mu t}\right)^2 \rho_0\big(\llbracket \boldsymbol{y} \rrbracket(\boldsymbol{x}, t)\big) \times$$

$$\times \left(\llbracket (y_1 - \llbracket y_1 \rrbracket)^2 \rrbracket \llbracket (y_2 - \llbracket y_2 \rrbracket)^2 \rrbracket - \llbracket (y_1 - \llbracket y_1 \rrbracket)(y_2 - \llbracket y_2 \rrbracket) \rrbracket^2 \right), \qquad (5)$$

where, as above, the square brackets signify the "spatial average" with respect to the distribution (7.1.3), and a calculation of the "thickness" of the distribution (3) is of principal importance. In particular, we have the following result:

Theorem 1. *If $t < t_1$ then the limit*

$$\lim_{\mu \to 0} \rho(\boldsymbol{x}, t) = \frac{\rho_0(\boldsymbol{y}(\boldsymbol{x}, t))}{j(\boldsymbol{x}, t)}, \qquad (6)$$

of the model density field (5) exists, where $\boldsymbol{y}(\boldsymbol{x}, t)$ are the Lagrangian coordinates and $j(\boldsymbol{x}, t)$ is the Jacobian of the transformation from Lagrangian to Eulerian coordinates of the continuous medium, the velocity field thereof satisfies the Riemann equation (7.1.9) with the initial condition (7.1.1).

PROOF. A direct calculation shows that the desired limit expression for the model density field is described by the right hand side of equation (6), where $\boldsymbol{y}(\boldsymbol{x}, t)$ are the coordinates of the minimum of function ϕ in formula (7.1.2) and $j(\boldsymbol{x}, t)$ is given by expression (4). Now, it suffices to prove that the fields $\boldsymbol{y}(\boldsymbol{x}, t)$ and $j(\boldsymbol{x}, t)$ defined in this fashion are indeed, respectively, the fields of Lagrangian coordinates and of the Jacobian. For that purpose, observe that the unique (for $t < t_1$) solution of the Riemann equation (7.1.9) can be written in the form

$$\boldsymbol{v}(\boldsymbol{x}, t) = \boldsymbol{v}_0(\boldsymbol{y}(\boldsymbol{x}, t)),$$

where $\boldsymbol{y}(\boldsymbol{x}, t)$ is the field of Lagrangian coordinates. Comparing the last equation with (7.1.2), and (7.1.2) with equality (7.1.8), we realize that the coordinates of the minimum of function (7.1.2) are indeed simultaneously the Lagrangian coordinates.

Furthermore, note that the Riemann equation describes the field of velocities in the continuous medium all the particles thereof move with uniform speed on straight lines. This means that the Eulerian and Lagrangian coordinates of such a medium are connected by equality

$$x = y + v_0(y)t, \qquad v_0(y) = \nabla S_0(y).$$

The Jacobian

$$J(y,t) = \left\| \frac{\partial x_i}{\partial y_j} \right\|$$

coincides with the expression (4) in the 2-D case under consideration. Thus the Theorem is proved.   ∎

Remark 1. The proof of Theorem 1 can be also easily carried out in the case of $x \in \mathbf{R}^d$ for arbitrary dimension d.

Remark 2. The time instant t_1, which appeared in the above considerations, is the infimum of positive roots of the equation $J(y,t) = 0$ over all y.

Remark 3. The right hand side of equation (6) is familiar for the physicists and has a transparent physical meaning: The density at a given point x is equal to the initial density at this point of the continuous medium divided by the Jacobian $j(x,t)$ which describes the influence of squeezing and stretching of the fluid.

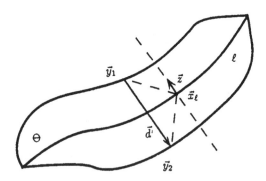

Fig. 7.2.1. A typical example of curve ℓ—a 2-D analog of 3-D "pancake structures" and corresponding region Θ bounded by curves $y_1(x_\ell, t)$ and $y_2(x_\ell, t)$. In the inviscid limit, the region Θ has a vivid physical interpretation: All particles inside it at the initial time $t = 0$, and driven by the Burgers velocity field, end up on the "pancake" curve ℓ at time t.

At this point we will suppose, for simplicity, that the initial density is the same at all points $\boldsymbol{x}$, i.e., that

$$\rho_0(\boldsymbol{x}) \equiv \rho_0 = \text{const}, \tag{7}$$

and, under these circumstances, we will analyze evolution of the density fields. This case corresponds, for instance, to the initial stage of development of the gravitation instability of matter in the Universe (cf. e.g., Shandarin and Zeldovich (1989), Weinberg and Gunn (1990), Kofman et al. (1992), Vergassola et al. (1994)).

For $0 < t < t_1$, the auxiliary field $J(\boldsymbol{y}, t)$ defined in (4) is positive for any $\boldsymbol{y}$. For $t > t_1$, there appear regions in the $\boldsymbol{y}$ plane, where $J(\boldsymbol{y}, t)$ is negative. At this stage, there appear on $\boldsymbol{x}$-plane curves ℓ, at all points $\boldsymbol{x}_\ell$ thereof function $\phi(\boldsymbol{y}, \boldsymbol{x}, t)$ defined in (2.2) has two minima of equal value. Let's mark the coordinates of these two minima as $\boldsymbol{y}_1(\boldsymbol{x}_\ell, t)$ and $\boldsymbol{y}_2(\boldsymbol{x}_\ell, t)$. As point $\boldsymbol{x}$ travels over curve ℓ, points $\boldsymbol{y}_1, \boldsymbol{y}_2$ draw a closed curve in the $\boldsymbol{y}$-plane which bounds a certain region Θ (see Fig. 7.2.1).

The curve ℓ shown on Fig. 7.2.1 represents a 2-D analog of 3-dimensional "pancake" structures which appear in the large-scale distribution of matter in the Universe (see Gurbatov et al. (1991) and the astrophysical papers quoted above).

If μ is sufficiently small, the density in a small vicinity of these curves is very large and we can calculate the density field there via a bimodal approximation. In this approximation, function $f_\mu(\boldsymbol{y}; \boldsymbol{x}, t)$ from (7.1.3) is represented by a superposition of two unimodal distributions (3)

$$f_\mu(\boldsymbol{y}; \boldsymbol{x}, t) = \frac{f_1(\boldsymbol{y}; \boldsymbol{x}_\ell, t) + R f_2(\boldsymbol{y}; \boldsymbol{x}_\ell, t)}{1 + R}, \tag{4.8}$$

where

$$R = \sqrt{\frac{J_1}{J_2}} \exp\left(\frac{\vec{dz}}{2\mu t}\right), \quad \vec{d}(\boldsymbol{x}_\ell, t) = \boldsymbol{y}_2(\boldsymbol{x}_\ell, t) - \boldsymbol{y}_1(\boldsymbol{x}_\ell, t), \tag{9}$$

and where $\vec{z} = \boldsymbol{x} - \boldsymbol{x}_\ell$ is a vector located on the dashed straight line r perpendicular to $\boldsymbol{x}_\ell$ (see Fig. 7.2.1). Functions f_1, J_1, f_2, J_2 are now obtained from (3) and (4), after substituting for $\boldsymbol{y}(\boldsymbol{x}, t)$ vectors $\boldsymbol{y}_1(\boldsymbol{x}_\ell, t)$ and $\boldsymbol{y}_2(\boldsymbol{x}_\ell, t)$, respectively.

Let's calculate, with the help of bimodal distribution (8), the density field

$$\rho(\boldsymbol{x}, t) = \rho_0 \|a_{ij}\| \tag{10}$$

in the vicinity of curve ℓ . The elements of matrix $[a_{ij}]$, after some calculations, can be expressed in the form

$$a_{ij} = \frac{1}{2\mu t}\left(\frac{b_{ij}^1 + Rb_{ij}^2}{1+R} + \frac{R}{(1+R)^2}d_id_j\right),\tag{11}$$

where

$$b_{11}(\boldsymbol{y},t) = \frac{2\mu t}{J(\boldsymbol{y},t)}\left(1 + t\frac{\partial^2 S_0(\boldsymbol{y})}{\partial y_2^2}\right),$$

$$b_{22}(\boldsymbol{y},t) = \frac{2\mu t}{J(\boldsymbol{y},t)}\left(1 + t\frac{\partial^2 S_0(\boldsymbol{y})}{\partial y_1^2}\right),$$

$$b_{12}(\boldsymbol{y},t) = -\frac{2\mu t^2}{J(\boldsymbol{y},t)}\frac{\partial^2 S_0(\boldsymbol{y})}{\partial y_1 \partial y_2}.$$

The superscripts in (11) indicate that in formulas for b_{ij}, components of vector $\boldsymbol{y}$ must be replaced by components of vectors $\boldsymbol{y}_1(\boldsymbol{x}_\ell, t)$ and $\boldsymbol{y}_2(\boldsymbol{x}_\ell, t)$, respectively.

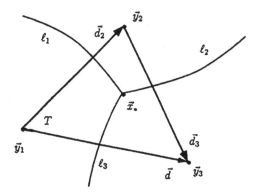

Fig. 7.2.2. Pancakes intersect at points $\boldsymbol{x}_*$, and nodes of very high density are created.

Substituting (11) into (10) we obtain that

$$\rho(\boldsymbol{x},t) = \frac{\rho_1}{(1+R^2)} + \frac{\rho_2}{(1+1/R)^2} + \frac{\rho_{12}}{(\sqrt{R}+1/\sqrt{R})^2}$$

$$+\frac{\rho_0}{2\mu}\frac{R}{(1+R)^3}\left((\vec{d}\cdot\nabla_1)^2\phi(\boldsymbol{y}_1,t) + R(\vec{d}\cdot\nabla_2)^2\phi(\boldsymbol{y}_2,t)\right),\tag{12}$$

with

$$\phi(\boldsymbol{y},t) = \phi(\boldsymbol{y},0,t) = \frac{y^2}{2t} + S_0(\boldsymbol{y}).$$

Here ρ_1 and ρ_2 are values of the density for different sides of pancake ℓ in the immediate neighborhood of the pancake, and ρ_{12} represents a certain additional "mixed" density:

$$\rho_{12} = \frac{\rho_0}{4\mu t^2} \left(b_{11}^1 b_{22}^2 + b_{11}^2 b_{22}^1 - 2b_{12}^1 b_{21}^2 \right).$$

The last summand in (12) is the "main" component of the density, which describes the high density of matter getting stuck in the vicinity of the "pancake rim". It can also be conveniently estimated by the "double delta-function" approximation

$$f_\mu(\boldsymbol{y}; \boldsymbol{x}, t) = \frac{\delta(\boldsymbol{y} - \boldsymbol{y}_1(\boldsymbol{x}_\ell, t)) + R\delta(\boldsymbol{y} - \boldsymbol{y}_2(\boldsymbol{x}_\ell, t))}{1 + R}, \qquad (13)$$

which is simpler than (8). It corresponds to neglecting lower order densities $\rho_1, \rho_2, \rho_{12}$.

The next stage of the density field evolution is associated with appearance of points $\boldsymbol{x}_*$, where pancakes intersect and nodes of very high density are created (see Fig. 7.2.2). For a rough calculation of the density in a vicinity of these nodes we may use a "triple delta-function" approximation:

$$f_\mu = \frac{\delta(\boldsymbol{y} - \boldsymbol{y}_1) + R_2\delta(\boldsymbol{y} - \boldsymbol{y}_2) + R_3\delta(\boldsymbol{y} - \boldsymbol{y}_3)}{1 + R_2 + R_3}, \qquad (14)$$

where

$$R_2 = \sqrt{\frac{\rho_2}{\rho_1}} \exp\left(\frac{\boldsymbol{d}_2 \boldsymbol{z}}{2\mu t}\right),$$

$$R_3 = \sqrt{\frac{\rho_3}{\rho_1}} \exp\left(\frac{\boldsymbol{d}_3 \boldsymbol{z}}{2\mu t}\right),$$

with

$$\boldsymbol{d}_2 = \boldsymbol{y}_2 - \boldsymbol{y}_1, \quad \boldsymbol{d}_3 = \boldsymbol{y}_3 - \boldsymbol{y}_1, \quad \boldsymbol{d} = \boldsymbol{y}_3 - \boldsymbol{y}_2, \quad \boldsymbol{z} = \boldsymbol{x} - \boldsymbol{x}_*.$$

It follows from (14) that

$$a_{ij} = \frac{R_2 d_{2i} d_{2j} + R_3 d_{3i} d_{3j} + R_2 R_3 d_i d_j}{2\mu t (1 + R_2 + R_3)^2}. \qquad (15)$$

Substituting this expression into (10) we get, after simple calculations, that

$$\rho(\boldsymbol{x}, t) = \frac{\rho_0 |T|^2 R_2 R_3}{\mu^2 t^2 (1 + R_2 + R_3)^3}. \qquad (16)$$

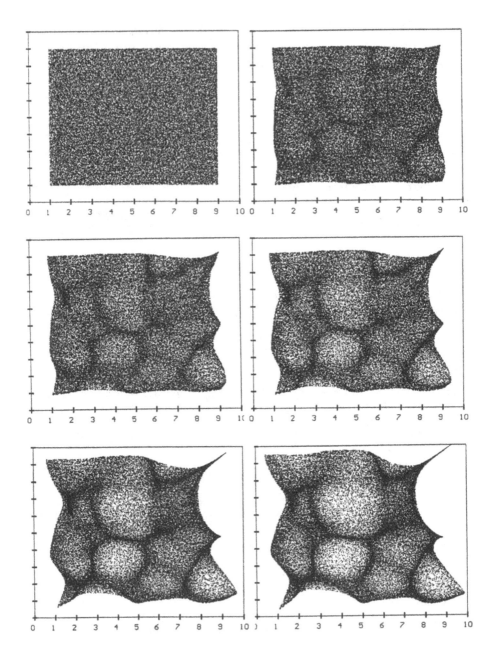

Fig. 7.2.3. Evolution of passive tracer density in 2-D Burgers' velocity field with shot-noise initial velocity data. The simulation was performed for 100,000 particles. The initial distribution of mass was chosen to be uniform on the rectangle $[1,9] \times [1,9]$. The consecutive frames show the location of passive tracer particles at t = 0.0, 0.3, 0.6, 1.0, 2.0 , 3.0 (from Janicki et al. (1995)).

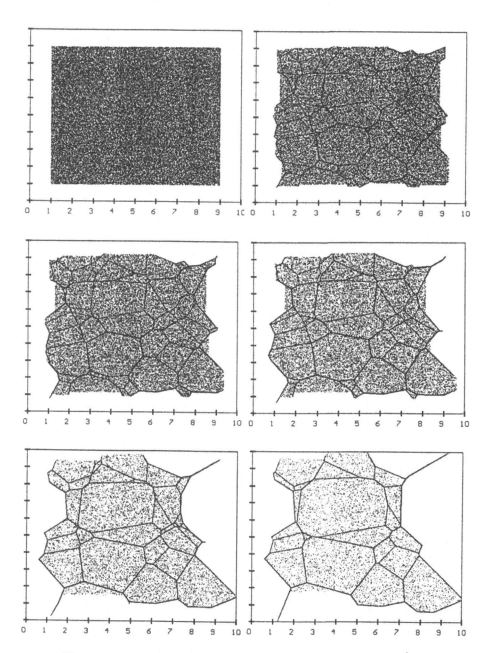

Fig. 7.2.4. Evolution of passive tracer density in the *inviscid* 2-D Burgers' velocity field with shot-noise initial velocity data. The simulation was performed for 100,000 particles. The initial distribution of mass was chosen to be uniform on the rectangle $[1, 9] \times [1, 9]$. The consecutive frames show the location of passive tracer particles at t = 0.0, 0.3, 0.6, 1.0, 2.0 , 3.0 (from Janicki et al. (1995)).

Here $|T|$ is the area of triangle T with vertices at points $\boldsymbol{y}_1, \boldsymbol{y}_2$, and $\boldsymbol{y}_3$ (see Fig. 7.2.2).

The total mass concentrated at the node is obtained by integration of density field (16) over $\boldsymbol{x}$:

$$m = \int \rho(\boldsymbol{x}, t) d\boldsymbol{x} = \rho_0 |T|. \tag{17}$$

This indicates that the passive tracer mass, which at time $t = 0$ was uniformly distributed of triangle T, eventually concentrates at such a node.

Figs. 7.2.3-4 compare evolution of cellular structures of the true density field in Burgers velocity field with analogous evolution in the inviscid limit. The lines actually indicate the location of shock fronts in the inviscid velocity field at different epochs. These figures should also be compared with N-body simulations in Kofman et al. (1992).

7.3   Evolution of 1-D model density

The laws of evolution of the model density field (7.1.14) of the passive tracer are easiest to understand in the 1-D case, and this section provides their comprehensive discussion. It follows from (7.1.14) and (7.1.12) that in the 1-D case the model density field is of the form:

$$\rho(x, t) = \rho_0([\![y]\!](x, t)) \frac{\partial [\![y]\!](x, t)}{\partial x}, \tag{1}$$

where the auxiliary field $[\![y]\!](x, t)$ satisfies equation

$$\frac{\partial [\![y]\!]}{\partial t} + v \frac{\partial [\![y]\!]}{\partial x} = \mu \frac{\partial^2 [\![y]\!]}{\partial x^2}, \tag{2}$$

with the initial condition $[\![y]\!](x, t = 0) = x$. If we introduce a new function

$$\rho_1(x, t) = \frac{\partial [\![y]\!]}{\partial x}, \tag{3}$$

then the density field (1) can be rewritten in the form:

$$\rho(x, t) = \rho_0 \cdot \rho_1. \tag{4}$$

First of all, let us observe that the equation for model density field (4) has a divergence form. Indeed, it follows from (2) and (3) that the auxiliary fields ρ_0 and ρ_1 satisfy equation

$$\frac{\partial \rho_0}{\partial t} + v \frac{\partial \rho_0}{\partial x} = \frac{\mu}{\rho_1} \frac{\partial \rho_0}{\partial x} \frac{\partial \rho_1}{\partial x}, \tag{5}$$

and it can be clearly seen from (2-5) that the model density field (4) satisfies equation

$$\frac{\partial \rho}{\partial t} + \frac{\partial}{\partial x}(\rho v) = \mu \frac{\partial}{\partial x}\left(\rho_0 \frac{\partial \rho_1}{\partial x}\right) \tag{6}$$

which has a divergence form. This implies the validity of a global conservation law

$$\int_{-\infty}^{\infty} \rho(x, t) dx = \text{const.} \tag{7}$$

It follows from (6) that, in addition to a global conservation law (7), we also have a local conservation law

$$\frac{\partial}{\partial t} \int_{x_1}^{x_2} \rho(x, t) dx + \left(\rho v - \mu \rho_0 \frac{\partial \rho_1}{\partial x}\right)\Big|_{x_1}^{x_2} = 0. \tag{8}$$

Notice, that in the limit $\mu \to 0$, when the field weakly converges to a generalized solution of equations

$$\frac{\partial v}{\partial t} + v \frac{\partial v}{\partial x} = 0, \qquad \frac{\partial \rho}{\partial t} + \frac{\partial}{\partial x}(v\rho) = 0, \tag{9}$$

and when x_1 and x_2 are not singularity points of that generalized solution, the conservation law (5.8) acquires a hydrodynamic-type form:

$$\frac{\partial}{\partial t} \int_{x_1}^{x_2} \rho(x, t) dx + v\rho \Big|_{x_1}^{x_2} = 0. \tag{10}$$

Finally, let's consider a special case of constant initial density $\rho_0 = \text{const.}$ Then it is customary to write the expression for model density field in the form

$$\rho(x, t) = \rho_0 \left(1 - t \frac{\partial v(x, t)}{\partial x}\right). \tag{11}$$

Here $v(x, t)$ is a solution of the 1-D Burgers equation. It is clear from (11) that if the velocity field satisfies boundary conditions $v(-\infty, t) = v(+\infty, t) = 0$, then

$$\int_{-\infty}^{\infty} \rho v \, dx = \rho_0 \int_{-\infty}^{\infty} v(x, t) dx = \text{const.}$$

Thus, in this case, in addition to the above mentioned mass conservation law, the momentum conservation law is also valid.

It is instructive to substitute in expression (11), a sample explicit solution

$$v(x,t) = \frac{1}{t}\left(x - y\tanh\left(\frac{xy}{2\mu t}\right)\right),\qquad(12)$$

of the Burgers equation, which describes the evolution of a shock wave in the inertial

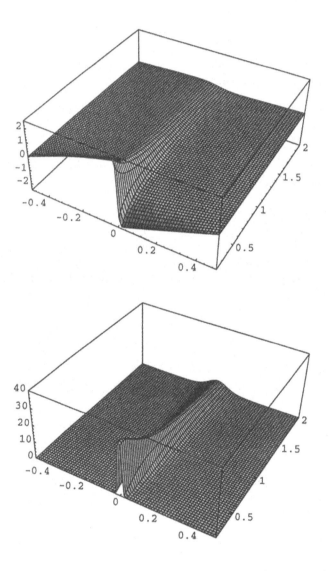

Fig. 7.3.1. *Top:* The special Burgers velocity field $v(x,t)$ from (12) with $\mu = 0.01, y = 0.5$. *Bottom:* The corresponding density field $\rho(x,t)$ from (13).

coordinate system, moving together with the shock, and which has been chosen in such a way that the shock's center coincides with the origin of the coordinate system (see Fig. 7.3.1). This solution gives for the density field ρ the formula

$$\rho(x,t) = \frac{\rho_0 y^2}{2\mu t \cosh^2((x-Vt)y/2\mu t)}, \tag{13}$$

which is now written in the stationary coordinate system, relative to which the shock moves with a velocity V. Formula (13) describes the tracer density, the total mass thereof

$$M = \int \rho(x,t)\,dx = 2\rho_0 y$$

is constant. For $\mu \to 0+$ the density weakly converges to the delta-function:

$$\lim_{\mu \to 0+} \rho(x,t) = M\delta(x-Vt). \tag{14}$$

It means that in the inviscid limit of the above sample solution, the tracer particles, initially uniformly distributed over the interval $x \in [y_-, y_+]$ (in our case $y_\mp = \mp y$) of length $l = 2y$, coalesce into a single "macroparticle" with singular density and mass $M = \rho_0 l$.

7.4 KdV-Burgers transport in 1-D compressible gas

Up to this point the assumption was that the velocity field satisfies the Burgers equation that, in the 1-D case has the form

$$\frac{\partial v}{\partial t} + v\frac{\partial v}{\partial x} = \mu\frac{\partial^2 v}{\partial x^2}, \tag{1}$$

and that doesn't allow certain important effects that accompany the motion of hydrodynamic waves in a compressible gas and, for example, the dispersion of waves on the ocean surface. To escape these constraints, in the present section we consider a more general equation

$$\frac{\partial v}{\partial t} + v\frac{\partial v}{\partial x} + \frac{1}{\rho}\frac{\partial P}{\partial x} = \mathcal{L}v + F(x,t), \tag{2}$$

where P denotes pressure, and $\mathcal{L}$ stands for a linear operator, the typical example thereof would be

$$\mathcal{L} = \mu\frac{\partial^2}{\partial x^2} + \nu\frac{\partial^3}{\partial x^3},\tag{3}$$

where μ represent the viscosity and ν—dispersion factors. We couple equation (2) with the following model equation for the density field:

$$\frac{\partial \rho}{\partial t} + \frac{\partial}{\partial x}(v\rho) = \mathcal{L}\rho,\tag{4}$$

which extends an idea introduced and justified in Section 2, where $\nu = 0$.

To close the system (2-4) we suppose that the gas is polytropic $(P \sim \rho^\delta)$ with index $\delta = 3$, in which case the pressure

$$P = \frac{k^2}{3}\rho^3,$$

and equation (2) is transformed into

$$\frac{\partial v}{\partial t} + v\frac{\partial v}{\partial x} + k^2\rho\frac{\partial \rho}{\partial x} = \mathcal{L}v + F(x,t).\tag{5}$$

For convenience, introduce the local sound velocity $c = k\rho$, evolution thereof is described, in view of (4), by equation

$$\frac{\partial c}{\partial t} + \frac{\partial}{\partial x}(cv) = \mathcal{L}c.\tag{6}$$

It follows from (5) and (6), that the set of equations for $v(x,t)$ and $c(x,t)$ acquires the form

$$\frac{\partial v}{\partial t} + v\frac{\partial v}{\partial x} + c\frac{\partial c}{\partial x} = \mathcal{L}v + F(x,t),$$

$$\frac{\partial c}{\partial t} + v\frac{\partial c}{\partial x} + c\frac{\partial v}{\partial x} = \mathcal{L}c,$$

$$v(x,t=0) = v_0(x), \qquad c(x,t=0) = c_0(x).$$

Now, notice that the above discussion and an introduction of the auxiliary fields defined by formulas

$$u_\pm(x,t) = v(x,t) \pm c(x,t)$$

immediately leads to the following particular result:

Theorem 1. *The velocity v and density ρ fields of a polytropic gas with index $\delta = 3$ satisfying equations*

$$\frac{\partial v}{\partial t} + v\frac{\partial v}{\partial x} + \frac{1}{\rho}\frac{\partial P}{\partial x} = \mu\frac{\partial^2 v}{\partial x^2} + \nu\frac{\partial^3 v}{\partial x^3} + F(x,t),$$

$$\frac{\partial \rho}{\partial t} + \frac{\partial}{\partial x}(v\rho) = \mu\frac{\partial^2 \rho}{\partial x^2} + \nu\frac{\partial^3 \rho}{\partial x^3}$$

are given by formulas

$$v = \frac{u_+ + u_-}{2}, \qquad \rho = \frac{u_+ - u_-}{2k}, \tag{7}$$

where the auxiliary fields $u_\pm$ satisfy the same non-homogeneous KdV-Burgers equation

$$\frac{\partial u_\pm}{\partial t} + u_\pm\frac{\partial u_\pm}{\partial x} = \mu\frac{\partial^2 u_\pm}{\partial x^2} + \nu\frac{\partial^3 u_\pm}{\partial x^3} + F(x,t), \tag{8}$$

with (different) initial conditions $u_\pm(x, t = 0) = v_0(x) \pm c_0(x)$.

Remark 1. Equation (2) with operator $\mathcal{L}$ of the form (3) and zero dissipation $\mu = 0$, becomes the well-known Korteweg-de Vries equation, and the question of what happens in the limit $\nu \to 0$ and its relation to the problem of turbulence has been discussed before (see, e.g., Lax and Levermore (1983), and Lax (1991)). Also, the polytropic compressible gas assumption can be satisfied in certain situations on the ocean-atmosphere interface which are of interest to physical oceanography (Gossard and Khuk (1978)).

In the case of nondispersive media ($\nu = 0$), and absence of external forces ($F \equiv 0$), equation (8) reduces to the Burgers equation discussed previously. This means, in particular, that system (2-4) of model adiabatic gas equations, which takes into account the pressure forces, admits a general analytic solution via the Hopf-Cole formula (6).

Remark 2. The idea of the homogeneous KdV-Burgers equation

$$\frac{\partial u}{\partial t} + u\frac{\partial u}{\partial x} = \mu\frac{\partial^2 u}{\partial x^2} + \nu\frac{\partial^3 u}{\partial x^3}, \tag{9}$$

as a normal equation for turbulence has been pursued by Liu and Liu (1992) (also, see references therein). In particular, they obtained traveling wave solutions

$$u(x,t) = u(x - ct)$$

of the following forms (see Fig. 7.4.1):

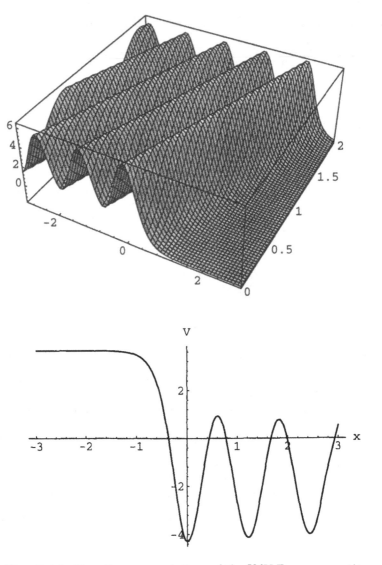

Fig. 7.4.1. Traveling wave solutions of the KdV-Burgers equation. *Top:* The full evolution corresponding to case (i); *Bottom:* The solution field corresponding to case (ii) at $t = 0$.

(i) For positive viscosity μ and negative dispersion ν, and $\mu^2 < -4\nu\sqrt{c^2 + 2A}$, we have $u(x,t) =$

$$\begin{cases} u_1 + \frac{u_1-u_2}{2} e^{-\mu(x-ct)/2\nu} \cos\sqrt{\frac{u_2-u_1}{2\nu} - \left(\frac{\mu}{2\nu}\right)^2}(x-ct), & \text{for } x - ct \le 0; \\ u_2 + \frac{3(u_1-u_2)}{2} \operatorname{sech}^2 \sqrt{\frac{u_2-u_1}{8\nu}}(x-ct), & \text{for } x - ct \ge 0; \end{cases}$$

and

(ii) For positive viscosity μ and positive dispersion ν, $\mu^2 < 4\nu\sqrt{c^2 + 2A}$, we have $u(x, t) =$

$$\begin{cases} u_1 + \frac{3(u_1-u_2)}{2}\operatorname{sech}^2\sqrt{\frac{u_1-u_2}{8\nu}}(x-ct), & \text{for } x - ct \leq 0; \\ u_2 + \frac{u_1-u_2}{2}e^{-\mu(x-ct)/2\nu}\cos\sqrt{\frac{u_2-u_1}{2\nu} - \left(\frac{\mu}{2\nu}\right)^2}(x-ct), & \text{for } x - ct \geq 0; \end{cases}$$

where

$$u_1 = c + \sqrt{c^2 + 2A}, \qquad u_2 = c - \sqrt{c^2 + 2A},$$

correspond to the two steady states in the phase plane, and A is a constant such that $c^2 + 2A > 0$.

In case (i), the upper part of the trajectory in the phase plane $(u, \dot{u})$ is close to a soliton, while the lower part consitutes an attenuating oscillatory tail. Analyzing the related cascading down process, Liu and Liu arrived at the conclusion that the energy spectrum slope (in the log-log coordinates) is between -1.76 and -1.97, and the corresponding fractal dimensions lie between 2.09 and 2.72.

Let $\rho_0 = $ const. Then, the model density field satisfying equations (3-4) is described by expression (7.3.11). Consequently, the results of Liu and Liu quoted above indicate that the spectrum of the model density field is almost flat and $\propto k^{-n}, 0.03 < n < 0.24$. In the next section we will show that for strongly nonlinear waves, and in a certain frequency band, a similar effect of appearance of the "quasi-white" piece of the density fluctuations spectrum takes place also when the dispersion is absent.

Spectral properties of the field of nonlinear fluctuations of density. In the special case of the above KdV-Burgers' model which assumes absence of dispersion and of external forces, equations for velocity and density fields take the form

$$\frac{\partial v}{\partial t} + v\frac{\partial v}{\partial x} + \kappa^2\rho\frac{\partial\rho}{\partial x} = \mu\frac{\partial^2 v}{\partial x^2}, \tag{10a}$$

$$\frac{\partial\rho}{\partial t} + \frac{\partial}{\partial x}(v\rho) = \mu\frac{\partial^2\rho}{\partial x^2}, \tag{10b}$$

$$v(x, t = 0) = v_0(x), \quad \rho(x, t = 0) = \rho_0(x).$$

This model provides an excellent opportunity for the analytic study of how nonlinearity, pressure and viscosity compete to influence the spectral properties of the density and velocity fields. In this section we will follow, at the physical level of rigorousness, the time evolution of statistical properties of the density field assuming, for the sake of simplicity, that the initial density is everywhere the same, i.e., that

$$\rho_0(x) = \rho_0 = \text{const},$$

and that the initial velocity field $v_0(x)$ is a statistically homogeneous stochastic function, with Gaussian distributions, zero means $\langle v \rangle = 0$, and a known correlation function

$$b_0(s) = \langle v_0(x)v_0(x+s) \rangle, \quad \sigma_0^2 = \langle v_0^2 \rangle = b_0(0).$$

Here, and elsewhere in this section, the angle brackets $\langle . \rangle$ denote the statistical averaging over the ensemble of realizations of the initial random velocity field $v_0(x)$. The density field satisfying the system of equations (10) can be written in the form

$$\rho(x,t) = \rho_0 + \tilde{\rho}(x,t), \tag{11}$$

where the fluctuation part, of principal interest to us, has a representation

$$\tilde{\rho}(x,t) = \rho_0 \frac{u(x-ct,t) - u(x+ct,t)}{2c}, \tag{12}$$

where $c = \kappa\rho_0$, and $u(x,t)$ is an auxiliary field which satisfies the Burgers equation

$$\frac{\partial u}{\partial t} + u\frac{\partial u}{\partial x} = \mu\frac{\partial^2 u}{\partial x^2}, \quad u(x,t=0) = v_0(x). \tag{13}$$

The spectral distribution

$$G_\rho(k,t) = \frac{1}{2\pi}\int \langle \tilde{\rho}(x,t)\tilde{\rho}(x+s,t) \rangle e^{iks}ds$$

of density fluctuations can be expressed through the analogous spectral distribution $G_u(k,t)$ of the auxiliary field $u(x,t)$ as follows:

$$G_\rho(k,t) = \frac{\rho_0^2}{2c^2}(1 - \cos(2kct))G_u(k,t). \tag{14}$$

Now, notice that for $k = 0$ and $G_u(k,t)$ bounded in k in the neighborhood of $k = 0$, we obtain an invariant

$$G_\rho(k=0,t) = 0, \tag{15}$$

which is equivalent to the well-known in statistical hydrodynamics invariant

$$\int \langle \tilde{\rho}(x,t)\tilde{\rho}(x+s,t)\rangle ds = 0,$$

which itself is a statistical consequence of the dynamic mass conservation law.

Let us analyze expression (15) for the spectral density in more detail. It has a vivid physical interpretation. Here, factor $1 - \cos(2kct)$ represents the interference of the acoustic waves propagating in the opposite directions of the medium, and $G_u(k,t)$ describes the evolution of their spectrum caused by nonlinear distortions. Moreover, the behavior of the spectrum of density (15) is qualitatively different for $k < k_p(t)$ and $k > k_p(t)$, where

$$k_p(t) = \frac{2\pi}{l(t)}, \quad l(t) = 2ct, \tag{16}$$

is the characteristic wavenumber which separates spectral components of the density field into two types: the first (for $k > k_p(t)$) which takes into account the influence of pressure forces, and the second (for $k < k_p(t)$) which behaves as if there was no pressure in the medium. For $k \ll k_p(t)$ the density spectrum "does not feel" pressure forces, and it can be replaced by a simpler expression

$$G_\rho(k,t) = (\rho_0 kt)^2 G_u(k,t). \tag{17}$$

It corresponds to the formula

$$\tilde{\rho}(x,t) = -\rho_0 t \frac{\partial u(x,t)}{\partial x},$$

which follows from (12) in the limit $c \to 0+$, and which connects fluctuations of the density with the divergence of the auxiliary field u.

In the remainder of this section we restrict our discussion to an analysis of the density fluctuation field in the inviscid limit $\mu \to 0$. In this case, in the initial nonlinear stage of evolution of the velocity and density fields, which until now had no discontinuities in the profile of the auxiliary field $u(x,t)$, equation (13) can be replaced by the Riemann equation

$$\frac{\partial u}{\partial t} + u\frac{\partial u}{\partial x} = 0, \quad u(x,t=0) = v_0(x).$$

The spectrum of the solution of the Riemann equation is well known (see, e.g., Gurbatov et al. (1991)) and has the form

$$G_u(k,t) = \frac{e^{-(\sigma_0 kt)^2}}{2\pi(kt)^2}\int_{-\infty}^{\infty}\left(e^{b_0(s)k^2t^2} - 1\right)e^{iks}ds. \tag{18}$$

Here, we also can introduce the characteristic wavenumber

$$k_n(t) = \frac{1}{\sigma_0 t},$$ (19)

which separates the harmonic contents of field $u(x,t)$ into relevant and irrelevant nonlinear distortions. The sufficiently large-scale harmonic components, for which $k < k_n(t)$, are in practice insensitive to nonlinear distortions. Indeed, for $k \ll k_n(t)$ the exponent inside integral (18) can be expanded into a Taylor series in b_0, and—with good accuracy—one can restrict oneself to the first two terms of the expansion. Then, it turns out that

$$G_u(k,t) = G_0(k) = \frac{1}{2\pi} \int_{-\infty}^{\infty} b_0(s) e^{iks} ds,$$ (20)

or that, in other words, the spectrum of field $u(x,t)$ is identical with the spectrum of the initial density field. For $k \ll k_n(t)$, the density spectrum (14) takes the form

$$G_\rho(k,t) = \frac{\rho_0^2}{2c^2} \big(1 - \cos(2kct)\big) G_0(k)$$ (21)

and describes an evolution of the density field spectrum for chaotic linear acoustic waves propagating in opposite directions.

In what follows, we will assume that the initial velocity field $v_0(x)$ has a unique characteristic spatial scale l_0, so that the spectrum (20) of $v_0(x)$ is concentrated in the neighborhood of $k = 0$ with the bandwidth $k_0 = 2\pi/l_0$. In this case, if $k_0 < k_n(t)$, that is if $t < t_n$, where

$$t_n = l_0/2\pi\sigma_0$$ (22)

is the characteristic time of appearance of nonlinear effects, then formula (21) is valid for all k and describes the linear regime of propagation of chaotic waves in the model gas.

Finally, we briefly describe behavior of the spectrum for times $t \gg t_n$, that is at times much larger than the characteristic time (22) of appearance of nonlinear effects. At such times, realizations of the auxiliary field $u(x,t)$ assume a quasiregular sawtooth-like shape, and as a result of multiple coalescence of shock fronts, they form a self-similar regime (Gurbatov et al. (1991)) such that the form of the spectral density does not vary with time and

$$G_u(k,t) = \frac{L^3(t)}{t^2} \tilde{g}(L(t)k).$$

Here, $L(t)$ is the external scale of the Burgers turbulence and the dimensionless function $\tilde{g}(p)$ determines the form of self-similar spectrum. The shocks in the sawtooth-like sample paths of the auxiliary field $u(x,t)$ lead, for $k \gg 2\pi/L(t)$ to a universal power asymptotics

$$G_u(k,t) \sim L(t)t^{-2}k^{-2}$$

of the spectrum of the strongly nonlinear Burgers turbulence. When the pressure forces are absent then, for any k, formula (17) is valid, and this leads to a flat spectrum

$$G_\rho(k,t) \sim \rho_0^2 L(t) \qquad (23)$$

of the density field. It has a clear cut physical interpretation: In the absence of pressure forces, in the epoch $t \gg t_n$, almost all the mass of gas concentrates at the points of discontinuity of the velocity field, creating thus a gas of "macroparticles" with density similar to that of (7.3.14), and the realizations of the density field take the form

$$\rho(x,t) = \sum_k M_k \delta(x - x_k(t)),$$

where $x_k(t)$ are the coordinates of shock fronts and M_k are full masses of matter concentrated at those discontinuites.

In presence of pressure forces, the independent of k spectrum (23) changes, for $k > k_p$, into a spectrum which decays at $k \to \infty$ as k^{-2}.

7.5 Exact formulas for 1-D Burgers turbulent diffusion

In this section we will discuss in detail another important consequence of the model gas equations (7.4.5-6). An analysis of their solution (7.4.7) in absence of dispersion ($\nu = 0$), and for vanishing pressure forces ($\kappa \to 0$), permits to find general explicit solutions for the density field equation

$$\frac{\partial \rho}{\partial t} + \frac{\partial}{\partial x}(v\rho) = \mu \frac{\partial^2 \rho}{\partial x^2}, \qquad (1)$$

$$\rho(x, t = 0) = \rho_0(x),$$

in the case of the velocity field $v(x,t)$ satisfying the 1-D Burgers equation

$$\frac{\partial v}{\partial t} + v\frac{\partial v}{\partial x} = \mu\frac{\partial^2 v}{\partial x^2}, \qquad (2)$$

$$v(x, t = 0) = v_0(x),$$

It should be emphasized that, from a physicist's point of view, equation (1), in contrast to the model density field equation (7.3.6), provides a completely adequate description of the occuring physical processes. For that reason, an acceptable for physicists exact solution of equation (1) is of independent interest. They also give us the ability to compare the model density (7.3.1) with the "real density" satisfying equation (1). This, in turn, permits a more detailed response the question of validity of the proposed density field model (7.3.1) in the general case of nonhomogeneous initial density field $\rho_0(x) \neq const$.

Theorem 1. *Let the density field $\rho(x, t)$ be a solution of equation (1) with the velocity field $v(x, t)$ satisfying the Burgers equation (2). Then*

$$\rho(x, t) = \frac{\partial}{\partial x}[\![M(y)]\!], \tag{3}$$

where

$$M(x) = \int^x \rho_0(z) \, dz$$

is the cumulative initial passive tracer mass distribution function, and the double brackets $[\![.]\!]$ denotes the "spatial" averaging introduced in (7.1.2-4) with $S_0(y) = \int^y v_0(z) \, dz$.

PROOF. In view of (7.4.7) and (7.1.5), the desired solution is equal to the limit

$$\rho(x, t) = \lim_{\kappa \to 0} \frac{u_+ - u_-}{2\kappa} = \frac{1}{2t} \lim_{\kappa \to 0} \frac{[\![y]\!]_- - [\![y]\!]_+}{\kappa},$$

where

$$[\![y]\!]_\pm = \frac{\int y \exp[\phi(y, x, t) \pm \kappa M(y))/2\mu] \, dy}{\int \exp[(\phi(y, x, t) \pm \kappa M(y))/2\mu] \, dy}.$$

A direct calculation of the limit gives

$$\rho(x, t) = \frac{1}{2\mu t} \Big([\![yM(y)]\!] - [\![y]\!][\![M(y)]\!]\Big). \tag{4}$$

Now, in view of the properties of function (7.1.4), we obtain (3), which ends the proof of Theorem 1. ∎

Remark 1. As expected, in the case of the homogeneous initial density $\rho_0 = const$ (i.e., $M(y) = \rho_0 y$), equation (4) becomes a 1-D analog of the model density field (7.1.13-14):

$$\rho(x, t) = \frac{\rho_0}{2\mu t} \Big([\![y^2]\!] - [\![y]\!]^2\Big),$$

and the relation (3), up to replacement of $\langle y \rangle$ by $[\![y]\!]$, coincides with formula (7.1.35a).

Remark 2. Observe that solution (3) of the density field equation (1) can also be found by a more "physical" method based on the stochastic intepretation of function

$$f_\mu(y; x, t) = \mathcal{Q}(y; t|x) \tag{5}$$

(see Remark 3.3), and applying the general formula (7.1.20) for the density field:

$$\rho(x, t) = \int \rho_0(y) \mathcal{P}(x; t|y) \, dy. \tag{6}$$

This is due to the fact that, in the 1-D case, the relationship between the probability distribution $\mathcal{Q}(y; t|x)$ of the Lagrangian coordinate of the passive tracer particle and the probability distribution $\mathcal{P}(x; t|y)$ of its Eulerian coordinate, becomes especially simple. Indeed, from (7.1.25a) and (7.1.35) applied in the 1-D case, one can get the formula

$$\mathcal{P}(x; t|y) = \left\langle \frac{\partial y(x, t)}{\partial x} \delta(y - y(x, t)) \right\rangle.$$

Differentiating the above equality with respect to y, we obtain

$$\frac{\partial \mathcal{P}}{\partial y} + \frac{\partial \mathcal{Q}}{\partial x} = 0.$$

Taking into account (5), we can solve the above equation for $\mathcal{P}$ to obtain

$$\mathcal{P}(x; t|y) = -\frac{\partial}{\partial x} \int_{-\infty}^{y} f_\mu(z; x, t) \, dz.$$

Substituting this expression into (6) it is easy to see that the obtained expression is equivalent with the solution (3).

Let us return to the analysis of the evolution of the density field. Note, that in the 1-D case under consideration in this section, the expression (7.1.14) for the model density becomes

$$\rho_m(x, t) = \rho_0([\![y]\!](x, t)) \frac{\partial [\![y]\!](x, t)}{\partial x}. \tag{7}$$

Throughout the remainder of this section the model density will be denoted by ρ_m to distinguish it from the "real" density (3).

Suppose that $[\![y]\!](x, t)$ is an everywhere continuously differentiable function (which is the case if, for example, $v_0(x)$ is everywhere continuous and bounded and $\mu > 0$). Then, equation (7) can be rewritten in the form

$$\rho_m(x, t) = \frac{\partial}{\partial x} M\big([\![y]\!](x, t)\big), \tag{8}$$

which, in the inviscid limit $\mu \to 0+$, converges weakly to

$$\rho_m(x, t) = \frac{\partial}{\partial x} M\big(y(x, t)\big), \tag{9}$$

where $y(x, t)$ is the coordinate of the absolute minimum of the function (7.1.2), and the derivative is understood in the distribution-theoretic sense. In particular, this means that if x^* is an isolated discontinuity of function $y(x, t)$ such that

$$\lim_{x \to x^*-0} y(x, t) = y_- < \lim_{x \to x^*+0} y(x, t) = y_+$$

then, inside a neighborhood of point x^* containing no other discontinuities, the model density (9) is of the form

$$\rho(x, t) = M(y_-, y_+)\delta(x - x^*) + \rho_c(x, t),$$

where the first summand is a singular density of "macroparticles" like (7.3.14) and $\rho_c(x, t)$ is the nonsingular component of the density described by the classical expression (8), and $M(y_-, y_+)$ is the passive tracer mass concentrated in the interval (y_-, y_+) at $t = 0$.

Observe, that both, the density (3), and the model density (9) converge in the inviscid limit to the same weak limit

$$\rho(x, t) = \frac{\partial}{\partial x} M(y(x, t)). \tag{10}$$

In this fashion, at least in the 1-D case, the "real" and model density fields coincide in the most interesting for physical applications inviscid limit.

Remark 3. The obtained relations are likely to be useful in efforts to solve, in the Burgers equation context, the central problem of the turbulence theory: understanding of the evolution of statistical properties of the density $\rho(x, t)$ of the passive tracer in the hydrodynamic random velocity field $v(x, t)$. This broad and attractive theme deserves a separate detailed treatment. Here we restrict ourselves to only first steps in that direction.

First of all, note that if the initial velocity field $v_0(x)$ in (2) is random, then the density field (8.3) is also random. Consequently, the mean density field is given by

$$\langle \rho(x,t) \rangle = \frac{\partial}{\partial x} \langle [\![M(y)]\!] \rangle, \tag{11}$$

where the angled bracked stand for the statistical averaging over the ensemble of realizations of the initial velocity field $v_0(x)$. Since the averaged expression is a complex nonlinear functional of the field $v_0(x)$, the exact calculation of the statistical average in (11) is not possible. Nevertheless, in certain physically interesting cases, one can obtain for $\langle \rho(x,t) \rangle$ some asymptotic formulas. In particular, in the inviscid limit, the expression (11) becomes

$$\langle \rho(x,t) \rangle = \frac{\partial}{\partial x} \langle M(y(x,t)) \rangle, \tag{12}$$

where the statistical averaging is over the ensemble of random coordinates $y(x,t)$ of the absolute minimum of the random field (7.1.2). Clearly, if $v_0(x)$ is a statistically homogeneous random field, then the probability distribution of $y(x,t)$

$$\mathcal{Q}(y;t|x) = \mathcal{Q}(y-x;t)$$

depends only on the difference of coordinates x and y and the expression (12) is transformed into

$$\langle \rho(x,t) \rangle = \int \rho_0(y) \mathcal{Q}(y-x;t) \, dy. \tag{13}$$

Gurbatov et al. (1991), have demonstrated that at the stage of multiple coalescence of shock fronts ($t \gg t_n$, where t_n is given by (7.4.22)), the probability distribution $\mathcal{Q}(y;t)$ becomes self-similar:

$$\mathcal{Q}(y;t) = \frac{1}{L(t)} q \left(\frac{y}{L(t)} \right),$$

where $L(t)$ is the external scale of the Burgers turbulence (also, see Lectures 4-5, and Funaki et al. (1995), Molchanov et al. (1995) and Surgailis and Woyczynski (1994) for a detailed probabilistic classification of the related parabolic and hyperbolic scaling limits). Substituting the last expression into (13) and taking, for example, $\rho_0(x) = \delta(x)$, we obtain the following asymptotic law of the Burgers turbulent diffusion

$$\langle X^2(t) \rangle = \int x^2 \langle \rho(x,t) \rangle dx \sim L^2(t).$$

It is known (see Gurbatov and Saichev (1993)) that the rate $L(t)$ of growth of the external scale of the Burgers turbulence depends on the behavior near $k = 0$ of the spectral density (7.4.20) of the initial velocity field. For typical asymptotics

$$G_0(k) \sim \beta_p^2 |k|^p, \qquad k \to 0, \quad -1 < p < \infty,$$

of the velocity field spectrum, one obtains the following rates of growth of the external scale of turbulence:

$$L(t) \sim \begin{cases} (\beta_p t)^{2/(3+p)}, & \text{for } p < 1; \\ (\sigma_s t)^{1/2}, & \text{for } p > 1, \end{cases}$$

where σ_s^2 is the dispersion of the (statistically homogeneous for $p > 1$) initial random velocity potential $S_0(x)$. Consequently, we obtain the following rates for the Burgers turbulent diffusion:

$$\langle X^2(t) \rangle \sim \begin{cases} (\beta_p t)^{4/(3+p)}, & \text{for } p < 1; \\ \sigma_s t, & \text{for } p > 1. \end{cases}$$

Notice that, for $p > 1$, this gives the Brownian rate of growth $\langle X^2(t) \rangle \sim t$ (see (7.1.2a)). The similarity between the molecular and turbulent diffusion is further reinforced by the fact that the probability distribution $Q(y, t)$ is in this case Gaussian (see Gurbatov and Saichev (1993)). However, that is where the similarities end since the molecular diffusion is generated by a stationary in time forcing process $\xi(t)$ with the corellation function of the form (7.1.17), whereas the turbulent diffusion is generated by the nonstationary field of Burgers turbulence, the dispersion thereof decays in time to zero at the rate

$$\langle v^2(x, t) \rangle \sim \frac{L^2(t)}{t^2} \sim \begin{cases} \beta_p^{4/(3+p)} t^{-2(1+p)/(3+p)}, & \text{for } p < 1; \\ \sigma_s^2 t^{-1}, & \text{for } p > 1. \end{cases}$$

7.6 Fourier-Lagrangian representation for non-smooth Lagrangian-Eulerian maps

In this section, considering just the 1-D case, we will discuss some problems of description of the density field $\varrho(x, t)$ when mapping

$$x = X(y, t) \tag{1}$$

of Lagrangian coordinates y into Eulerian coordinates x does not satisfy the standard assumptions of strict monotonicity and differentiability. Notice, that these assumptions are violated even in the

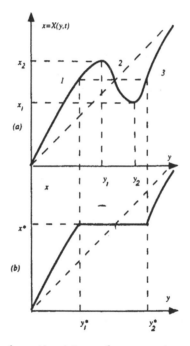

Fig. 7.6.1. A schematic picture of nonmonotone and nondifferentiable Lagrangian-to-Eulerian maps.

simplest physical situations. Indeed, for a gas of noninteracting particles with initial velocities $v_0(y)$, equation (1) takes the form

$$x = y + v_0(y)t. \tag{2}$$

It is obvious that, for a general (not monotonically increasing) function $v_0(y)$, one can find a $t_1 < \infty$ such that for $t > t_1$ mapping (2) is not monotone. A typical graph of mapping (2) which is not monotonically increasing is shown on Fig. 7.6.1 (top). On the other hand, if the velocity field $v(x,t)$ satisfies the Burgers equation

$$\frac{\partial v}{\partial t} + v\frac{\partial v}{\partial x} = \mu\frac{\partial^2 v}{\partial x^2},$$

$$v(x, t = 0) = v_0(x),$$

in the inviscid limit ($\mu \to 0+$), mapping (1) is monotone (not strictly though) but not differentiable. This is clearly seen on Fig. 7.6.1 (bottom), where a typical segment of the Lagrangian-to-Eulerian map for this case is pictured.

Also, observe that even more pathological (from the viewpoint of classical analysis) situations, where mapping (1) has a fractal character and a non-trivial Hausdorff dimension, are of physical interest. We just quote here Sinai's (1992) result on solutions of the Burgers equation with Brownian motion initial data (see Lecture 5), related studies by Avellaneda and Weinan E (1994) on the Burgers equation with the white noise initial data, and the work of Vergassola, Dubrulle, Frisch and Noullez (1994), Molchan (1998), and Reade (1998), on the Burgers equation with fractional Brownian motion initial data . In all these (and other) situations we encounter a mathematical problem of finding a physically acceptable description of the density field for a known (in the dynamical and statistical sense) nonsmooth Lagrangian-to-Eulerian map. One such possible generalized description is given in this section.

To begin with, recall that in the distributional notation (see, e.g., Saichev and Woyczynski (1997)), for a smooth and strictly increasing in y function $X(y,t)$, the density field can be written in the form

$$\varrho(x,t) = \int \varrho_0(y)\delta\big(x - X(y,t)\big)dy, \qquad (3)$$

where $\varrho_0(x)$ is the initial velocity field. If function $\varrho_0(y)$ is absolutely integrable, then the Fourier transform

$$\tilde{\varrho}(k,t) = \frac{1}{2\pi}\int \rho_0(y)e^{-ikX(y,t)}dy, \qquad (4)$$

of the density field (3) is well defined. However, the above integral may exist even in cases when functional (3) looses its distributional meaning. For that reason, a more general description of the density field is obtained if one defines it as the inverse Fourier transform

$$\varrho(x,t) = \int \tilde{\rho}(k,t)e^{ikx}dk \qquad (5)$$

of the integral (4).

We shall explain the mechanics of the above Fourier-Lagrangian representation for the density field on simple, but characteristic, examples.

Example 1. Suppose that, as on Fig. 7.6.1 (bottom), the Lagrangian-to-Eulerian map has a flat segment, that is,

$$X(y,t) = x^*, \quad \text{for} \quad y \in (y_1^*, y_2^*),$$

and that it is a smooth, strictly increasing function for $y < y_1^*$ and $y > y_2^*$. Then, in view of (4), the Fourier transform of the density field is

$$\tilde{\varrho}(k,t) = \frac{1}{2\pi} m(y_1^*, y_2^*) e^{-ikx^*} + \tilde{\varrho}_c(k,t), \tag{6a}$$

where

$$m(y_1^*, y_2^*) = \int_{y_1^*}^{y_2^*} \varrho_0(y) dy$$

is the total mass of particles contained in the interval (y_1^*, y_2^*) at the initial time, and where

$$\tilde{\varrho}_c(k,t) = \frac{1}{2\pi} \left[\int_{-\infty}^{y_1^*} + \int_{y_2^*}^{\infty} \right] \rho_0(y) e^{ikX(y,t)} dy.$$

Taking the inverse Fourier transform of (6a), we obtain the generalized density field

$$\varrho(x,t) = m(y_1^*, y_2^*)\delta(x - x^*) + \varrho_c(k,t), \tag{6b}$$

which has a clear-cut physical meaning. The first summand describes the density of particles accumulated at point x^*, and the second summand

$$\varrho_c(x,t) = \int \tilde{\varrho}_c(k,t) e^{ikx} dk$$

describes the continuous part of the density field on either side of point x^*. A typical picture of the density field corresponding to the Lagrangian-to-Eulerian map from Fig. 7.6.1b, is shown on Fig. 7.6.2b.

Example 2. Consider the density of a multistream gas of noninteracting particles. Suppose that, as on Fig. 7.6.1.a, the Lagrangian-to-Eulerian map (2) monotonically decreases in the interval (y_1, y_2), and monotonically increases on the remainder of the Lagrangian axis y. Physically, this means that in the corresponding interval (x_1, x_2) of the Eulerian coordinate axis (see Fig. 7.6.2a), a 3-stream regime of motion of particles occurs. Each stream has its own density field

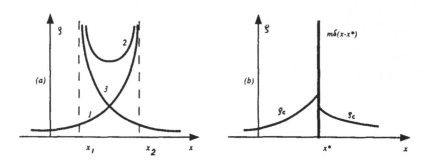

Fig. 7.6.2. Density fields for multistream and nondifferentiable regimes.

$\varrho_n(x,t)$, where the subscript $n = 1, 2, 3$, indicates the number of the stream (see Fig. 7.6.1a). In the general case, at each point x and time t, there exist $N(x,t)$ streams of the gas of noninteracting particles, each with its own density $\varrho_n(x,t)$. It is not difficult to show that in such a situation the Fourier-Lagrangian representation (4-5) considered above, leads to the density field

$$\varrho(x,t) = \sum_{n=1}^{N(x,t)} \varrho_n(x,t). \qquad (7)$$

Notice, that the generalized density field obtained above as a result of formal mathematical operations reflects the physical reality of the density field equal to the sum of densities of all streams of particles found at a given point.

The Fourier-Lagrangian representation also provides satisfactory answers in more complex cases where, for example, the initial data $v_0(y)$ in (2) form a Brownian motion process with the structure function

$$\langle [v_0(y) - v_0(y+s)]^2 \rangle = 2q|s|, \quad v_0(0) = 0. \qquad (8)$$

Although in this case it is practically impossible to find values of a realization of the field $\varrho(x,t)$ which corresponds to a given realization of the initial Brownian motion $v_0(x)$, it is relatively easy to calculate arbitrary moment functions

$$B_M(x_1, t_1, \ldots, x_M, t_M) = \left\langle \prod_{m=1}^{M} \varrho(x_m, t_m) \right\rangle. \qquad (9)$$

Let us point out another, purely mathematical, aspect of the above Fourier-Lagrangian representation. Notice, that the integral in (3) represents a functional of a composition of the delta-function with another function. In the more usual distribution-theoretical form this functional can be written as

$$T_{\delta(f)}[\phi] = \int \phi(y)\delta(f(y))dy = \phi(\bar{y})/f'(\bar{y}), \qquad (10)$$

where $\phi(y)$ is a test function and $\bar{y}$ is a root of equation $f(y) = 0$. If $f(y)$ is a smooth, strictly monotone function, then the theory assigns to this functional the value that appears on the right hand side of equality (10). On the other hand, the Fourier-Lagrangian representation, which gives functional (10) the value

$$T_{\delta(f)}[\phi] = \int C(k)dk,$$

where

$$C(k) = \frac{1}{2\pi} \int \phi(y)e^{-ikf(y)}dy,$$

permits a rigorous definition of the functional $T_{\delta(f)}[\phi]$ for a very broad class of functions $f(y)$, including functions that are not monotone, not differentiable, or that have a fractal character.

The Fourier-Lagrangian representation was first considered by Fournier and Frisch (1983), and also has been introduced independently in a paper by Gurbatov and Saichev (1993), where it was applied to solutions of the Riemann equation

$$\frac{\partial v}{\partial t} + v\frac{\partial v}{\partial x} = 0, \qquad (11)$$

$$v(x, t = 0) = v_0(x),$$

describing the velocity field $v(x,t)$ of a gas of noninteracting particles. We shall briefly describe these papers' results that are related to our main topic.

If $v_0(x)$ is a smooth function with the derivative bounded from below, that is such that

$$\min_{x \in \mathbf{R}} v'(x) = -u_0, \quad 0 < u_0 < \infty,$$

then, for times t such that

$$0 < t < t_1 = 1/u_0,$$

there exists a unique solution $v(x,t)$ of the initial value problem (11). Its Fourier transform

$$\tilde{v}(k,t) = -\frac{1}{2\pi i k t} \int \left(e^{-ikX(y,t)} - e^{-iky} \right) dy, \qquad (12)$$

where $X(y,t)$ is given by the right hand side of (2). The integral in (12) exists in the usual sense for any $t > 0$ whenever function $v_0(y)$ decreases sufficiently fast to 0 as $|y| \to \infty$. However, for $t > t_1$, the solution of problem (11) is no longer single-valued and we face the problem: Which single-valued function is represented by the Fourier transform (12)?

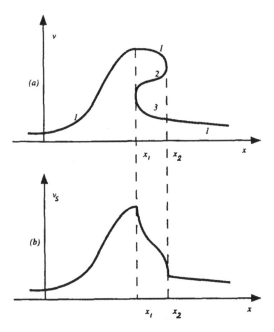

Fig. 7.6.3. Multi-stream velocity regime and the corresponding single-stream function v_s.

A detailed analysis of integral (12) leads to the following answer. Let $v(x,t)$ be the velocity field of a gas of noninteracting particles. For $t > t_1$, at each point x we find $N(x,t) \geq 1$ particles of gas with different velocities

$$v_n(x,t), \quad n = 1, 2, \ldots, N(x,t).$$

Let us list these velocities in the decreasing order

$$v_1 \geq v_2 \geq \ldots \geq v_N,$$

and form an alternating sum

$$v_s(x, t) = \sum_{n=1}^{N(x,t)} (-1)^{n-1} v_n(x, t). \tag{13}$$

Here, subscript s stands for "single-stream". It turns out that the integral in (12) is the Fourier transform of the above function v_s.

Figs. 7.6.3a and b show the typical graph of a multi-stream function $v(x, t)$, and the corresponding single-stream function $v_s(x, t)$ constructed according to the above recipe.

In order to study evolution of the density field corresponding to motion of a gas of noninteracting particles it is necessary to complement the Riemann equation (11) by the continuity equation

$$\frac{\partial \varrho}{\partial t} + \frac{\partial}{\partial x}(v\varrho) = 0. \tag{14}$$

In what follows we will restrict our observations to the simplest case of the uniform initial density

$$\varrho(x, t = 0) = \varrho_0 = \text{const.}$$

In this case, as can be verified by a direct substitution, solution ϱ of equation (14) can be expressed in terms of a solution v of the Riemann equation as follows:

$$\varrho(x, t) = \varrho_0 \left[1 - t \frac{\partial}{\partial x} v(x, t) \right]. \tag{15}$$

It turns out that, essentially, the same expression remains valid also for $t > t_1$, where the uniqueness condition of the solution of equation (11) is violated. The only adjustment that is required is the replacement of the multistream function $v(x, t)$ in (66) by the single-stream function $v_s(x, t)$ constructed above (see (13). In this case equation (15) leads to the same result as equation (7).

It follows from the above discussion and from formula (15), that the deviation

$$\delta\varrho(x, t) = \varrho(x, t) - \varrho_0,$$

of the current density field of the gas of noninteracting particles from its initial value is described by the formula

$$\delta\varrho(x, t) = -t\varrho_0 \frac{\partial}{\partial x} v_s(x, t), \tag{16}$$

which remains valid for any $t > 0$. In this fashion, function $v_s(x,t)$ which, for $t > t_1$, looses its physical meaning as a velocity field, preserves its physical meaning as a measure of amount of matter of the initially uniformly distributed gas. In particular, since

$$t\varrho_0[v_s(b,t) - v_s(a,t)] = \int_a^b \delta\rho(x,t)\,dx,$$

the increment of function v_s over interval (a,b) is proportional to the deviation of gas mass in the interval (a,b) from the initial mass $\varrho_0(b-a)$ of particles in this interval.

Next, let us consider the case where the initial velocity field $v_0(x)$ is a stationary, zero-mean Gaussian stochastic process with the structure function

$$D(\theta) = \langle[v_0(x+\theta) - v_0(x)]^2\rangle.$$

Then, using (12), one can show that the spectrum

$$G_v(k,t) = \frac{1}{2\pi} \int \langle v_s(x,t)v_s(x+\theta,t)\rangle e^{-ik\theta}\,d\theta$$

of the velocity field $v_s(x,t)$ is expressed by the formula

$$G_v(k,t) = \tag{17}$$

$$\frac{1}{2\pi(kt)^2} \int \left[\exp\left(-\frac{1}{2}D(\theta)k^2t^2\right) - \exp\left(-\frac{1}{2}D(\infty)k^2t^2\right)\right] e^{-ik\theta}\,d\theta.$$

Random motion of particles generates a random density field $\varrho(x,t)$. Obviously, its statististical mean

$$\langle\varrho(x,t)\rangle = \varrho_0,$$

and its fluctuations around the mean value are given by expression (16). Consequently, the spectrum of density fluctuations is of the form

$$G_\varrho(k,t) = \varrho_0^2 t^2 k^2 G_v(k,t). \tag{18}$$

The inertial nonlinearity in the Riemann equation (11) leads to the appearance of power tails (as $k \to \infty$) in spectra (17-18). However, the power character of these spectra and the mechanism of their formation is qualitatively different in the cases when $v_0(x)$ is a smooth function, and when $v_0(x)$ is an nowhere differentiable function of the fractal Brownian motion type. In the case of a smooth random function $v_0(x)$, its structure function enjoys the universal asymptotics

$$D(s) \sim ds^2 \quad (s \to 0), \tag{19}$$

which leads to the following decay laws for spectra (17) and (18):

$$G_v(k, t) \sim \frac{d}{\sqrt{2\pi}(|k|\tau)^3} \exp\left(-\frac{1}{2\tau^2}\right), \qquad (20a)$$

$$G_\varrho(k, t) \sim \frac{\varrho_0^2}{\sqrt{2\pi}|k|\tau} \exp\left(-\frac{1}{2\tau^2}\right). \qquad (20b)$$

Variable $\tau = t\sqrt{d}$ stands for the nondimensionalized time.

As an example of a nowhere differentiable initial velocity field $v_0(x)$ consider a Brownian motion with structure function (8). In this case the integral in (17) can be evaluated exactly which gives

$$G_v(k, t) = \frac{q}{\pi k^2} \cdot \frac{1}{\varepsilon^2 k^2 + 1}, \quad G_\varrho(k, t) = \frac{\varrho_0^2}{\pi} \cdot \frac{\varepsilon}{\varepsilon^2 k^2 + 1}, \qquad (21)$$

where $\varepsilon = qt^2$.

It is worthwhile to observe a peculiar inversion of properties of the initial field $v_0(x)$, and the resulting fields $v_s(x, t)$ and $\varrho(x, t)$, in the case of smooth and of nowhere smooth initial fields $v_0(x)$.

For $v_0(x)$ with smooth realizations, the spectrum of field v_s decays according to the power law $G_v \sim k^{-3}$, which follows from the nondifferentiability of the field $v_s(x, t)$ at points on different sides thereof the number of streams of the multi-stream field $v(x, t)$ are different (points x_1 and x_2 on Fig. 7.6.3). Following the geometric optics terminology we will call these points the points of caustics. The corresponding asymptotics $G_\varrho \sim k^{-1}$ of the fluctuations of density spectrum reflects the singularity of realizations of the density field in the neighborhood of caustic points (see Fig. 7.6.2a). This very slow decay law of density fluctuations' spectrum (as $k \to \infty$) implies that the variance of the random density field is infinite for any $t > 0$: $\langle(\delta\varrho)^2\rangle = \infty$.

On the other hand, in the case of the nowhere differentiable Brownian motion process $v_0(x)$, the spectrum of field $v_s(x, t)$ decays faster, according to the power law $G_v \sim k^{-4}$, and that implies smoothness of realizations of field $v_s(x, t)$ for any $t > 0$. As a result, the density spectrum decays as $G_\varrho \sim k^{-2}$ (more rapidly than in the case of a smooth field $v_0(x)$), and the variance of density fluctuations remains bounded: $\langle(\delta\varrho)^2\rangle = \varrho_0^2$.

The described above effect of creation, in the case of smooth initial conditions, of infinite singularities of the density field, and an absence of similar singularities in the realizations of the field $\varrho(x, t)$ for nowhere differentiable initial data, can be easily explained from the physical point of view. The point is that for a nowhere differentiable initial field $v_0(x)$, we observe right away, beginning with time $t = 0+$, creation of a complex, infinite-stream regime of motion of gas particles. It is similar to the regime of motion of a "warm" gas, where each particle has its own, independent of other particles, thermal component of velocity. As a result, the isolated singularities of the density field, which are characteristic for a gas of noninteracting particles with smooth initial velocity $v_0(x)$, are "washed out" and do not solidify.

7.7 Concentration field in reacting Burgers flows

In this section we will go beyond the passive tracer model and address a number of issues related to *reacting Burgers flows*. Physical characteristics and other features of solutions of the concentration field equation are discussed. In addition, an exactly solvable model of a nonlinear reaction-diffusion equation is studied. The material is taken from Saichev and Woyczynski (1997b).

The basic equation is the multidimensional Burgers equation of the form

$$\frac{\partial \boldsymbol{v}}{\partial t} + (\boldsymbol{v} \cdot \boldsymbol{\nabla}) \boldsymbol{v} = \mu \Delta \boldsymbol{v} + \vec{F}(\boldsymbol{x}, t), \tag{1}$$

$$\boldsymbol{v}(\boldsymbol{x}, t = 0) = \boldsymbol{v}_0(\boldsymbol{x}).$$

In the case of potential velocity field $\boldsymbol{v}(\boldsymbol{x}, t)$, the Hopf-Cole substitution reduces this nonlinear equation to a linear Schrödinger-type equation. We studied this problem in Lecture 6.

The solutions of the Burgers equation describe evolution of the velocity field $\boldsymbol{v}(\boldsymbol{x}, t)$. However, in many physical applications, the more interesting object is not the velocity field itself but the related concentration field $C(\boldsymbol{x}, t)$ or the density field $\rho(\boldsymbol{x}, t)$, and relatively little has been done in terms of the rigorous study of this problem (see the last section of the lecture). The concentration field of passive tracer transported in the velocity field $\boldsymbol{v}$ obeys the equation

$$\frac{\partial C}{\partial t} + (\boldsymbol{v} \cdot \boldsymbol{\nabla}) C = 0, \tag{2}$$

(see, e.g., Csanady (1980) for important environmental applications, and Saichev and Woyczynski (1997) for a general discussion of concentration and density fields), while the density field obeys the continuity equation

$$\frac{\partial \rho}{\partial t} + \text{div}\,(v\rho) = 0, \tag{3}$$

which describes, e.g., the evolution of matter distribution in the Universe. It is worthwhile to remember that the concentration measures not the absolute, as the density does, but relative proportion of the tracer in the physically infinitesimal unit of the medium; while the density increases when the medium is compressed and decreases when the medium expands, the concentration preserves its value in the neighborhood of an arbitrary fixed particle. For reacting systems it is the concentration that matters. For the Burgers velocity field v, exact solutions of these equations are not known and approximate methods for (3) were discussed earlier.

In this section we obtain novel exact general solutions of hydrodynamic type equations related to (2-3) which, in particular, include the equation

$$\frac{\partial C}{\partial t} + (v \cdot \nabla)\,C = \mu \Delta C + QC + e \tag{4}$$

$$C(x, t = 0) = C_0(x),$$

in space of arbitrary dimension ($x \in \mathbf{R}^d$, $d \geq 1, t > 0$), where the potential velocity field $v(x, t)$ satisfies the nonhomogeneous multidimensional Burgers equation (1) with the external force field

$$\vec{F}(x, t) = \vec{\nabla} R(x, t), \qquad v_0(x) = \vec{\nabla} S_0(x),$$

of potential type. The initial conditions $C_0(x), S_0(x)$, and the functions $e(x, t), Q(x, t)$ and $R(x, t)$, are arbitrary, the only assumption being that the solutions of equations (1) and (4) exist in the classical sense for any value of the viscosity parameter $\mu > 0$. The term $\mu \Delta C$ in the concentration equation (4) takes into account the molecular diffusion process of the passive tracer (see, e.g., discussion below and Csanady (1980)). The term $e = e(x, t)$ models creation (or annihilation) of tracer particles due to chemical reactions, and the term QC describes chain reaction-like effects (for further discussion of this subject, see Smoller (1994)). Observe, that in the inviscid limit and in the special case $Q \equiv e \equiv 0$, the equation (4) coincides with the usual concentration field equation (2), and the corresponding inviscid limit of the above mentioned exact solution defines a "generalized" solution of the equation (2).

To begin with, the solution of the coupled system of the *forced Burgers equation* (1) and the *advection-diffusion-reaction equation* (4) is expressed in terms of a solution of a pair of coupled linear diffusion equations with variable coefficients. This novel observation, the basis of much of what is proposed in this lecture, permits us to use the classical Feynman-Kac formula (see Lecture 6) to write the exact solution of the system (1, 4) as an expression involving paths integrals. I the case of absent chain reactions ($Q \equiv 0$) and no external forces ($R \equiv 0$) these solutions can be expressed in terms of the usual, finite-dimensional integrals.

Then we will discuss the molecular diffusion effects in the concentration equation (4). Finally, we shall obtain exact solutions of a nonlinear reaction-diffusion equation coupled with the Burgers-like velocity equation in which the velocity also depends on the concentration. The reaction-diffusion equation contains a quadratic nonlinearity modeling binary reactions. It is shown that the behavior of solutions of this system is qualitatively different for spaces of different dimensions.

Exact solution of the mean concentration field equation. In the present section we find an explicit expression for the concentration field $C(\boldsymbol{x}, t)$ discussed before. The proposed approach will be also used in other contexts in the following sections.

To begin with, let us consider a pair of auxiliary linear, variable coefficient, diffusion equations

$$\frac{\partial a}{\partial t} = \mu \Delta a + \eta(\boldsymbol{x}, t)a + e(\boldsymbol{x}, t)b, \tag{5}$$

$$\frac{\partial b}{\partial t} = \mu \Delta b + \theta(\boldsymbol{x}, t)b, \tag{6}$$

for unknown functions $a = a(\boldsymbol{x}, t)$ and $b = b(\boldsymbol{x}, t)$. Functions $\eta(\boldsymbol{x}, t)$, $e(\boldsymbol{x}, t)$, and $\theta(\boldsymbol{x}, t)$ are given. Their form, as well as the initial conditions in (5-6), will be provided later on.

Our first observation is that the ratio

$$C(\boldsymbol{x}, t) = \frac{a(\boldsymbol{x}, t)}{b(\boldsymbol{x}, t)} \tag{7}$$

of solutions a and b of the system (5-6) satisfies the equation

$$\frac{\partial C}{\partial t} = \mu \left(\frac{\Delta a}{b} - \frac{a \Delta b}{b^2} \right) + QC + e, \tag{8}$$

where

$$\eta - \theta = Q. \tag{9}$$

This fact is easily verified by differentiating (7) with respect to t, and taking into account equations (5-6). For the term in parenthesis we have an obvious identity

$$\frac{\Delta a}{b} - \frac{a\Delta b}{b^2} \equiv \Delta C + \frac{2}{b}\left[\frac{\nabla b \cdot \nabla a}{b} - \frac{a(\nabla b)^2}{b^2}\right]. \tag{10}$$

Since

$$\nabla C = \frac{\nabla a}{b} - \frac{a\nabla b}{b^2},$$

we can rewrite (10) in the form

$$\frac{\Delta a}{b} - \frac{a\Delta b}{b^2} \equiv \Delta C - \frac{1}{\mu}(v \cdot \nabla)C, \tag{11}$$

where

$$v(x, t) = -2\mu\nabla \log b(x, t). \tag{12}$$

Replacing the expression in parentheses in (8) by the right-hand side of the identity (11) we discover that the ratio $C(x, t)$ (7) of the solutions of the system (5-6) satisfies equation (8).

It remains to note that the substitution (12) is nothing but the Hopf-Cole formula expressing solutions of the Burgers equation (1) through solutions of the diffusion equation (2) with

$$\theta = -\frac{1}{2\mu}R. \tag{13}$$

Solving the system of algebraic equations (9) and (13) with respect to η and θ, we obtain the following final result:

Ansatz 1. *Assume that the potential vector velocity field* $v(x, t)$ *is a solution of the Burgers equation (1). Then the solution* $C(x, t)$ *of the concentration equation (4) is given by the formula (7), where* $a(x, t)$ *and* $b(x, t)$ *are solutions of the following coupled initial-value problem:*

$$\frac{\partial a}{\partial t} = \mu\Delta a - \frac{1}{2\mu}(R - 2\mu Q)a + eb, \tag{14}$$

$$a(x, t = 0) = C_0(x)\exp\left[-\frac{1}{2\mu}S_0(x)\right],$$

$$\frac{\partial b}{\partial t} = \mu \Delta b - \frac{1}{2\mu} R b, \tag{15}$$

$$b(\boldsymbol{x}, t = 0) = \exp\left[-\frac{1}{2\mu} S_0(\boldsymbol{x})\right],$$

where $S_0(\boldsymbol{x})$ is the initial velocity potential. Solutions of these initial-value problems are expressed via the following well-known Feynman-Kac type formulas (see, Lecture 6)

$$a(\boldsymbol{x}, t) = \mathbf{E}\left[C_0(\boldsymbol{x} - \boldsymbol{\omega}(t)) \exp\left(-\frac{1}{2\mu} S_0(\boldsymbol{x} - \boldsymbol{\omega}(t))\right. \right. \tag{16}$$

$$\left. -\frac{1}{2\mu} \int_0^t (R - 2\mu Q)(\boldsymbol{x} - \boldsymbol{\omega}(t) + \boldsymbol{\omega}(\tau), \tau) \, d\tau\right)\right]$$

$$+ \int_0^t d\tau \, E\left[e(\boldsymbol{x} - \boldsymbol{\omega}(t) + \boldsymbol{\omega}(\tau), \tau) \exp\left(-\frac{1}{2\mu} S_0(\boldsymbol{x} - \boldsymbol{\omega}(t))\right.\right.$$

$$\left.\left. -\frac{1}{2\mu} \int_0^t R(\boldsymbol{x} - \boldsymbol{\omega}(t) + \boldsymbol{\omega}(s), s) \, ds + \int_\tau^t Q(\boldsymbol{x} - \boldsymbol{\omega}(t) + \boldsymbol{\omega}(s), s) \, ds\right)\right],$$

$$b(\boldsymbol{x}, t) = \mathbf{E}\left[\exp\left(-\frac{1}{2\mu} S_0(\boldsymbol{x} - \boldsymbol{\omega}(t)) - \frac{1}{2\mu} \int_0^t R(\boldsymbol{x} - \boldsymbol{\omega}(t) + \boldsymbol{\omega}(\tau), \tau) \, d\tau\right)\right]. \tag{17}$$

The statistical averaging (expectation) $\mathbf{E}[\,.\,]$ is performed over trajectories of the vector Wiener process $\boldsymbol{\omega}(t) = (\omega_1(t), \ldots \omega_d(t))$ determined by the conditions

$$\boldsymbol{\omega}(0) = 0, \quad E[\boldsymbol{\omega}(t)] = 0, \quad E[\omega_l(t)\omega_m(t)] = 2\mu \delta_{lm} t, \quad l, m = 1, 2, \ldots, d. \tag{18}$$

Let us take a look at expressions (17-18) in the simple, but important, no external field and no reaction special case

$$R = Q = e \equiv 0. \tag{19}$$

Then the paths integrals (16-17) are expressed via the usual d - dimensional integrals

$$a_0(\boldsymbol{x}, t) = \left(\frac{1}{4\pi\mu t}\right)^{d/2} \int C_0(\boldsymbol{y}) \exp\left[-\frac{1}{2\mu} \Phi(\boldsymbol{y}, \boldsymbol{x}, t)\right] d\boldsymbol{y}, \tag{20}$$

$$b_0(\boldsymbol{x}, t) = \left(\frac{1}{4\pi\mu t}\right)^{d/2} \int \exp\left[-\frac{1}{2\mu} \Phi(\boldsymbol{y}, \boldsymbol{x}, t)\right] d\boldsymbol{y}, \tag{21}$$

where
$$\Phi(\boldsymbol{y}, \boldsymbol{x}, t) := \frac{(\boldsymbol{y} - \boldsymbol{x})^2}{2t} + S_0(\boldsymbol{y}). \qquad (22)$$

So, in the case (19), the exact solution to the system (1) and (4) is given explicitly in the form

$$C(\boldsymbol{x}, t) = \frac{\int C_0(\boldsymbol{y}) \exp[-\Phi(\boldsymbol{y}, \boldsymbol{x}, t)/2\mu]\, d\boldsymbol{y}}{\int \exp[-\Phi(\boldsymbol{y}, \boldsymbol{x}, t)/2\mu]\, d\boldsymbol{y}}.$$

The physical interpretation of the mean concentration field equation. This subsection provides a physical justification for the exactly solvable form (4) of the concentration field equation. The reasoning here is based on a fairly standard, Langevin-type equation model (see, e.g. van Kampen (1985), pp. 237-252).

Initially, let us consider a hydrodynamic velocity field $\boldsymbol{v}(\boldsymbol{x}, t)$ of a continuous medium. Assume that $\boldsymbol{v}(\boldsymbol{x}, t)$ is a twice continuously differentiable function with bounded first spatial derivatives. Then, the solution $\boldsymbol{X} = (X_1, \ldots, X_d)$ of the initial value problem

$$\frac{d\boldsymbol{X}}{dt} = \boldsymbol{v}(\boldsymbol{X}, t), \quad \boldsymbol{X}(\boldsymbol{y}, t = 0) = \boldsymbol{y}, \qquad (23)$$

generates a continuously differentiable and one-to-one mapping $\boldsymbol{x} = \boldsymbol{X}(\boldsymbol{y}, t)$, of $\boldsymbol{y} \in \mathbf{R}^d$ onto $\boldsymbol{x} \in \mathbf{R}^d$, such that the Jacobian of this mapping

$$J(\boldsymbol{y}, t) = \|\partial X_i / \partial y_j\| \qquad (24)$$

is continuous, finite everywhere and strictly positive: $0 < J(\boldsymbol{y}, t) < \infty$. Consequently, there exists an inverse mapping $\boldsymbol{y} = \boldsymbol{Y}(\boldsymbol{x}, t)$ with the same properties

Suppose, that the passive tracer particles (of unit mass) have been released at time $t = 0$ in the above continuous medium and consider their density field $\rho(\boldsymbol{x}, t)$ and concentration field $C(\boldsymbol{x}, t)$. Recall, that the density of particles is proportional to their number in a physically infinitesimal (unit) volume, while their concentration is proportional to the ratio of the number of passive tracer particles in relation to the number of medium particles in the same volume. The singular density of the tracer's n-th particle located at $\boldsymbol{y}_n$ at time $t = 0$ is

$$\rho_s(\boldsymbol{x}, t | \boldsymbol{y}_n) = \delta(\boldsymbol{x} - \boldsymbol{X}(y_n, t)). \qquad (25)$$

The singular concentration $C_s(\boldsymbol{x}, t | \boldsymbol{y}_n)$ which, at $t = 0$, is $C_s(\boldsymbol{x}, t = 0 | \boldsymbol{y}_n) = \rho_s(\boldsymbol{x}, t = 0 | \boldsymbol{y}_n) = \delta(\boldsymbol{x} - \boldsymbol{y}_n)$, differs for positive times from its singular density by the Jacobian factor (24):

$$C_s(\boldsymbol{x}, t | \boldsymbol{y}_n) = J(\boldsymbol{y}_n, t)\delta(\boldsymbol{x} - \boldsymbol{X}(y_n, t)). \qquad (25a)$$

This equality, in view of the rules of the distributional calculus, can be rewritten as follows:

$$C_s(\boldsymbol{x}, t | \boldsymbol{y}_n) = \delta(\boldsymbol{y}_n - \boldsymbol{Y}(\boldsymbol{x}, t)). \tag{25b}$$

In realistic situations, the equation of motion of the n-th particle, subject to the Brownian motion caused by molecular motions of adjacent particles, is different from (23) and has the form

$$\frac{d\boldsymbol{X}_n}{dt} = \boldsymbol{v}(\boldsymbol{X}_n, t) + \boldsymbol{\xi}_n(t), \quad \boldsymbol{X}_n(\boldsymbol{y}, t = 0) = \boldsymbol{y}_n, \quad n = 1, 2, \ldots, N, \tag{26}$$

where N is the total number of the passive tracer particles, and $\boldsymbol{\xi}_n(t)$ are mutually independent, zero-mean, Gaussian stochastic (white noise) vector processes with identical correlation tensors $\langle \xi_n^i(t)\xi_n^j(t + \tau) \rangle = 2\nu\delta_{ij}\delta(\tau)$, $i, j = 1, 2, \ldots, d$. The angled brackets indicate here the averaging over the ensemble of realizations of the vector processes $\boldsymbol{\xi}_1, \ldots, \boldsymbol{\xi}_N$. Consequently, instead of the deterministic density (25), it is necessary to consider a random singular density $\tilde{\rho}_s(\boldsymbol{x}, t | \boldsymbol{y}_n) = \delta(\boldsymbol{x} - \boldsymbol{X}_n(\boldsymbol{y}_n, t))$ and its statistical mean

$$\rho(\boldsymbol{x}, t | \boldsymbol{y}) = \langle \delta(\boldsymbol{x} - \boldsymbol{X}_n(\boldsymbol{y}, t)) \rangle. \tag{27}$$

The latter has a clear-cut probabilistic interpretation: it is the transition probability density of the vector Markov process $\boldsymbol{X}_n(\boldsymbol{y}, t)$, satisfying the stochastic equations (26). The density (27) itself satisfies the forward Kolmogorov equation

$$\frac{\partial \rho}{\partial t} + \operatorname{div}(\boldsymbol{v}\rho) = \nu\Delta\rho, \quad \rho(\boldsymbol{x}, t = 0 | \boldsymbol{y}) = \delta(\boldsymbol{x} - \boldsymbol{y}). \tag{28}$$

The full microscopic density $\tilde{\rho}(\boldsymbol{x}, t) = m\sum_{n=1}^{N}\delta(\boldsymbol{x} - \boldsymbol{X}_n(\boldsymbol{y}_n, t))$ of the passive tracer (m is the mass of each particle) has the statistical mean $\rho(\boldsymbol{x}, t) = \langle \tilde{\rho}(\boldsymbol{x}, t) \rangle = m\sum_{n=1}^{N}\rho(\boldsymbol{x}, t | \boldsymbol{y}_n)$ which, in the macroscopic description, is replaced by the integral

$$\rho(\boldsymbol{x}, t) = \int \rho_0(\boldsymbol{y})\rho(\boldsymbol{x}, t | \boldsymbol{y})\, d\boldsymbol{y}, \tag{29}$$

where $\rho_0(\boldsymbol{y})$ is the passive tracer's initial macroscopic density. The full mean density (29) satisfies the same equation (28) as the density (27) of the single particle, but with the initial condition $\rho(\boldsymbol{x}, t = 0) = \rho_0(\boldsymbol{x})$. Similarly, the mean concentration of the passive tracer

$$C(\boldsymbol{x}, t) = \int C_0(\boldsymbol{y})C(\boldsymbol{x}, t | \boldsymbol{y})\, d\boldsymbol{y}, \tag{30}$$

where $C_0(y)$ is the initial tracer concentration, and

$$C(x, t|y) = \langle \delta(y - Y_n(x, t)) \rangle \qquad (31)$$

is the average concentration of the single tracer particle.

To find equations satisfied by the mean concentrations (30-31), consider a system of $(d + 1)$ stochastic equations

$$\frac{dX}{dt} = v(X, t) + \xi(t), \qquad X(y, t = 0) = y, \qquad (32)$$

$$\frac{dJ}{dt} = J \operatorname{div} v, \qquad J(y, t = 0) = 1,$$

where, as usual, $\operatorname{div} v = \sum_{i=1}^{d} \partial v_i(X, t)/\partial X_i$. The solutions $(X(y, t), J(y, t))$ of equations (32) form a $(d + 1)$-dimensional Markov process with the joint transition probability density

$$\mathcal{P}(x, j; t|y) = \langle \delta(x - X(y, t)) \delta(j - J(y, t)) \rangle \qquad (33)$$

satisfying the forward Kolmogorov equation

$$\frac{\partial \mathcal{P}}{\partial t} + \operatorname{div}(v\mathcal{P}) + (\operatorname{div} v)\frac{\partial}{\partial j}(j\mathcal{P}) = \nu \Delta \mathcal{P}, \qquad (34)$$

$$\mathcal{P}(x, j; t = 0|y) = \delta(x - y)\delta(j - 1).$$

Note, that in view of the properties of the Dirac delta, (31) and (33), $C(y, t|x) = \int j\mathcal{P}(x, j; t|y)\, dj$. Consequently, multiplying equation (34) by j and integrating over all j's, we arrive at the backward Kolmogorov equation

$$\frac{\partial C}{\partial t} + (v \cdot \nabla)C = \nu C \qquad (35)$$

for the mean concentration (31) of a single particle. It is clear that the full concentration $C(x, t)$ (30) also satisfies the same equation (34), but with the initial condition $C(x, t = 0) = C_0(x)$.

The above process can be repeated if we add to equation (35) the extra terms $e(x, t)$ and QC, which describe effects of chain and chemical reactions, yielding (up to the replacement of ν by μ) the full equation (4) for the mean concentration of passive tracer subject to the Brownian motion of its particles. In the case when the molecular diffusion coefficient ν coincides with the viscosity coefficient μ appearing in the Burgers' equation (1), i.e., $\nu = \mu$, we can use the exact solutions of the system of equations (1), (4).

Reaction-diffusion equations. The method introduced above gives, in a particular case, an exact solution to the equation of *reaction-diffusion type* (see, e.g., Smoller (1994) for a general background on the reaction-diffusion equations). Consider the fields $a(\boldsymbol{x}, t)$ and $b(\boldsymbol{x}, t)$ which are solutions of the system of linear diffusion equations

$$\frac{\partial a}{\partial t} = \mu \Delta a \qquad (36)$$

$$\frac{\partial b}{\partial t} = \mu \Delta b - \kappa a \equiv \mu \Delta b - \kappa C b. \qquad (37)$$

Arguing as above, we obtain that the ratio (7) of the solutions of this system satisfies the nonlinear equation

$$\frac{\partial C}{\partial t} + (\boldsymbol{v} \cdot \boldsymbol{\nabla}) C = \mu \Delta C + \kappa C^2, \qquad (38)$$

where the field $\boldsymbol{v}$ is defined by the Cole-Hopf substitution (12), and as is clear from (37), satisfies the equation

$$\frac{\partial \boldsymbol{v}}{\partial t} + (\boldsymbol{v} \cdot \boldsymbol{\nabla}) \boldsymbol{v} = \mu \Delta \boldsymbol{v} + 2\mu\kappa \boldsymbol{\nabla} C. \qquad (39)$$

Thus we arrive at a system of nonlinear partial differential equations (38-39) which, as indicated above, has an exact analytic solution for arbitrary initial data

$$C(\boldsymbol{x}, t = 0) = C_0(\boldsymbol{x}), \qquad v(\boldsymbol{x}, t = 0) = \vec{\nabla} S_0(\boldsymbol{x}). \qquad (40)$$

Indeed, complementing equations (36-37) by the initial conditions of the equations (14-15), and solving them, we get

$$a(\boldsymbol{x}, t) = a_0(\boldsymbol{x}, t), \qquad b(\boldsymbol{x}, t) = b_0(\boldsymbol{x}, t) - \kappa t a_0(\boldsymbol{x}, t), \qquad (41)$$

where functions a_0 and b_0 are given by the expressions (2-22). Substituting these functions into (7) and (12), we obtain the sought explicit solution of equation (38-39). In particular, the concentration field is given by the expression

$$C(\boldsymbol{x}, t) = \frac{a_0(\boldsymbol{x}, t)}{b_0(\boldsymbol{x}, t) - \kappa t a_0(\boldsymbol{x}, t)}. \qquad (42)$$

Although the system (38-39) is of independent interest as an elegant example of exactly solvable pair of quasilinear equations of parabolic type, our main interest in it was driven by the fact that similar equations arise in mathematical modeling of chemical reactions and combustion processes in which the diffusive and convective behavior is significant. Indeed, the last term in (38) describes the growth of concentration due to a binary chemical reaction, and the last term in (39) reflects the force of a "negative pressure" which attracts the surrounding medium into the region of higher concentration of the chemical reaction products. The tracer here is no longer passive as its concentration affects the medium's motion itself. The existence of the above explicit solutions makes it possible to discover several interesting and novel nonlinear effects displayed by the system (38-39).

Our main observation is that the behavior of the concentration field (42) depends qualitatively on the spatial dimension d. Indeed, suppose that $d \geq 2$, that the ambient medium is at rest at $t = 0$ (i.e., $S_0 \equiv 0$), and that the initial concentration is

$$C_0(\boldsymbol{x}) = C_m \exp\left(-\frac{x^2}{4l^2}\right). \tag{43}$$

In this case, the expression (4.7) takes the form

$$C(\boldsymbol{x}, t) = \frac{a_0(\boldsymbol{x}, t)}{1 - \kappa t a_0(\boldsymbol{x}, t)}, \tag{44}$$

where

$$a_0(\boldsymbol{x}, t) = C_m \left(\frac{l^2}{l^2 + \mu t}\right)^{d/2} \exp\left(-\frac{x^2}{4(l^2 + \mu t)}\right). \tag{45}$$

Let us rewrite (44) in terms of the dimensionless variables

$$c = C/C_m, \quad r = |\boldsymbol{x}|/2l, \quad \tau = \mu t/l^2. \tag{46}$$

As a result, we get

$$c(r, \tau) = \frac{1}{(1 + \tau)^{d/2} \exp(r^2/(1 + \tau)) - \delta\tau}, \tag{47}$$

where the dimensionless parameter

$$\delta = \kappa C_m l^2/\mu,$$

describes competitive influence of the nonlinearity and diffusion on the concentration behavior. The parameter has two critical values:

$$\delta_1 = d/2, \quad \text{and} \quad \delta_2 = (\delta_1 - 1)(1 - 1/\delta_1)^{-\delta_1},$$

which split the half-axis $\delta > 0$ into three intervals where the qualitative behavior of concentration is different.

For $\delta < \delta_1$, the maximal concentration $c(0, \tau)$ monotonically decreases with time τ, and the concentration field asymptotically tends to the solution

$$c(r, \tau) = (1 + \tau)^{-d/2} \exp(-r^2/(1 + \tau))$$

of the linearized equation (38).

For $\delta_1 < \delta < \delta_2$, the maximal concentration initially increases as a result of the chemical reaction, and then, beginning with the time instant

$$\tau^* = \left(\frac{\delta}{\delta_1}\right)^{1/(\delta_1 - 1)} - 1,$$

the diffusion processes take over and the evolution of concentration follows the scenario for the case $\delta < \delta_1$ discussed above.

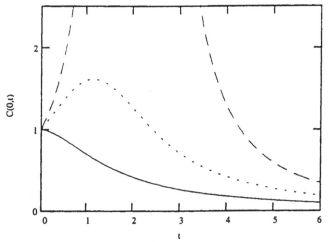

Fig. 7.7.1. The graphs of time evolution of the 3-D concentration field for $r = 0$ and three different values of δ. The case of $\delta = 1.4 < \delta_1$ is indicated by the solid line, that of $\delta = 2.2 \in (\delta_1, \delta_2)$—by the dotted line, and that of $\delta = 2.6 > \delta_2$—by the dashed line.

Finally, for $\delta > \delta_2$, the combustion of the chemical reaction leads, at a certain time $t_1 < \infty$, to an explosion, that is, to the divergence of the concentration to infinity. In this case, there exists a time-interval (τ_1, τ_2) such that, for $\tau_1 < \tau < \tau_2$, the solution (47) is negative for $r < r(\tau)$, where

$$r^2(\tau) = (1 + \tau) \ln \left(\frac{\delta \tau}{(1 + \tau)^{d/2}}\right) > 0, \qquad \tau_1 < \tau < \tau_2.$$

As r tends to $r(\tau)$ from either side, the concentration field diverges to infinity. Let us call the region of negative values of the exact solution of equation (38) the *conflagration region*.

The time evolution of the conflagration region's boundary is qualitatively different in spaces of different dimension:

For $d = 2$, when $\delta_2 = \delta_1 = 1$, for any $\delta > 1$ the conflagration region monotonically expands and gradually engulfs the entire plane $\boldsymbol{x} \in \mathbf{R}^2$. In this case $\tau_2 = \infty$.

For $d = 3$, when $\delta_1 = 3/2$, $\delta_2 = 3\sqrt{3}/2$, for any $\delta > \delta_2$, the conflagration lasts a finite time $\tau_2 < \infty$. The radius $r(\tau)$ of the conflagration region initially increases, and then decreases to zero, the conflagration is extinguished and, subsequently, the concentration field evolves as in the case $\delta < \delta_1$.

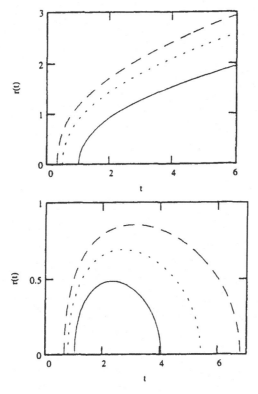

Fig. 7.7.2. *Top:* The graphs of time evolution of the conflagration region boundary $r(\tau)$ in the 2-D case for $\delta = 2$ (solid line), $\delta = 3$ (dotted line), and $\delta = 4$ (dashed line). *Bottom:* Analogous graphs in the 3-D case for $\delta = 2.8$ (solid line), $\delta = 3$ (dotted line), and $\delta = 3.2$ (dashed line). Note that, in 2-D, the conflagration area grows indefinitely, whereas in 3-D it eventually shrinks and disappears.

Fig. 7.7.1 provides graphs of $c(0, \tau)$ (4.12) for different δ and $d = 3$, and Fig. 7.7.2 shows the conflagration region's boundaries in the 2-D and 3-D spaces, for different values of δ.

7.8 Concentration field in potential and rotational flows

The method of reduction of systems of nonlinear model equation to systems of linear diffusion equations proposed in Section 7.7 and based on substitutions (7.7.7) and (7.7.12) is quite flexible and applies in several other cases such as, up-to-now unsolved, problems concerning evolution of Burgers' velocity fields in presence of external potential, as well as rotational flows. The latter case has been consipuously absent in the literature. The reason is obvious: the Hopf-Cole formula is then not applicable.

Consider the system of equations:

$$\frac{\partial C}{\partial t} + (\boldsymbol{V} + \boldsymbol{v} \cdot \boldsymbol{\nabla})C = \mu \Delta C,$$

$$\frac{\partial \boldsymbol{v}}{\partial t} + (\boldsymbol{v}\boldsymbol{\nabla})\boldsymbol{v} + \boldsymbol{\nabla}(\boldsymbol{V} \cdot \boldsymbol{v}) = \mu \Delta \boldsymbol{v}, \tag{1}$$

$$C(\boldsymbol{v}, t = 0) = C_0(\boldsymbol{x}), \qquad \boldsymbol{v}(\boldsymbol{x}, t = 0) = \boldsymbol{\nabla} S_0(\boldsymbol{x}),$$

where $\boldsymbol{V}(\boldsymbol{x}, t)$ is a known "external" velocity field. The system takes into account, for example, a general stretching and deformation of the continuous medium similar to, e.g., Hubble expansion of the Universe in the astrophysical large-scale structure problem discussed in the Lecture 1, or an expansion of the cloud of gas following an explosion, or an external eddy motion forming the background against which the potential component $\boldsymbol{v}(\boldsymbol{x}, t)$ of the velocity field evolves. It is easy to check that the initial-value problem for linear diffusion equations

$$\frac{\partial a}{\partial t} = \mu \Delta a - (\boldsymbol{V} \cdot \boldsymbol{\nabla})a, \qquad a(\boldsymbol{x}, t = 0) = C_0(\boldsymbol{x}) \exp\left[-\frac{1}{2\mu} S_0(\boldsymbol{x})\right],$$
$$\tag{2}$$
$$\frac{\partial b}{\partial t} = \mu \Delta b - (\boldsymbol{V} \cdot \boldsymbol{\nabla})b, \qquad b(\boldsymbol{x}, t = 0) = \exp\left[-\frac{1}{2\mu} S_0(\boldsymbol{x})\right],$$

is reducible, via substitutions (7.7.7),(7.7.12), to the initial-value problem (1) .

In order to solve the system of quasilinear equations (1), it suffices to solve the linear diffusion equations (2). Unfortunately, in the general case, these equations do not have explicit analytic solutions. Nevertheless, for some simple but sufficiently interesting from the physical viewpoint cases, general exact solutions of the equations (2) can be found and we discuss them in what follows.

Let us utilize results of Section 7.7 to analyze solutions of the first equation in (2)

$$\frac{\partial a}{\partial t} + (\boldsymbol{V} \cdot \boldsymbol{\nabla})a = \mu \Delta a, \qquad a(\boldsymbol{x}, t = 0) = a_0(\boldsymbol{x}), \qquad (3)$$

where $a_0(\boldsymbol{x})$ is the initial field (which will be assumed sufficiently smooth), and $\vec{V}(\boldsymbol{x}, t)$ is the above mentioned known velocity field (which will be assumed to be everywhere twice differentiable with bounded first derivatives). In the case when the external velocity field $\vec{V}$ depends linearly on $\boldsymbol{x}$ and is given by the formula

$$\boldsymbol{V} = A\boldsymbol{x}, \qquad (3)$$

where A is an arbitrary time-independent matrix, the equation can be solved explicitly by standard tools.

Consider, for example, the 2-D case and simplest matrices

$$A_p = \begin{pmatrix} h & 0 \\ 0 & h \end{pmatrix}, \qquad A_r = \begin{pmatrix} 0 & \omega \\ -\omega & 0 \end{pmatrix}, \qquad (4)$$

where subscript p stands for "potential" and r—for "rotational". The first matrix corresponds to the potential field $\vec{V} = h\boldsymbol{x}$ describing the homogeneous and isotropic expansion of the medium (like the Hubble expansion of the Universe mentioned earlier). The second corresponds to the velocity field $\boldsymbol{v}_r = \boldsymbol{\omega} \times \boldsymbol{x}$ of a rotating medium. Here, $\boldsymbol{\omega}$ is the angular velocity vector perpendicular to the (x_1, x_2)-plane and giving the left-handed orientation to the coordinate system $(1, 0, 0), (0, 1, 0), \boldsymbol{\omega}$.

In the first, potential, case the corresponding exact solution for the concentration equation (1) is

$$C_p(\boldsymbol{x}, t) = \frac{\int C_0(\boldsymbol{y}) \exp[-\Phi_p(\boldsymbol{y}, \boldsymbol{x}, t)/2\mu] \, d\boldsymbol{y}}{\int \exp[-\Phi_p(\boldsymbol{y}, \boldsymbol{x}, t)/2\mu] \, d\boldsymbol{y}}, \qquad (5)$$

where

$$\phi_p(\boldsymbol{y}, \boldsymbol{x}, t) = h \frac{(\boldsymbol{y} - \boldsymbol{x}e^{-ht})^2}{1 - e^{-2ht}} + S_0(\boldsymbol{y}). \qquad (6)$$

Notice that if one introduces the coordinate system $z = xe^{-ht}$ moving with the the expanding medium and the "effective" time

$$t_e = \frac{1}{2h}(1 - e^{-2ht}),\tag{7}$$

then the expression (6) will coincide with $\Phi(y, z, t_e)$ from (7.7.22). This means that, as time t increases to infinity, the velocity field $v(z, t)$ and concentration field $C(z, t)$ replicate the evolution of solutions of the standard Burgers equation (with $V \equiv 0$), and the corresponding concentration field in the time interval $t \in (0, 1/2h)$. In other words, the expansion of the medium dominates the nonlinear and dissipative processes.

For a rotational motion, the concentration field is described by the same expression (5), with Φ_p, however, replaced by

$$\begin{aligned}\Phi_r(y, x, t) \;=\; & \frac{1}{2t}\Big[(y_1 - x_1 \cos \omega t + x_2 \sin \omega t)^2 \\ & + (y_2 - x_1 \sin \omega t - x_2 \cos \omega t)^2\Big] + S_0(y).\end{aligned}$$

7.9 Burgers' density field revisited

Analysis of the evolution of the large scale structure of the mass distribution in the Universe requires analysis of not the concentration field $C(x, t)$ but the mean density field $\rho(x, t)$ of the passive tracer driven by the Burgers' velocity field $v(x, t)$ satisfying equation (7.7.34). In other words, it is necessary to solve the system of equations

$$\begin{aligned}\frac{\partial \rho}{\partial t} + \operatorname{div}(v\rho) &= \mu\Delta\rho, & \rho(x, t = 0) &= \rho_0(x), \\[2mm] \frac{\partial v}{\partial t} + (v \cdot \nabla)v &= \mu\Delta v, & v(x, t = 0) &= \nabla S_0(x).\end{aligned}\tag{1}$$

After substitutions (7.7.7) and (7.7.12), the system (1) reduces to a system of linear diffusion equations (7.7.14-15) with $R = e = 0$, $Q = -2\mu\Delta \log b$, and with $C_0(x)$ in the initial condition of (7.7.14) replaced by $\rho_0(x)$. Unfortunately, as is clear from (7.7.16), even without the external forcing term in the Burgers equation ($R \equiv 0$) the corresponding solution for the density field is expressed via the full Feynman-Kac path integral which is not easy to study analytically.

In this section, we will present an alternative approximate expression for such density fields which is significantly simpler and also happens to have a clear-cut physical interpretation. Notice, that in contrast to equation (7.7.34), we deliberately put $\nu = \mu$, since only in this special case we can obtain convenient analytic approximations for solutions of the system (1). A comparison of this approximation with the exact 1-D solution shows that, in the most physically interesting inviscid case ($\mu \to 0+$), both weakly converge to the same limit.

To construct the above-mentioned approximate solution we return to the interpretations of the concentration and density fields discussed in detail in Section 7.7. Let, initially, $\boldsymbol{X}(\boldsymbol{y}, t)$ be a solution of the fully determined initial-value problem (7.7.23). Then, the exact expression for the density field of the continuous medium in the velocity field $\boldsymbol{v}(\boldsymbol{x}, t)$, with the initial ($t = 0$) density field $\rho_0(\boldsymbol{x})$, is of the form

$$\rho(\boldsymbol{x}, t) = \int \rho_0(\boldsymbol{y})\delta(\boldsymbol{x} - \boldsymbol{X}(\boldsymbol{y}, t))\, d\boldsymbol{y}. \tag{2}$$

Expressing the singular concentration via the Dirac delta we get

$$C_s(\boldsymbol{x}, t|\boldsymbol{y}) = \delta(\boldsymbol{y} - \vec{Y}(\boldsymbol{x}, t)), \tag{3}$$

so that, in view of (7.7.25a-b), we have the equality

$$\delta(\boldsymbol{x} - \boldsymbol{X}(\boldsymbol{y}, t)) = \frac{C_s(\boldsymbol{x}, t|\boldsymbol{y})}{\int C_s(\boldsymbol{x}, t|\boldsymbol{y})\, d\boldsymbol{x}}. \tag{4}$$

We shall obtain the exact solution of the first equation in (1) if we take as $\boldsymbol{X}(\boldsymbol{y}, t)$ the solution of the stochastic equation (7.7.26) for $\boldsymbol{y}_n = \boldsymbol{y}$, and then average (2) over the statistical ensemble of realizations of the white noise $\boldsymbol{\xi}(t)$:

$$\rho(\boldsymbol{x}, t) = \int \rho_0(\boldsymbol{y}) \left\langle \frac{\tilde{C}_s(\boldsymbol{x}, t|\boldsymbol{y})}{\int \tilde{C}_s(\boldsymbol{x}, t|\boldsymbol{y})\, d\boldsymbol{x}} \right\rangle d\boldsymbol{y}. \tag{5}$$

Here, $\tilde{C}_s(\boldsymbol{x}, t|\boldsymbol{y})$ is the singular random concentration of the single tracer particle defined by (3) where, now, $\vec{Y}(\boldsymbol{x}, t)$ is a random field which, for each realization of $\boldsymbol{v}(\boldsymbol{y}, t)$ provides a mapping (7.7.26), inverse to the mapping (7.7.24). Replacing in (5) the mean of the ratio by the ratio of the means we arrive at the promised approximate expression for the density field:

$$\rho^a(\boldsymbol{x}, t) = \int \rho_0(\boldsymbol{y})R(\boldsymbol{x}, t|\boldsymbol{y})\, d\boldsymbol{y}, \tag{6}$$

where

$$R(\boldsymbol{x}, t|\boldsymbol{y}) = \frac{C(\boldsymbol{x}, t|\boldsymbol{y})}{\int C(\boldsymbol{x}, t|\boldsymbol{y})\, d\boldsymbol{x}} \tag{7}$$

and where $C(\boldsymbol{x}, t|\boldsymbol{y})$ (7.7.40) is known and equal to

$$C(\boldsymbol{x}, t|\boldsymbol{y}) = \frac{\exp[-\Phi(\boldsymbol{y}, \boldsymbol{x}, t)/2\mu]}{\int \exp[-\Phi(\boldsymbol{y}, \boldsymbol{x}, t)/2\mu]\, d\boldsymbol{y}}. \qquad (8)$$

Note that $R(\boldsymbol{x}, t|\boldsymbol{y})$, as a function of variable $\boldsymbol{x}$, has all the properties of a probability distribution. Therefore, in analogy with angled brackets, we will utilize braces to denote the integral of the form

$$\{g\}(\boldsymbol{y}, t) = \int g(\boldsymbol{x}) R(\boldsymbol{x}, \boldsymbol{y}, t)\, d\boldsymbol{x}. \qquad (9)$$

Among numerous physical and mathematical arguments in favor of using formula (6) as an approximation for the density field, we will mention a couple based on the simplicity of their verification.

First, observe that in view of the physical meaning of the density field, expression (6) satisfies the mass conservation law

$$\int \rho^a(\boldsymbol{x}, t)\, d\boldsymbol{x} = \int \rho_0(\boldsymbol{y})\, d\boldsymbol{y} = \text{const.}$$

Furthermore, consider the inviscid limit

$$\rho^0(\boldsymbol{x}, t) = \lim_{\mu \to 0+} \rho^a(\boldsymbol{x}, t).$$

Let $S_0(\boldsymbol{y})$ be a thrice everywhere continuously differentiable function with bounded second derivatives. Then there exists a $t_1 > 0$ such that, for $0 \le t < t_1$, the function $\Phi(\boldsymbol{y}, \boldsymbol{x}, t)$ (7.7.22) has, for any fixed $\boldsymbol{x}$, a unique minimum $\boldsymbol{y} = \vec{Y}^0(\boldsymbol{x}, t)$ as a function of argument $\boldsymbol{y}$ which, as a mapping of $\boldsymbol{x}$ into $\boldsymbol{y}$, is twice continuously differentiable. Therefore,

$$R^0(\boldsymbol{x}, t|\boldsymbol{y}) = \lim_{\mu \to 0+} R(\boldsymbol{x}, t|\boldsymbol{y}) = \delta(\boldsymbol{x} - \boldsymbol{X}^0(\boldsymbol{y}, t)), \qquad (10)$$

where

$$\boldsymbol{x} = \boldsymbol{X}^0(\boldsymbol{y}, t) = \boldsymbol{y} + \boldsymbol{v}_0(\boldsymbol{y})t \qquad (11)$$

is the inverse mapping of $\boldsymbol{y}$ into $\boldsymbol{x}$, the Jacobian

$$J(\boldsymbol{y}, t) = \left\| \delta_{ij} + t\frac{\partial^2 v_0(\boldsymbol{y})}{\partial y_i \partial y_j} \right\| \qquad (12)$$

thereof is everywhere positive, continuously differentiable and bounded. In addition, as is easy to see,

$$\rho^0(\boldsymbol{x}, t) = \frac{\rho_0(\vec{Y}^0(\boldsymbol{x}, t))}{J(\vec{Y}^0(\boldsymbol{x}, t), t)} \qquad (13)$$

is an exact, in the classical sense, solution of the continuity equation

$$\frac{\partial \rho^0}{\partial t} + \text{div}\,(\boldsymbol{v}^0 \rho^0) = 0,$$

$$\rho^0(\boldsymbol{x}, t = 0) = \rho_0(\boldsymbol{x}),$$

$$(14)$$

where $\boldsymbol{v}^0(\boldsymbol{x}, t)$ is the classical solution of the Riemann equation

$$\frac{\partial \boldsymbol{v}^0}{\partial t} + (\boldsymbol{v}^0 \cdot \boldsymbol{\nabla})\boldsymbol{v}^0 = 0, \qquad \boldsymbol{v}^0(\boldsymbol{x}, t = 0) = \boldsymbol{\nabla} S_0(\boldsymbol{x}) = \boldsymbol{v}_0(\boldsymbol{x}). \qquad (15)$$

For $t > t_1$, the solutions of equations (14-15) cease to exist in the classical sense. Instead, there exists, obtained as a weak inviscid limit of the Cole-Hopf solution, the generalized solution

$$\boldsymbol{v}^0(\boldsymbol{x}, t) = \frac{\boldsymbol{x} - \boldsymbol{Y}^0(\boldsymbol{x}, t)}{t}, \qquad (16)$$

of equation (15), where $\boldsymbol{Y}^0(\boldsymbol{x}, t)$ is the coordinate vector of the absolute minimum of function $\Phi(\boldsymbol{y}, \boldsymbol{x}, t)$ (7.7.22). For $t > t_1$, $\boldsymbol{Y}^0(\boldsymbol{x}, t)$ is a piecewise smooth function of $\boldsymbol{x}$ with discontinuities located on hypersurfaces $\mathcal{D}$ which are the loci of points $\boldsymbol{x}$ where function $\Phi(\boldsymbol{y}, \boldsymbol{x}, t)$ has two or more minima of the same magnitude. Recall, that in the adhesion approximation for the large-scale mass distribution of the Universe, these surfaces correspond to the so called "pancake" regions of high matter density which were discussed earlier in this lecture. In the vicinity of these surfaces, in the inviscid limit, the density field has a complex singular structure. However, outside the surface $\mathcal{D}$, in the so-called "dark" regions, function $\boldsymbol{Y}^0(\boldsymbol{x}, t)$ is continuously differentiable and the density field is defined by the classical expression (13). So, our approximate expression for the density turns out to be exact in the inviscid limit also for $t > t_1$, at least in the dark regions.

The study of the full structure of the singular, in the inviscid limit and $t > t_1$, density field $\rho(\boldsymbol{x}, t)$ has received, up to now, only limited attention. The difficulty of the problem is exacerbated by the fact that, as is known from the Riemann equation example, there could be infinitely many generalized solution of the same initial-value problem. To extract the one that satisfies physical requirements additional information is needed. In our case, in analogy with (16), it is natural to take as the "honest" singular generalized density field the density field obtained as a weak limit for $\mu \to 0+$ of the exact classical (for $\mu > 0$) solution

$$\rho(\boldsymbol{x}, t) = \frac{\rho_0(\boldsymbol{Y}(\boldsymbol{x}, t))}{J(\boldsymbol{Y}(\boldsymbol{x}, t), t)} \qquad (17)$$

of the continuity equation (14). Here, in contrast to (13), $Y(x,t)$ is the inverse function of $X(y,t)$, where $X(y,t)$ is the solution of the initial-value problem (7.7.23) with everywhere smooth, for $\mu > 0$, Burgers' equation (1) solutions $v(x,t)$ appearing on the right-hand side of equation (7.7.23). Intuitively, it is clear (although the rigorous proof is quite difficult) that, for $\mu \to 0+$, the impact of the Brownian motion of the tracer particles on the density distribution on the pancake surface is negligibly small, and the weak limit of solution (17) of the continuity equation (14) should coincide with the weak limit of the solution to the first equation in (1). If this is the case, then it suffices to compare the weak limit of the exact solution of the density equation (1) with the weak limit of the approximate solution (6). It is shown in Saichev and Woyczynski (1997b) that, at least in the 1-D case, the above weak limits coincide.

Concluding this section we will derive an equation which has as its exact solution the approximate density field (6). Differentiating the equality (6) with respect to time, we get

$$\frac{\partial \rho}{\partial t} = \int \rho_0(y) \frac{\partial C(x,t|y)/\partial t}{\int C(x,t|y)\,dx}\,dy - \int \rho_0(y) R(x,t|y) \int \frac{\partial C(x,t|y)/\partial t}{C(x,t|y)\,dx}\,dy.$$

Taking into account the fact that $C(x,t|y)$ satisfies equation

$$\frac{\partial C}{\partial t} + (v(x,t)\cdot\nabla)C = \mu\Delta C,$$

and also that

$$\int (v(x,t)\cdot\nabla)\,C\,dx = -\int C(x,t|y)\,\mathrm{div}\,v(x,t)\,dx,$$

and

$$\int \Delta C(x,t|y)\,dx = 0,$$

we obtain that

$$\frac{\partial \rho}{\partial t} + (v\cdot\nabla)\rho = \mu\Delta\rho - I(x,t), \qquad (18)$$

where

$$I(x,t) = \int \rho_0(y) R(x,t|y)\,\{\,\mathrm{div}\,v(x,t)\}\,(y,t)\,dy.$$

For $\mu \to 0+$, outside the pancakes $\mathcal{D}$, where the Laplacian in equation (18) can be neglected and instead of $R(x,t|y)$ one can insert the function R^0 (6.10), the last integral converges to

$$I(x,t) = \rho(x,t)\,\mathrm{div}\,v(x,t),$$

and the equation (18) becomes the standard continuity equation.

7.10 Generalized variational principles for systems of conservation laws

In this section we report on the results of E, Rykov and Sinai (1996) on generalized variational principles, global weak solutions and behavior with random initial data for systems of conservation laws of the following form:

$$(\rho u)_t + (\rho u^2)_x = 0,$$
$$\rho_t + (\rho u)_x = 0, \tag{1}$$

and

$$(\rho u)_t + (\rho u^2)_x = -\rho g_x,$$
$$\rho_t + (\rho u)_x = 0, \tag{2}$$
$$g_{xx} = \rho.$$

The basic behavior can be described as follows: At any positive time $t > 0$, the density $\rho(.,t)$ becomes a purely singular measure which is supported on a dense set which can be considered as the shock set of u.

Observe that for smooth solutions, equation (1) is equivalent to the passive tracer problem for the Riemann equation

$$u_t + uu_x = 0, \qquad \rho_t + (\rho u)_x = 0. \tag{3}$$

As we have seen in Lecture 1, for small t the initial value problem for (3) can be solved via the method of characteristics :

$$u(x,t) = u_0(\phi_t^{-1}(x)), \qquad \rho(x,t) = \rho_0(\phi_t^{-1}(x)) \left| \frac{\partial x}{\partial y} \right|^{-1} \tag{4}$$

where

$$x = \phi_t(y) = y + tu_0(y), \tag{5}$$

connects the Lagrangian coordinate y with the Eulerian coordinate x. However, after a certain critical time t^*, the Lagrangian-to-Eulerian map ceases to be one-to-one : a whole interval can be mapped to a single point which is the location of a shock.

However, in all cases ϕ_t defines a partition ξ_t of R^1 where elements of the partition are given by

$$D_t(x) = \{\phi_t^{-1}(x), x \in \mathbf{R}\} \tag{6}$$

and are either single points or intervals. The solutions (continuous from the right) can be reconstructed from the two conservation laws:

$$\phi_t(y) = \frac{\int_{C_t(y)}(\eta + tu_0(\eta))dP_0(\eta)}{\int_{C_t(y)} dP_0(\eta)}, \quad u(x,t) = \frac{\int_{D_t(x)} u_0(\eta))dP_0(\eta)}{\int_{D_t(x)} dP_0(\eta)} \quad (7)$$

where $C_t(y)$ denotes the element of the partition ξ_t containing y, and $dP_0(\eta)(= \rho_0(\eta)d\eta)$ is the initial distribution of mass given by a nonnegative Borel measure.

The partitions $\xi_t, t > 0$, are defined according to the *generalized variational principle (GVP)* as a follows: $y \in \mathbf{R}^1$ *is the left endpoint of an element of ξ_t iff for any $y^-, y^+ \in \mathbf{R}^1$ such that $y^- < y < y^+$ the following inequality holds:*

$$\frac{\int_{[y^-,y)}(\eta + tu_0(\eta))dP_0(\eta)}{\int_{[y^-,y)} dP_0(\eta)} < \frac{\int_{[y,y^+)}(\eta + tu_0(\eta))dP_0(\eta)}{\int_{[y,y^+)} dP_0(\eta)} \quad (8)$$

Now, ϕ_t is defined via (7) and P_t and I_t, the density and momentum distributions at time t by

$$P_t(\Delta) = P_0(\phi_t^{-1}(\Delta)), \qquad I_t(\Delta) = I_0(\phi_t^{-1}(\Delta)), \quad (9)$$

For continuous $U_0(x)$ the mapping ϕ_t is continuous. Moreover I_t is absolutely continuous with respect to P_t with

$$u(x,t) = \frac{dI_t}{dP_t}(x). \quad (10)$$

Theorem 1. *Suppose that:*

(i) The initial distribution of mass P_0 is finite on compact sets and is either discrete or absolutely continuous with respect to the Lebesgue measure, with positive density in the support of P_0. If the support is unbounded then $\int_0^x sdP_0(s) \to +\infty$ as $|x| \to +\infty$.

(ii) The initial distribution of momentum I_0 is absolutely continuous with respect to P_0 with $u_0 = dI_0/dP_0$. If P_0 is absolutely continuous then we assume u_0 is continuous.

(iii) For any $z > 0$, $\sup_{|x| \le z} |u_0(x)| \le b_0(z)$ and $\lim b_0(z)/z \to 0$ as $|z| \to \infty$.

Then, the family $(P_t, I_t, u)_{t \ge 0}$ constructed using the generalized variational principle gives the weak solution of (1) with initial data (P_0, I_0) in the sense that $P_t \to P_0$ and $I_t \to I_0$ weakly as $t \to 0^+$.

The triple $(P_t, I_t, u)_{t \geq 0}$ is understood as a weak solution of (1) in the following sense: for any $f, g \in C^1(\mathbf{R})$ with compact support, and $0 < t_1 < t_2$,

$$\int f(\eta)dP_{t_2}(\eta) - \int f(\eta)dP_{t_1}(\eta) = \int_{t_1}^{t_2} d\tau \int f'(\eta)dI_\tau(\eta),$$

$$\int g(\eta)dI_{t_2}(\eta) - \int g(\eta)dI_{t_1}(\eta) = \int_{t_1}^{t_2} d\tau \int g'(\eta)u(\eta,\tau)dI_\tau(\eta).$$

As far as system (2) is concerned, note that its third equation can be interpreted as

$$-g_x = \left(\int_x^\infty \rho(\xi, t)d\xi - \int_{-\infty}^x \rho(\xi, t)d\xi \right), \tag{11}$$

that is the acceleration at a point is proportional to the difference betweeen the total masses from the right and from the left of that point. The characteristics of (2) are now given by quadratic functions of t

$$x(t) = y + u_0(y)t + a_0(y)t^2/2, \tag{12}$$

with (7) replaced by

$$\phi_t(y) = \frac{\int_{C_t(y)} (\eta + tu_0(\eta))dP_0(\eta)}{\int_{C_t(y)} dP_0(\eta)} + a(C_t(y))\frac{t^2}{2}, \tag{13a}$$

$$u(x, t) = \frac{\int_{D_t(x)} u_0(\eta))dP_0(\eta)}{\int_{D_t(x)} dP_0(\eta)} + a(D_t(x))t, \tag{13b}$$

where

$$a(C_t(y)) = P_0(I^+) - P_0(I^-), \tag{14}$$

and I^+, I^- are, respectively, the right and left component of $\mathbf{R} \setminus C_t(y)$.

In this case the generalized variational principle permits construction of the family of partitions $\xi_t, t \geq 0$, with the left endpoint y such that

$$\frac{\int_{[y^-, y)} (\eta + tu_0(\eta))dP_0(\eta)}{\int_{[y^-, y)} dP_0(\eta)} + \frac{t^2}{2}(P_0(y, +\infty) - P_0(-\infty, y^-))$$

$$< \frac{\int_{[y, y^+)} (\eta + tu_0(\eta))dP_0(\eta)}{\int_{[y, y^+)} dP_0(\eta)} + \frac{t^2}{2}(P_0(y^+, +\infty) - P_0(-\infty, y)) \tag{15}$$

Then ϕ_t and u can be constructed from (13), P_t as before and I_t from u and P_t by simple integration.

Theorem 2. *Suppose that in addition of (i)-(iii) of Theorem 1 we suppose that:*

(iv) The initial mass $P_0(\mathbf{R})$ is finite.

Then the family $(P_t, I_t, u)_{t \geq 0}$ constructed using the generalized variational principle gives the weak solution of (2) with initial data (P_0, I_0) in the sense that $P_t \to P_0$ and $I_t \to I_0$ weakly as $t \to 0^+$.

The triple $(P_t, I_t, u)_{t \geq 0}$ is understood as a weak solution of (2) in the following sense: for any $f, g \in C^1(\mathbf{R})$ with compact support, and $0 < t_1 < t_2$,

$$\int f(\eta) dP_{t_2}(\eta) - \int f(\eta) dP_{t_1}(\eta) = \int_{t_1}^{t_2} d\tau \int f'(\eta) dI_\tau(\eta),$$

$$\int g(\eta) dI_{t_2}(\eta) - \int g(\eta) dI_{t_1}(\eta) = \int_{t_1}^{t_2} d\tau \int g'(\eta) u(\eta, \tau) dI_\tau(\eta)$$

$$+ \int_{t_1}^{t_2} d\tau \int g(\eta) (P_\tau(\eta, -\infty) - P_\tau(-\infty, \eta)) \, dP_\tau(\eta).$$

Random initial data. It turns out that for continuous but nowhere differentiable random initial data (like the Brownian motion), almost surely, the solution $u(x, t)$ becomes discontinuous for any $t > 0$, the set of discontinuities (shocks) is dense and almost all masses are absorbed in shocks.

Theorem 3. *Suppose that:*

(i) The probability distribution Q of the initial velocity u_0 is defined on the Borel σ-algebra of the space of continuous functions on $\mathbf{R}$ and for Q-almost all u_0, $u_0(x)/|x| \to 0$ as $|x| \to \infty$, and, for any $\eta_0 \in \mathbf{R}$,

$$\lim_{h \to 0^+} \sup h^{-2} \int_{\eta_0 - h}^{\eta_0} (u_0(\eta) - u_0(\eta_0)) \, d\eta = \infty,$$

$$\lim_{h \to 0^+} \inf h^{-2} \int_{\eta_0}^{\eta_0 + h} (u_0(\eta) - u_0(\eta_0)) \, d\eta = -\infty.$$

(ii) The initial distribution of masses ρ_0 is either bounded and bounded away from zero with a bounded derivative, or positive only on a finite interval with a bounded derivative (for problem (1)) and positive only on a finite interval with a bounded derivative (for problem (2)).

Then, the measures P_t corresponding to the weak solutions of problems (1) and (2) are pure point measures, that is $P_t(x) = \sum_i m_i \delta(x - x_i)$, $m_i > 0$, and the closure of the set $\{x_i\}$ is equal to the support of ρ_0.

The proofs of the above three theorems can be found in E, Rykov and Sinai (1996).

Lecture 8
Fractal Burgers-KPZ Models

8.1 Existence and uniqueness problems

In the first three sections we will review local and global in time solutions to a class of multidimensional generalized Burgers-type equations with a fractional power of the negative Laplacian $(-\Delta)$ in $\mathbf{R}^d$ replacing the term u_{xx} of the usual Burgers' equation (1.1.13). Also, we replace the quadratic term $(u^2)_x$ of (1.1.13) by a more general algebraic power nonlinearity which allows for multiparticle interactions. Such equations naturally appear in continuum mechanics. The results include existence, uniqueness, regularity and asymptotic behavior of solutions to the Cauchy problem as well as a construction of self-similar solutions. The role of critical exponents is also explained. There are obvious connections to the multidimensional fractal (anomalous) diffusion related to the Lévy flights (see, e.g., Stroock (1975), Dawson and Gorostiza (1990), Shlesinger et al. (1995), Zaslavsky (1994), Zaslavsky and Abdullaev (1995), and the references quoted therein). The results of Sections 8.1-3 are due to Biler, Funaki and Woyczynski (1998a) (see also Biler and Woyczynski (1998)).

Consider equations

$$u_t = -\nu(-\Delta)^{\alpha/2}u - a \cdot \nabla(u^r), \tag{1}$$

where $x \in \mathbf{R}^d$, $d = 1, 2, \ldots$, $t \geq 0$, $u : \mathbf{R}^d \times \mathbf{R}^+ \to \mathbf{R}$, $\alpha \in (0, 2]$, $r \geq 1$, and $a \in \mathbf{R}^d$ is a fixed vector. For noninteger r, by u^r we mean $|u|^r$. In the sequel we assume $\nu \equiv 1$, without loss of generality. The case $\alpha = 2$ and $r = 1$ corresponds to the standard (Gaussian) *linear* diffusion equation with a drift, see Section 2.1.

For the equation (1), the Hopf-Cole formula is not longer available and another major difference with the classical Burgers equation

(1.1.13) is the presence in (1) of the singular integro-differential operator $(-\Delta)^{\alpha/2}$. The equations are no longer local.

There is ample physical motivation justifying consideration of the nonlocal Burgers equation (1), one of them being the eventual goal of studying the Navier-Stokes problem

$$u_t = -\nu(-\Delta)^{\alpha/2}u - (u \cdot \nabla)u - \nabla p$$

$$\nabla \cdot u = 0$$

with modified dissipativity as suggested by Frisch and his collaborators (see, e.g., Frisch et al. (1974) and Bardos et al. (1979)). A large variety of physically motivated (linear) fractal differential equation can be found in Shlesinger et al. (1995), including applications to hydrodynamics, statistical mechanics, physiology and molecular biology. Fractal relaxation models are described in Saichev and Woyczynski (1996) (the book also contains a pedestrian introduction to fractal calculus). Linear fractional kinetic equations have been studied by Saichev and Zaslavsky (1997). Models of several other hydrodynamical phenomena (including hereditary and viscoelastic behavior and propagation of nonlinear acoustic waves in a tunnel with an array of Helmholtz resonators) employing the Burgers equation involving the fractional Laplacian have also been developed (Sugimoto and Kakutani (1986), Sugimoto (1989, 1991, 1992)). An additional motivation was an extended fractional KPZ model of surface growth which includes surface trapping effects and was suggested in Woyczynski (1997) and Mann and Woyczynski (1997), and the needs of the theory of nonlinear Markov processes and propagation of chaos associated with fractal Burgers equation, see Sections 8.4-5, Funaki and Woyczynski (1998) and Biler, Funaki and Woyczynski (1998b).

Throughout this lecture we will use the standard notation: $|u|_p$ for the Lebesgue $L^p(\mathbf{R}^d)$-norms of functions, $\|u\|_{\beta,p}$ for the Sobolev $W^{\beta,p}(\mathbf{R}^d)$-norms, and $\|u\|_\beta \equiv \|u\|_{\beta,2}$ for the most frequent case of Hilbert Sobolev space $H^\beta(\mathbf{R}^d)$. The constants independent of solutions considered will be denoted by the same letter C, even if they may vary from line to line. For various interpolation inequalities we refer to Adams (1975), Ladyženskaja et al. (1988), Triebel (1983, 1992), Mikhlin and Prössdorf (1986) and Henry (1982).

1-D case: a direct approach via a priori inequalities. Consider the equation

$$u_t = -D^\alpha u - uu_x, \qquad 0 < \alpha \le 2, \qquad (2)$$

where

$$D^\alpha \equiv (-\partial^2/\partial x^2)^{\alpha/2}.$$

Using simplest *a priori* estimates we will prove some results on local and global in time solvability of the Cauchy problem for (2). This will show the role of dissipative operator $-D^\alpha$ and, in particular, its strength compared to the nonlinearity uu_x. We define D^α as

$$(D^\alpha v)(x) = \mathcal{F}^{-1}\left(|\xi|^\alpha \hat{v}(\xi)\right)(x),$$

where $\hat{} \equiv \mathcal{F}$ denotes the Fourier transform and $\mathcal{F}^{-1}$ its inverse.

We look for *weak* solutions of (2) supplemented with the initial condition

$$u(x,0) = u_0(x), \tag{3}$$

i.e., functions

$$u \in V_2 \equiv L^\infty((0,T); L^2(\mathbf{R})) \cap L^2((0,T); H^1(\mathbf{R}))$$

satisfying the integral identity

$$\int u(x,t)\phi(x,t) - \int_0^t \int u\phi_t + \int_0^t \int \left(D^{\alpha/2}u\, D^{\alpha/2}\phi - \frac{1}{2}u^2\phi_x\right)$$

$$= \int u_0(x)\phi(x,0)$$

for a.e. $t \in (0,T)$ and each test function $\phi \in H^1(\mathbf{R} \times (0,T))$; all integrals with no integration limits are understood as $\int_{\mathbf{R}} \cdot\, dx$.

Observe that we assume $u(t) \in H^1(\mathbf{R})$ a.e. in $t \in (0,T)$, instead of just $u(t) \in H^{\alpha/2}(\mathbf{R})$ a.e. in t, which could be expected from a straightforward generalization of the definition of the weak solution of a parabolic second order equation (see, e.g., Ladyženskaja et al. (1988)). We need this supplementary regularity to simplify slightly our construction; for the initial data $u_0 \in H^1(\mathbf{R})$ it is a consequence of the assumptions.

Theorem 1. *Let* $\alpha \in (3/2, 2]$, $T > 0$, *and* $u_0 \in H^1(\mathbf{R})$. *Then the Cauchy problem (2-3) has a unique weak solution* $u \in V_2$. *Moreover,* u *enjoys the following regularity properties:*

$$u \in L^\infty((0,T); H^1(\mathbf{R})) \cap L^2((0,T); H^{1+\alpha/2}(\mathbf{R})),$$

and

$$u_t \in L^\infty((0,T); L^2(\mathbf{R})) \cap L^2((0,T); H^{\alpha/2}(\mathbf{R}))$$

for each $T > 0$. For $t \to \infty$, this solution decays so that

$$\lim_{t \to \infty} |D^{\alpha/2} u(t)|_2 = \lim_{t \to \infty} |u(t)|_\infty = 0.$$

PROOF. We begin with formal calculations to obtain *a priori* inequalities for various norms of (sufficiently regular) solutions to (2-3). Given these *a priori* estimates, the proof of the theorem will proceed in a rather routine fashion. First, we introduce spatial truncations of (2) to $(-R, R) \subset \mathbf{R}$, $R > 0$. Then we consider k-dimensional approximations to (2) with the homogeneous Dirichlet boundary conditions for $x = \pm R$ via the Galerkin procedure (note that $\partial/\partial x$ commutes with D^β). Finally, the *a priori* estimates permit us to pass to the limit $k \to \infty$ and with $R \to \infty$ (by the diagonal choice of subsequences).

Suppose that u is a weak solution of (2-3). Multiplying (2) by u, after applying the definition of the diffusion operator D^α we arrive at

$$\frac{d}{dt} |u|_2^2 + 2|D^{\alpha/2} u|_2^2 = 0. \tag{4}$$

Similarly, differentiating (2) with respect to x and multiplying by u_x we obtain

$$\frac{d}{dt} |u_x|_2^2 + 2|D^{1+\alpha/2} u|_2^2 \leq |u_x|_3^3, \tag{5}$$

since

$$- \int (u u_x)_x u_x = \int u(u_x u_{xx}) = \frac{1}{2} \int u(u_x^2)_x = -\frac{1}{2} \int u_x^3.$$

The right-hand side of (5) can now be estimated by

$$|u_x|_3^3 \leq \|u\|_{1,3}^3 \leq C \|u\|_{1+\alpha/2}^{7/(2+\alpha)} |u|_2^{3-7/(2+\alpha)} \leq \|u\|_{1+\alpha/2}^2 + C|u|_2^m$$

for some $m > 0$; note that the assumption $\alpha > 3/2$ has been used in the interpolation of the $W^{1,3}$-norm of u by the norms of its fractional derivatives to have $7/(2 + \alpha) < 2$. Indeed, this follows from Henry (1982, p. 99) with extensions for nonintegral order derivatives like in, e.g., Triebel (1983, 1992). Combining this with (4-5) we get

$$\frac{d}{dt} \|u\|_1^2 + \|u\|_{1+\alpha/2}^2 \leq C(|u|_2^2 + |u|_2^m),$$

and since (4) implies $|u(t)|_2 \leq |u_0|_2$ for $t \in [0, T]$, we arrive at

$$\|u(t)\|_1^2 + \int_0^t \|u(s)\|_{1+\alpha/2}^2 \, ds \leq C = C(T, \|u_0\|_1). \tag{6}$$

To get the estimate for the time derivative of the solution, let us differentiate (2) with respect to t and multiply by u_t. After elementary calculations we obtain

$$\frac{d}{dt}|u_t|_2^2 + |D^{\alpha/2}u_t|_2^2 = -\int u_x u_t^2, \tag{7}$$

because

$$-\int(uu_x)_t u_t = -\int u_x u_t^2 - \frac{1}{2}\int u(u_t^2)_x = -\frac{1}{2}\int u_x u_t^2.$$

The right-hand side of (7) is now estimated by

$$\frac{1}{2}\int|u_x|u_t^2 \leq C\|u_t\|_{\alpha/2}^{1/\alpha}|u_t|_2^{2-1/\alpha}|u_x|_2$$

$$\leq \frac{1}{2}\|u_t\|_{\alpha/2}^2 + C|u_t|_2^2,$$

as we applied the (locally) uniform in time estimate for $\|u(t)\|_1$. It is clear now that a standard application of the Gronwall inequality leads to

$$|u_t(t)|_2^2 + \int_0^t \|u_t(s)\|_{\alpha/2}^2\, ds \leq C(T). \tag{8}$$

The *a priori* estimates (6) and (8) are sufficient to apply the Galerkin approximation procedure as sketched above (note that C's in (6) and (8) are independent of the order of approximation and of the interval $(-R, R)$). Thus, the existence and regularity part of the conclusions of Theorem 1 have been established.

To resolve the uniqueness problem, let us consider two weak solutions u, v of (2-3). Then their difference $w = u - v$ satisfies

$$\frac{d}{dt}|w|_2^2 + 2|D^{\alpha/2}w|_2^2 = 2\int(vv_x - uu_x)w$$

$$= -2\int(vww_x + w^2u_x) = 2\int w^2(v_x/2 - u_x). \tag{9}$$

Now, the right-hand side of (9) can be estimated from above

$$|w|_4^2|v_x - 2u_x|_2 \leq C\|w\|_{\alpha/2}^{1/\alpha}|w|_2^{2-1/\alpha}(|u_x|_2 + |v_x|_2)$$

$$\leq \|w\|_{\alpha/2}^2 + C|w|_2^2,$$

since, in view of (6), the factor $|u_x|_2 + |v_x|_2$ is bounded. An application of the Gronwall lemma implies that $w(t) \equiv 0$ on $[0, T]$.

Remarks. We proceeded formally with differential inequalities. A rigorous proof is obtained by rewriting them as integral inequalities, like (6) and (8), which are direct consequences of the definition of the weak solution. Note that the proof of *uniqueness* also works for weak solutions of (2) with $\alpha > 1/2$. Indeed, the crucial estimate of (9) only requires that $1/\alpha < 2$. Moreover, a slight modification of the uniqueness proof can give the (local in time) *continuous dependence of solutions on the initial data.*

The proof of Theorem 1 gives also the local in time existence of solutions to (2) for $\alpha > 1/2$. If $\alpha \in (1/2, 3/2]$ they may loose regularity after some time $T > 0$, and can be considered only as a kind of weaker solutions. We discuss this issue in remarks to the next section which deals with such a generalization of weak solutions studied here. The proof of the asymptotic estimates

$$\lim_{t \to \infty} |D^{\alpha/2}u(t)|_2 = \lim_{t \to \infty} |u(t)|_\infty = 0$$

can be accomplished as follows. Multiplying (2) by $D^\alpha u$ and integrating, we obtain

$$\frac{1}{2}\frac{d}{dt}|D^{\alpha/2}u|_2^2 + |D^\alpha u|_2^2 = -\int uu_x D^\alpha u \le \frac{1}{2}|D^\alpha u|_2^2 + \frac{1}{2}\int u^2 u_x^2.$$

Since $|u|_\infty^2 \le 2|u_x|_2|u|_2$ holds and $|u(t)|_2$ is bounded, we have

$$\frac{d}{dt}|D^{\alpha/2}u|_2^2 + |D^\alpha u|_2^2 \le C|u_x|_2^3 \le C|D^\alpha u|_2^{3/\alpha}|u|_2^{3-3/\alpha}$$

$$\le \frac{1}{2}|D^\alpha u|_2^2 + C$$

(again $\alpha > 3/2$ is applied). In particular, $(d/dt)|D^{\alpha/2}u|_2^2$ is bounded, which together with the integrated form of (4)

$$|u(t)|_2^2 + 2\int_0^t |D^{\alpha/2}u(s)|_2^2 \, ds \le |u_0|_2^2 = C,$$

for all $t \ge 0$, implies that $\lim_{t \to \infty} |D^{\alpha/2}u(t)|_2^2 = 0$. Obviously, $t \mapsto |D^{\alpha/2}u(t)|_2^2$ is positive and continuous, and it is clear that: $0 \le \phi, \phi' \le C, \int_0^\infty \phi < \infty$ implies $\lim_{t \to \infty} \phi(t) = 0$. The second asymptotic relation follows from the Sobolev imbedding $H^{\alpha/2} \subset L^\infty$ valid for $\alpha > 1$.   ∎

Parabolic regularization.   The above approach to the Cauchy problem (2-3) produces, in fact, regularity of the solutions; for $\alpha > 3/2$, the diffusion operator D^α is strong enough to control the nonlinear term uu_x. When $\alpha \leq 3/2$, a direct construction of weak global in time solutions is no longer possible for initial data (3) of arbitrary size, and we will review another technique to obtain candidate weak solutions; the construction will be done by the method of parabolic regularization, i.e., by first studying the initial-value problem

$$u_t = -D^\alpha u - uu_x + \epsilon u_{xx}, \qquad u(x,0) = u_0(x), \qquad (10)$$

with $u = u_\epsilon$, $\epsilon > 0$. This method is, of course, standard (see, e.g., Bardos et al. (1979), Saut (1979)). In particular, solutions of the inviscid Burgers (Riemann) equation $u_t = -uu_x$ can be obtained as limits of solutions to $u_t = -uu_x + \epsilon u_{xx}$ when $\epsilon \to 0$. Surely, we cannot expect to prove the uniqueness of solutions constructed in such a manner without proving their regularity; for the inviscid Burgers equation, besides the unique viscosity solution there are many others which are less regular (see, e.g., Smoller (1994)).

Theorem 2. *Let $0 < \alpha \leq 2$, and $u = u_\epsilon$, $\epsilon > 0$, be a solution to the Cauchy problem (10), with $u_0 \in L^1 \cap H^1$, $(u_0)_x \in L^1$. Then, for all $t \geq 0$,*

$$|u(t)|_2 \leq |u_0|_2, \qquad (11)$$

$$|u(t)|_1 \leq |u_0|_1, \qquad (12)$$

$$|u_x(t)|_1 \leq |(u_0)_x|_1. \qquad (13)$$

SKETCH OF THE PROOF. The existence of solutions to the regularized equation (10) is standard; as in Theorem 1 before, the main ingredients include the Galerkin method and the compactness argument.

The inequality (11) is a straightforward consequence of the differential inequality

$$\frac{d}{dt}|u|_2^2 + 2|D^{\alpha/2}u|_2^2 + 2\epsilon|u_x|_2^2 \leq 0,$$

which is a counterpart of (4). The L^1-contraction property (12) of solutions to (10) follows from the same property for the linear equation

$$u_t = -D^\alpha u + \epsilon u_{xx},$$

and from the structure of the nonlinear term. The details are in Biler, Funaki and Woyczynski (1996). The estimate (3.4) is obtained in a similar manner (see also Bardos et al. (1979), (2.14)) ∎

Having established the *a priori* estimates (11-13) we can pass to the limit $\epsilon \to 0$ in the regularized equation (10).

In this subsection by a weak solution of (2) we understand $u \in L^\infty((0,T); L^2(\mathbf{R}))$ satisfying the integral identity

$$\int u(x,t)\phi(x,t) - \int_0^t \int u\phi_t + \int_0^t \int \left(uD^\alpha\phi - \frac{1}{2}u^2\phi_x \right) = \int u_0(x)\phi(x,0)$$

for a.e. $t \in (0,T)$ and each test function $\phi \in C^\infty(\mathbf{R} \times [0,T])$ with compact support in x. Note that we do not assume $u(t) \in H^{\alpha/2}$ a.e. in t as in the first subsection.

Corollary 1. *Let* $0 < \alpha \leq 2$. *Given* $u_0 \in L^1 \cap H^1$ *with* $(u_0)_x \in L^1$, *there exists a weak solution* u *of (2.1) obtained as a limit of a subsequence of* u_ϵ*'s such that*

$$u \in L^\infty((0,\infty); L^\infty(\mathbf{R})) \cap L^\infty((0,\infty); H^{1/2-\delta}(\mathbf{R}))$$

for each $\delta > 0$. *Moreover,*

$$\|u(t); BV(\mathbf{R})\| \leq |(u_0)_x|_1,$$

where $BV(\mathbf{R})$ *denotes the space of functions on* $\mathbf{R}$ *with bounded variation.*

Remarks. Supplementary regularity properties of u can be read from the estimates (11-13) and the counterparts of (4-5) and (7) for the equation (10), i.e.,

$$\frac{d}{dt}|u|_2^2 + 2|D^{\alpha/2}u|_2^2 + 2\epsilon|u_x|_2^2 \leq 0,$$

$$\frac{d}{dt}|u_x|_2^2 + 2|D^{1+\alpha/2}u|_2^2 + 2\epsilon|u_{xx}|_2^2 \leq |u_x|_3^3,$$

$$\frac{d}{dt}|u_t|_2^2 + 2|D^{\alpha/2}u_t|_2^2 + 2\epsilon|u_{xt}|_2^2 \leq \int |u_x||u_t|^2.$$

For instance, if $\alpha > 1/2$ then weak solutions of (10) (they are unique by the proof of Theorem 1), constructed by the method of parabolic regularization, remain in $H^1(\mathbf{R})$ for $t \in [0,T)$ with some $T > 0$. Moreover, if $\|u_0\|_1$ is small enough, then these regular solutions are global in time. The crucial estimate to obtain this reads

$$\|u\|_{1,3}^3 \leq C\|u\|_{1+\alpha/2}^{1/\alpha} \|u\|_1^{3-1/\alpha} \leq \|u\|_{1+\alpha/2}^2 + C\|u\|_1^{2(3\alpha-1)/(2\alpha-1)}.$$

Then from the inequality

$$\frac{d}{dt}\|u\|_1^2 + \|u\|_{1+\alpha/2}^2 \leq C\left(|u|_2^2 + \|u\|_1^m\right)$$

with some $m > 2$, we may conclude either the boundedness of $\|u(t)\|_1$ on some time interval $[0, T)$, or the global smallness of $\|u(t)\|_1$ under a smallness assumption on $\|u_0\|_1$. Indeed, the solutions $\Psi(t) = \|u(t)\|_1$ of the differential inequality

$$\frac{d}{dt}\Psi + \Psi \leq C(|u_0|_2^2 + \Psi^m)$$

remain bounded (and small) whenever $\Psi(0)$ is sufficiently small.

For $\alpha < 1$ those weak solutions, regular on a finite time interval only, may exhibit shocks, see Section 8.3.

Mild solutions. In this subsection we review an alternative, *mild solution* approach to the fractal Burgers-type equation (1) with a general power nonlinearity $a \cdot \nabla(u^r)$, $r > 1$, and $\alpha \in (1, 2]$. It replaces the partial differential equation (1.2) by the integral equation

$$u(t) = e^{tA}u_0 - \int_0^t (\nabla e^{(t-s)A}) \cdot (au^r(s))\, ds, \qquad (14)$$

which is a consequence of the variation of parameters formula. Here $A = -(-\Delta)^{\alpha/2}$ is the infinitesimal generator of an analytic semigroup (e^{tA}), $t \geq 0$, called the *Lévy semigroup*, on $L^p(\mathbf{R}^d)$ (and on other functional spaces), and the commutativity $\nabla A = A\nabla$ permits changing the order of application of ∇ and e^{tA}. We restrict our attention to $\alpha > 1$ since, as we shall see later on ((18)), the derivative of e^{tA} contributes the factor $(t - s)^{-1/\alpha}$ which is integrable on $[0, t]$ only if $\alpha > 1$.

Remark. The operators e^{tA} act by convolution with the kernel

$$p_{\alpha,t} = \mathcal{F}^{-1}\left(\exp(-t|\xi|^\alpha)\right), \qquad (15)$$

or, in other words, e^{tA} is a Fourier multiplier with the symbol $\exp(-t|\xi|^\alpha))$. Explicit representation of the convolution kernel (15) of the Lévy semigroup is known for only a few values of α ($= \frac{1}{2}, 1, 2$).

The idea to replace the partial differential equation by an abstract evolution equation goes back (at least) to H. Fujita and T. Kato's early sixties work. An elegant approach in this spirit to semilinear parabolic equations is due to Weissler (1980). Our Theorem 3 below is close to the results of Avrin (1987, Theorems 2.1-2), who considered the case $\alpha \geq 2$ and L^p spaces. However, the functional framework developed in Theorems 6.1-2 is different. We employ Morrey, instead of Lebesgue spaces to get local and global time solvability for less regular initial data. This approach was motivated by Biler (1995, Section 2), where (nearly optimal) results had been proved for a parabolic problem arising in statistical mechanics, see also Taylor (1992), Kozono and Yamazaki (1994), Cannone (1995)) for the related work on the Navier-Stokes system.

Recall the definition of the Morrey spaces and basic properties of the semigroup e^{tA} (see, e.g., Taylor (1992), Triebel (1982, 1991), Biler (1995), Cannone (1995)).

$M^p = M^p(\mathbf{R}^d)$ denotes the *Morrey space* of locally integrable functions such that the norm

$$\|f; M^p\| \equiv \sup_{x \in \mathbf{R}^d, 0 < R \leq 1} R^{d(1/p-1)} \int_{B_R(x)} |f|$$

is finite. $\dot{M}^p = \dot{M}^p(\mathbf{R}^d)$ is the *homogeneous* Morrey space, where in the above definition the supremum is taken over all $0 < R < \infty$. More general 2-index Morrey spaces include $M_q^p = M_q^p(\mathbf{R}^d) =$

$$\left\{ f \in L_{\text{loc}}^q(\mathbf{R}^d) : \|f; M_q^p\|^q \equiv \sup_{x \in \mathbf{R}^d, 0 < R \leq 1} R^{d(q/p-1)} \int_{B_R(x)} |f|^q < \infty \right\},$$
(16)

where $1 \leq q \leq p \leq \infty$, as well as their homogeneous versions, $\dot{M}_q^p = \dot{M}_q^p(\mathbf{R}^d) =$

$$\left\{ f \in L_{\text{loc}}^q(\mathbf{R}^d) : \|f; \dot{M}_q^p\|^q \equiv \sup_{x \in \mathbf{R}^d, 0 < R < \infty} R^{d(q/p-1)} \int_{B_R(x)} |f|^q < \infty \right\}.$$
(17)

Note that the $\dot{M}_q^p(\mathbf{R}^d)$-norm has the same type of scaling as the $L^p(\mathbf{R}^d)$-norm:

$$\|f(\lambda \cdot); \dot{M}_q^p\| = \lambda^{-d/p} \|f; \dot{M}_q^p\|,$$

which explains the adjective "homogeneous".

The Morrey spaces are larger than the Lebesgue spaces L^p, and they also contain the Marcinkiewicz weak-L^p spaces. We note the inclusions

$$L_{\text{unif}}^p \equiv \{f : \sup_{x \in \mathbf{R}^d} \int_{B_1(x)} |f|^p < \infty\} = M_p^p \subset M_q^p \subset M^p \qquad (L_{\text{unif}}^p \subset L_{\text{loc}}^p).$$

The estimates for the Lévy semigroup e^{tA} and for the operator ∇e^{tA} can be obtained (in the Fourier representation) in a way parallel to those for the usual Gaussian heat semigroup ($\alpha = 2$); another method would be to apply the concept of subordination of analytic semigroups. Thus we can rewrite inequalities for the heat semigroup from Taylor (1992, Th. 3.8, (3.71), (3.75), (4.18)), in the form

$$\|e^{tA}f; M_{q_2}^{p_2}\| \le C\, t^{-d(1/p_1 - 1/p_2)/\alpha}\, \|f; M_1^{p_1}\|, \tag{18}$$

where $1 < p_1 < p_2 < \infty$, $q_2 < p_2/p_1$, and

$$\|\nabla e^{tA}f; M_{q_2}^{p_2}\| \le C\, t^{-d(1/p_1 - 1/p_2)/\alpha - 1/\alpha}\, \|f; M_{q_1}^{p_1}\|, \tag{19}$$

valid for $1 < p_1 < p_2 < \infty$, $q_1 > 1$, and $q_2/q_1 = p_2/p_1$ provided $p_1 \le d$, and $q_2/q_1 < p_2/p_1$ otherwise.

The limit case $p_1 = p_2$ in (18) also holds true: $e^{tA} : M_q^p \to M_q^p$ is a bounded operator. However, e^{tA} (like $e^{t\Delta}$) is not a strongly continuous semigroup on M_q^p. This makes impossible a direct application of the scheme of the existence proof in Weissler (1980), where spaces of vector-valued functions continuous with respect to time have been used.

A remedy for this is either to consider a subspace of M_q^p on which e^{tA} forms a strongly continuous semigroup, or to weaken the usual definition of mild solution to (14) (cf. a discussion in Biler (1995, Section 2)). In the first case one needs to study the subspace

$$\ddot{M}_q^p = \left\{ f \in M_q^p : \|\tau_y f - f; M_q^p\| \longrightarrow 0 \quad \text{as} \quad |y| \to 0 \right\},$$

$y \in \mathbf{R}^d$, $\tau_y f(x) = f(x - y)$. $\ddot{M}_q^p$ is the maximal closed subspace of M_q^p on which the family of translations forms a strongly continuous group and, simultaneously, the maximal closed subspace on which e^{tA} is strongly continuous semigroup. Notice that

$$L^p \subset \ddot{M}_q^p \subset \mathring{M}_q^p \subset M_q^p,$$

where

$$\mathring{M}_q^p = \left\{ f \in M_q^p : \limsup_{R \to 0} R^{d(q/p-1)} \int_{B_R(x)} |f|^q = 0 \right\},$$

see Taylor (1992, (4.14)). The second possibility is to replace the space $C([0,T]; \mathcal{B})$ of norm continuous functions with values in a Banach space $\mathcal{B}$ of tempered distributions on $\mathbf{R}^d$ by the space $\mathcal{C}([0,T]; \mathcal{B})$ of weakly continuous (in the sense of distributions) functions which are bounded in the norm of $\mathcal{B}$, i.e., the subspace of $C([0,T]; \mathcal{S}'(\mathbf{R}^d))$ such that $u(t) \in \mathcal{B}$ for each $t \in [0,T]$ and $\{u(t) : t \in [0,T]\}$ is bounded in $\mathcal{B}$. When $\mathcal{B} = X^*$ is the dual of a Banach space X then $C([0,T]; \mathcal{B})$ coincides with the space of $\mathcal{B}$-valued functions that are continuous in the weak* topology of $\mathcal{B}$, cf. Cannone (1995).

Theorem 3. *Let $\alpha > 1$, $u_0 \in M_q^p$, with $p > \max(d(r-1)/(\alpha-1), r)$, $1 \le r \le q \le p$. Then there exists $T = T(u_0) > 0$ and a unique solution of the integral equation (14) in the space $\mathcal{X} = C([0,T]; M_q^p)$.*

A proof based on the contraction argument can be found in Biler, Funaki and Woyczynski (1996).

Remarks. For the usual Burgers equation with a quadratic nonlinear term, Theorem 6.1 applies with $p > \max(d/(\alpha - 1), 2)$, so, e.g., for $d = 1$, $\alpha \in (1, 3/2)$, $p > 1/(\alpha - 1)$. Evidently, for $d \ge 2$, $r \ge 2$, the inequality $p > d(r-1)/(\alpha - 1)$ is a sufficient condition. Note that for $p = q$ we recover an extension of the local existence result in $L^p(\mathbf{R}^d)$ spaces from Avrin (1987, Th. 2.2), where the assumption was $\alpha > 2$.

As far as the global existence of solutions is concerned we have

Theorem 4. *Let $\alpha > 1$. There exist $\epsilon > \tilde{\epsilon} > 0$ such that given $u_0 \in M_q^p$ with $p = d(r-1)/(\alpha - 1)$, $1 \le r < q \le p$, and*

$$l(u_0) \equiv \limsup_{t \to 0} t^\beta \|e^{tA} u_0 : M_{rq}^{rp}\| < \epsilon, \qquad \beta = (1 - 1/\alpha)/r,$$

there exist $T > 0$, and a local in time solution u of (6.1), $0 \le t \le T$, which is unique in the space

$$\mathcal{X} = C\left([0,T]; M_q^p\right) \cap \left\{u : [0,T] \to M_q^p; \sup_{0 < t \le T} t^\beta \|u(t); M_{rq}^{rp}\| < \infty\right\}.$$

Moreover, if

$$\sup_{t > 0} t^\beta \|e^{tA} u_0; \dot{M}_{rq}^{rp}\| < \tilde{\epsilon},$$

then this solution can be extended to a global one.

Notice that the second condition defining the space $\mathcal{X}$ in Theorem 4 is the more important one. In particular, the proof via contraction arguments is not sensitive to the size of the initial data in M_q^p, i.e., the length R of the N-invariant box $B_{R,\epsilon} \subset \mathcal{X}$.

The second assumption on u_0 is a sort of (rather weak) supplementary regularity of an element of M_q^p. Indeed, if $u_0 \in L^p$ or $u_0 \in \overset{\circ}{M}_q^p$, then $l(u_0) = 0$. So the assumption $u_0 \in \overset{\circ}{M}{}^p_q$ yields the local existence of solution *independently* of the size of $\|u_0; M^p_q\|$.

In the limit case when $p = d(r-1)/(\alpha-1) = 1$ (the method actually works, e.g., either for $r = 2, \alpha = 1, d = 1$, or when $\alpha \in (1, 2), 1 < r \leq (\alpha + 1)/2, d = 2$) we may obtain a similar conclusion replacing the Morrey space M^p_q by the space $\mathcal{M}(\mathbf{R}^d)$ of finite Borel measures on $\mathbf{R}^d$; see Biler (1995, Th. 2) for a similar situation.

Theorem 5. *Let* $\alpha > 1$, $d(r - 1) = \alpha - 1$, $\beta = (1 - 1/\alpha)/r$. *There exist* $\epsilon > \tilde{\epsilon} > 0$ *such that given* $u_0 \in \mathcal{M}(\mathbf{R}^d)$ *satisfying the condition*

$$l(u_0) \equiv \limsup_{t \to 0} t^\beta |e^{tA} u_0|_r < \epsilon,$$

there exists a local in time solution u *of (6.1) belonging to (and unique in) the space*

$$\mathcal{X} = \mathcal{C}\left([0, T] : \mathcal{M}(\mathbf{R}^d)\right) \cap \left\{ u : [0, T] \to \mathcal{M}(\mathbf{R}^d);\ \sup_{0 < t \leq T} t^\beta |u(t)|_r < \infty \right\}.$$

Moreover, if $\sup_{t>0} t^\beta |e^{tA} u_0|_r < \tilde{\epsilon}$, *then this solution can be continued to a global one.*

Remarks. (a) If u_0 is an integrable function then for the functional l in Theorem 5 we have $l(u_0) = 0$; hence a local solution starting from u_0 can be constructed. If the initial measure u_0 is sufficiently small, then the global existence holds.

(b) Note that the same functional-analytic tools permit us to study the space-periodic problem for the equation (1) (or (14)) for solutions satisfying the condition

$$u(x + e_j, t) = u(x, t)$$

for the unit coordinate vectors e_j, $1 \leq j \leq d$, and all $x \in \mathbf{R}^d$, $t \geq 0$. We consider then spaces of space-periodic functions with zero average over the cube $[0, 1]^d$: $\int_{[0,1]^d} u\, dx = 0$. Evidently, the long time behavior of solutions is then subexponential—instead of the algebraic decay rate $t^{-d/(2\alpha)}$ for the L^2-norms of solutions.

(c) We would like to stress once more importance of the assumption $\alpha > 1$ in subsection dealing with regularized and mild solutions. The parabolic regularization effect requires the strength of the linear diffusion operator to be above a certain threshold.

(d) Note that our linear operator is nonlocal while in Biler (1995) the nonlinear term has been defined by a singular integral.

(e) Let us remark that recently Dix (1996) studied the local in time solvability of the classical Burgers equation (1.1) with the initial data in Sobolev spaces of negative order: $u_0 \in H^\sigma$, $\sigma > -1/2$. His approach involves also mild solutions, and the space he is working with is also determined by the behavior of solutions to the linear diffusion equation, as in our Theorems 4-5. His presentation is based on Fourier transform arguments, so it works efficiently only for the quadratic nonlinearity. Since he deals with *a priori* very irregular objects, his uniqueness result (Theorem 5.1) is particularly delicate and based on a novel approach.

8.2   Time decay of solutions

This section is devoted to the problem of large-time decay of solutions to the fractal Burgers equation. We formulate the result in the form of an *a priori* estimate, in order to have a more versatile tool (to be applied later). It is an extension of a result (Theorem 2a)) from Biler (1984).

    Theorem 1. *Let $0 < \alpha \le 2$. Suppose u is a sufficiently regular solution of the (multidimensional) equation (8.1.1) with $u_0 \in L^1(\mathbf{R}^d)$. Then the L^2-norm of u decays, as $t \to \infty$, at the rate estimated by the inequality*

$$|u(t)|_2 \le C(1+t)^{-d/(2\alpha)}. \tag{1}$$

PROOF. The reasoning is based on the L^1-contraction estimate as in Theorem 8.1.2, and on the Fourier splitting method introduced in Schonbek (1980) and then successfully applied to various evolution equations including the Navier-Stokes system. Let us begin with the inequality

$$\frac{d}{dt}|u(t)|_1 \le 0$$

which follows from $\int (Au)\,\mathrm{sgn}\,u \le 0$ and $\int a \cdot \nabla(u^r)\,\mathrm{sgn}\,u = 0$. Next, rewrite the energy equality (8.1.4) in the Fourier representation

$$\frac{d}{dt}\int |\hat{u}(\xi,t)|^2\,d\xi = -2\int |\xi|^\alpha |\hat{u}(\xi,t)|^2\,d\xi,$$

and estimate the right-hand side by

$$\le -2\int_{B^c} |\xi|^\alpha |\hat{u}|^2 \le (1+t)^{-1}\int_{B^c} |\hat{u}|^2.$$

Here $B^c = B(t)^c = \{|\xi| > [2(1+t)]^{-1/\alpha}\}$ is the complement of a ball in $\mathbf{R}_\xi^d$. In other words,

$$\frac{d}{dt}\left((1+t)\int|\hat{u}|^2\right) \leq \int_{\mathbf{R}^d\setminus B^c}|\hat{u}|^2 \leq |\hat{u}|_\infty^2\int_B d\xi \leq C[2(1+t)]^{-d/\alpha},$$

so we get the desired estimate

$$|u|_2 = C\left(\int|\hat{u}|^2\right)^{1/2} \leq C(1+t)^{-d/(2\alpha)}. \qquad \blacksquare$$

Remarks. In particular, the above result applies to the solution u of (8.1.2-3) constructed in Theorem 8.1.1 whenever $u_0 \in L^1 \cap H^1$. Notice that the algebraic decay rate (1) in Theorem 1 is identical to that for solutions of the linear fractal diffusion equation $u_t + (-\Delta)^{\alpha/2}u = 0$. However, this kind of estimates for solutions of the latter equation follows straightforwardly from the properties of the fundamental solution $p_{\alpha,t}$ of the above linear equation. Indeed, for

$$p_{\alpha,t}(x) = C\int\exp\left(-t|\xi|^\alpha\right)\exp(ix\cdot\xi)\,d\xi,$$

we have, for all $t \geq 0$, $|p_{\alpha,t}|_1 = 1$, $|t^{d/\alpha}p_{\alpha,t}|_\infty \leq C < \infty$.

Having at our disposal (1), various decay estimates of u in other norms can be proved. For instance, if u solves the Cauchy problem (8.1.3-4) with $u_0 \in L^1 \cap H^1$, then $|u(t)|_\infty \leq C(1+t)^{-1/(4\alpha)}$ which follows from the interpolation $|u|_\infty^2 \leq 2\|u\|_1|u|_2$. Of course, this is far from being the decay rate for the linear equation when $|u(t)|_\infty \leq C(1+t)^{-1/\alpha}$. Results on the identical decay rates for solutions to both linear and nonlinear equations (in the spirit of a "nonlinear scattering of low energy solutions") can be proved under the assumption that the nonlinear term, say $f'(u)u_x$ in (8.1.2) instead of $-uu_x$, is described by a function f sufficiently flat at the origin. See, e.g., Theorem 2c) in Biler (1984), where under the hypothesis $|f'(u)| \leq C|u|^q$, with some $q > 2\alpha - 2$ and all $u \in [-1, 1]$, the bound $|u(t)|_\infty \leq C(1+t)^{-1/\alpha}$ is proved.

8.3 Traveling wave and self-similar solutions

Traveling wave solutions. Consider equation (8.1.2)

$$u_t + uu_x = -D^\alpha u, \qquad 0 < \alpha \leq 2.$$

A more general polynomial nonlinearity is also considered at the end
of this section. We look for solutions in the form

$$u(x,t) = U(x + vt). \tag{1}$$

Substituting (1) into (8.1.2) leads to a differential equation

$$vU' + UU' = -D^\alpha U, \tag{2}$$

with fractal derivative in one variable for $U = U(y)$. Assuming that
$U(-\infty) = 0$, the integration of (2) results in the equation

$$vU + \frac{1}{2}U^2 = -D^{(\alpha-2)}U'. \tag{3}$$

If we are looking for the solutions with finite $U(\infty)$, then for $\alpha > 1$ it
is natural to assume that $D^{(\alpha-2)}U'(y) \to 0$ as $y \to \infty$. Then, clearly,
$U(\infty) = -2v$. However, we try to determine $U(\infty)$ in a more general
situation. Multiplying (3) by U' and integrating over $(-\infty, y]$ gives

$$\frac{v}{2}U^2(y) + \frac{1}{6}U^3(y) = -\int_{-\infty}^{y} U'(z)D^{\alpha-2}U'(z)\,dz.$$

Passing to the limit $y \to \infty$ and taking into account the Parseval identity, this yields

$$\frac{v}{2}U^2(\infty) + \frac{1}{6}U^3(\infty) = C\int |\xi|^\alpha|\hat{U}(\xi)|^2\,d\xi. \tag{4}$$

On the other hand, multiplying (2) by U, integrating, and again taking
the limit $y \to \infty$, we get

$$\frac{v}{2}U^2(\infty) + \frac{1}{3}U^3(\infty) = -C\int |\xi|^\alpha|\hat{U}(\xi)|^2\,d\xi. \tag{5}$$

Comparing (4) and (5) we conclude that

$$U^3(\infty) = -12C\int |\xi|^\alpha|\hat{U}(\xi)|^2\,d\xi. \tag{6}$$

Now, assuming smoothness of U, the integral

$$\int |\xi|^\alpha|\hat{U}(\xi)|^2\,d\xi = \int |\xi|^{\alpha-2}|\widehat{U'}(\xi)|^2\,d\xi \tag{7}$$

converges for $|\xi| \to \infty$. On the other hand, if U' is integrable, then
$\widehat{U'} \not\equiv 0$ is bounded in a neighborhood of $\xi = 0$. Hence the integral (7)
is finite for $\alpha \in (1, 2]$ but (in general) infinite for $\alpha \in (0, 1]$. Thus we
arrive at the following

Proposition 1. *Let U be a solution of (2) such that $U(-\infty) = 0$ and $U', U'', U''' \in L^1(\mathbf{R})$.*

(i) If $1 < \alpha \leq 2$, then $U(\infty)$ is finite (and equal to $-2v$);

(ii) If $0 < \alpha \leq 1$, then $U(\infty) = -\infty$. Consequently, in this case, the fractal Burgers equation (8.1.2) does not admit bounded traveling wave solutions.

PROOF. Essentially, the proof has been sketched above. The only calculation needed is the degree of smoothness of U required to get the convergence of the integral (4). For instance, an application of the Tauberian theorem indicates that the integrability of the derivative U' of order $\lceil \alpha + 1 + \epsilon \rceil$, for some $\epsilon > 0$, suffices. ∎

A similar analysis can be carried out for a slightly more general equation

$$u_t + aru^{r-1}u_x = -D^\alpha u$$

with $r > 1$ and $a \in \mathbf{R}$. Namely, if $u(x,t) = U(x + vt)$, then

$$vU + aU^r = -D^{\alpha-1}U.$$

So if $\alpha > 1$, $U(-\infty) = 0$, then $U(\infty) = (-v/a)^{1/(r-1)}$, since we expect $D^{\alpha-2}U'(y)$ to decay as $y \to \infty$. Note that we put here a parameter a to study simultaneously two equations $u_t \pm ru^{r-1}u_x = -D^\alpha u$ with different properties when r is an odd integer. Similarly, we have

$$\frac{v}{2}U^2(\infty) + a(r+1)^{-1}U^{r+1}(\infty) = C \int |\xi|^\alpha |\hat{U}(\xi)|^2 \, d\xi,$$

as well as

$$\frac{v}{2}U^2(\infty) + ar(r+1)^{-1}U^{r+1}(\infty) = -C \int |\xi|^\alpha |\hat{U}(\xi)|^2 d\xi,$$

so

$$a\frac{r-1}{r+1}U^{r+1}(\infty) = -2C \int |\xi|^{\alpha-2} |\widehat{U'}(\xi)|^2 \, d\xi.$$

The remainder of the reasoning follows as before.

Concluding, we see that $\alpha = 1$ is a critical exponent of diffusion in (8.1.2) (see also Sugimoto (1989)). In the sequel we will see other phenomena related to the loss of regularity of solutions below a critical diffusion exponent α.

Self-similar solutions. In this section we study special solutions of the fractal Burgers-type equation (8.1.1) which enjoy certain invariance properties. Note that if a function u solves (8.1.1) then, for each $\lambda > 0$, the rescaled function

$$u_\lambda(x, t) = \lambda^\gamma u(\lambda x, \lambda^\alpha t), \qquad \gamma = (\alpha - 1)/(r - 1),$$

is also a solution of (1.2). The solutions satisfying the scaling invariance property

$$u_\lambda \equiv u, \qquad \forall \lambda > 0,$$

are called *forward self-similar solutions*. By the very definition they are global in time. It is expected that they describe large-time behavior of general solutions (see Lectures 3-5, and also Sinai (1992), Molchanov et al. (1995, 1997), Funaki et al. (1995), for stochastic analogs of this property for the usual Burgers equation). Indeed, if

$$\lim_{\lambda \to \infty} \lambda^\gamma u(\lambda x, \lambda^\alpha t) = U(x, t)$$

exists in an appropriate sense, then

$$t^{\gamma/\alpha} u(x t^{1/\alpha}, t) \to U(x, 1)$$

as $t \to \infty$ (to see this, take $t = 1$, $\lambda = t^{1/\alpha}$), and U satisfies the invariance property $U_\lambda \equiv U$. U is therefore a self-similar solution and

$$U(x, t) = t^{-\gamma/\alpha} U(x t^{-1/\alpha}, 1) \qquad (1)$$

is completely determined by a function of d variables $U(y) \equiv U(y, 1)$.

Let us observe that if

$$u_0(x) = \lim_{t \to 0} t^{-\gamma/\alpha} U(x t^{-1/\alpha})$$

exists, then u_0 is necessarily homogeneous of degree $-\gamma$. For $\gamma \neq d$, such $u_0 \not\equiv 0$ cannot have finite mass. A direct approach to these solutions via an elliptic equation with variable coefficients obtained from (8.1.1) by substituting the particular form (1) seems to be very hard. An analogous difficulty that appears for the Navier-Stokes system has been overcome by Y. Meyer and his collaborators (see, e.g., Cannone (1995)). Our techniques (and also those in Biler (1995)) are motivated by their results.

Of course, self-similar solutions to (8.1.1) and (8.1.2) can be obtained directly from Theorem 8.1.4 by taking suitably small u_0, homogeneous of degree $-\gamma$. Indeed, the Morrey space M_q^p with $p =$

$d(r-1)/(\alpha-1)$ does contain such u_0's since $p\gamma = d$, by the uniqueness, the solution obtained in Theorem 8.1.4 satisfies the scaling property. However, we are also interested in function spaces other than the Morrey spaces, e.g., Besov or symbol spaces. The purpose of such a generalization is that sufficient size conditions on u_0 might be weaker than those for the global existence part in Theorem 8.1.4.

As in the subsection of Section 8.1 about mild solutions we shall deal with solutions that are not necessarily positive.

Consider a Banach space $\mathcal{B} \subset \mathcal{S}'(\mathbf{R}^d)$ whose elements are tempered distributions and let $v \in \mathcal{X} = \mathcal{C}([0,T];\mathcal{B})$. Define the nonlinear operator $\mathcal{N} : \mathcal{X} \to \mathcal{X}$ by

$$\mathcal{N}(v)(t) = \int_0^t \left(\nabla e^{(t-s)A}\right) \cdot (av^r(s)) \, ds. \tag{2}$$

We are looking for (mild) self-similar solutions of (8.1.1), i.e., U of the form (1) satisfying the integral equation

$$U = V_0 + \mathcal{N}(U), \tag{3}$$

where $V_0 = e^{tA}u_0$. The crucial observation is that the equation (3) is well adapted to a study of self-similar solutions via an iterative algorithm.

Lemma 1. *(i) If $u_0 \in \mathcal{S}'(\mathbf{R}^d)$ is homogeneous of degree $-\gamma$: $u_0(\lambda x) = \lambda^{-\gamma}u_0(x)$, then*

$$V_0 \equiv e^{tA}u_0 = t^{-\gamma/\alpha}U_0(x/t^{1/\alpha})$$

for some U_0.

(ii) If U is of the form (1), i.e., $U = t^{-\gamma/\alpha}U(x/t^{1/\alpha})$, and $\mathcal{N}(U) \in \mathcal{S}'(\mathbf{R}^d)$ is well defined, then $\mathcal{N}(U)$ is again of the form (7.1):

$$\mathcal{N}(U) = t^{-\gamma/\alpha}V(x/t^{1/\alpha})$$

for some V.

Thus, it suffices to consider the equation (3) in $\mathcal{X}$ for $t = 1$ only, i.e., the study of (3) is reduced to the space $\mathcal{B}$. If we wanted to solve (3) by the iterative application of the operator $\mathcal{N}$:

$$V_{n+1} = V_0 + \mathcal{N}(V_n), \tag{5}$$

then for u_0 homogeneous of degree $-\gamma$ all V_n's would be of the self-similar form (1). Hence the iterative algorithm is entrapped in the set of self-similar functions. If we showed the convergence of this algorithm, the limit would automatically be a self-similar solution of (8.1.1).

The existence of solutions to (3) is proved under natural assumptions on $\mathcal{N}$ which generalize those in Cannone (1995, I Lemma 2.3; IV Lemma 2.9).

Lemma 2. *Suppose that* $\mathcal{N} : \mathcal{B} \to \mathcal{B}$ *is a nonlinear operator defined on a Banach space* $(\mathcal{B}, \|\,.\,\|)$ *such that* $\mathcal{N}(0) = 0$,

$$\|\mathcal{N}(U) - \mathcal{N}(V)\| \leq K \left(\|U\|^{r-1} + \|V\|^{r-1} \right) \|U - V\|,$$

with some $r > 1$ *and* $K > 0$ *(i.e.,* $\mathcal{N}$ *is a locally Lipschitz mapping). If* $\|V_0\|$ *is sufficiently small, then the equation (3) has a solution which can be obtained as the limit of* V_n's *defined by the recursive algorithm (5).*

Remark. When $\mathcal{N}$ is defined by a bounded bilinear form $B : \mathcal{B} \times \mathcal{B} \to \mathcal{B}$, $\mathcal{N}(U) = B(U, U)$, and $\|B(U, V)\| \leq K_1 \|U\| \|V\|$, then $\|V_0\| < 1/(4K_1)$ is an explicit sufficient condition for the existence of solutions to (7.3), cf. Biler (1995, Lemma 2).

Hence, a good functional framework to study the equation (3) should satisfy the following conditions:

(i) u_0 (with $U_0 \in \mathcal{B}$) is a distribution homogeneous of degree $-\gamma$;

(ii) $\mathcal{N}$ defined by (2) and represented by (4) for $t = 1$ is locally Lipschitz continuous, as in the hypothesis of Lemma 2.

We give in Theorems 1-2 below a suitable choice of function spaces (necessarily different from the usual L^p or Sobolev spaces) which satisfy these conditions. These are homogeneous Besov spaces and spaces containing functions related to symbols of classical pseudo-differential operators (as in Cannone (1995) and Biler (1995, Section 3)). The proofs are in Biler, Funaki and Woyczynski (1996).

Concerning the interpretation of $\lim_{t \to 0} t^{-\gamma/\alpha} U(x/t^{1/\alpha})$, note that solutions of (3) enjoy the same continuity properties as mild solutions in Section 8.1. Thus, the initial condition in (3) is attained in the sense of distributions, and the curve $t \mapsto U$ in (1) is bounded in $\mathcal{B}$.

We can interpret V_0 in (3) as the main term (trend) and $\mathcal{N}(U)$ as a fluctuation around the drift of u_0 described by the trajectory $e^{tA} u_0$ of the Lévy semigroup.

Below we recall the definition of the homogeneous Besov spaces using the Littlewood-Paley decomposition, cf. Triebel (1982, 1991) and Biler (1995). The advantage of Besov spaces is an easy frequency analysis; the inconvenience is that they restrict us to the quadratic case

$r = 2$ in Theorem 1 (while $r > 1$ will be an arbitrary real number in Theorem 2).

Let $\mathcal{S} = \mathcal{S}(\mathbf{R}^d)$, $\mathcal{Z} = \{v \in \mathcal{S} : D^\beta \hat{v}(0) = 0$ for every multiindex $\beta\}$, and $\hat{\mathcal{D}}_0 = \{v \in \mathcal{S} : \hat{v} \in C_0^\infty(\mathbf{R}^d \setminus \{0\})\}$. Since $\hat{\mathcal{D}}_0$ is dense in $\mathcal{Z}$ and $\mathcal{Z}$ is a closed subspace of $\mathcal{S}$, the inclusion $\mathcal{Z} \subset \mathcal{S}$ induces a surjective map $\pi : \mathcal{S}' \to \mathcal{Z}'$ such that $\ker \pi = \mathcal{P}$, the space of polynomials, so $\mathcal{Z}' = \mathcal{S}'/\mathcal{P}$.

Let $\hat{\psi} \in C_0^\infty(\mathbf{R}^d)$ satisfy $0 \le \hat{\psi} \le 1$, $\hat{\psi}(\xi) = 1$ for $|\xi| \le 1$, $\hat{\psi}(\xi) = 0$ for $|\xi| \ge 2$, and define for any $k \in \mathbf{Z}$

$$\hat{\phi}_k(\xi) = \hat{\psi}(2^{-k}\xi) - \hat{\psi}(2^{-(k+1)}\xi).$$

Evidently,

$$\operatorname{supp} \hat{\phi}_k \subset A_k \equiv \{\xi : 2^{k-1} \le |\xi| \le 2^{k+1}\}, \qquad \sum_k \hat{\phi}_k(\xi) = 1,$$

for any $\xi \ne 0$, with at most two nonzero terms in the series. The convolutions $\phi_k * v$ are meaningful not only for $v \in \mathcal{S}'$ but also for all $v \in \mathcal{Z}'$. The *homogeneous Besov space* $\dot{B}_{p\infty}^s \subset \mathcal{Z}'$ is defined for $s \in \mathbf{R}$, $1 \le p \le \infty$, by the condition

$$\|v; \dot{B}_{p\infty}^s\| \equiv \sup_k 2^{ks}|\phi_k * v|_p < \infty,$$

where $|\,.\,|_p$ is the usual $L^p(\mathbf{R}^d)$ norm.

More general homogeneous Besov spaces include $\dot{B}_{pq}^s$ with

$$\|v; \dot{B}_{pq}^s\| \equiv \left(\sum_k 2^{ksq}|\phi_k * v|_p^q\right)^{1/q} < \infty.$$

Clearly, $\mathcal{F}(\phi_k * v) = \hat{\psi}_k \hat{v}$, so the Besov norms control the size of the Fourier transform $\hat{v}$ over the dyadic annuli A_k, and the parameter s measures the smoothness of function v.

We will be interested in the (nonseparable) Banach spaces $\mathcal{B} = \dot{B}_{2\infty}^{d/2-\gamma}(\mathbf{R}^d)$, $d > 2(\alpha - 1)$, whose elements can be realized as tempered distributions (hence simpler to interpret than elements of $\mathcal{Z}' = \mathcal{S}'/\mathcal{P}$, see Biler (1995), Section 3). It is easy to check that $|x|^{-\gamma} \in \mathcal{B}$, since $\mathcal{F}(|x|^{-\gamma}) = c_{\gamma,d}|\xi|^{\gamma-d}$. Moreover, functions homogeneous of degree $-\gamma$ belong to $\mathcal{B}$ provided they are smooth enough on the unit sphere of $\mathbf{R}^d$, cf. Cannone (1995, IV, Theorem 2.1).

Theorem 1. *Let* $r = 2$, $\alpha \in (1, 2]$, $d > 2\gamma = 2(\alpha - 1)$. *If* $u_0 \in \mathcal{B} = \dot{B}_{2\infty}^{d/2-\gamma}$ *is homogeneous of degree* $-\gamma$ *and the norm* $\|u_0\|$ *in* $\mathcal{B}$ *is small enough, then there exists a solution* U *of the equation (3). This solution is unique in the class of distributions satisfying the condition* $\|U\| \leq R$ *with* R *as in the proof of Lemma 2.*

Remark. It is quite easy to prove that $V_0 \in L^p(\mathbf{R}^d)$ for each $p > d/\gamma$ (u_0 being homogeneous of degree $-\gamma$) but $V_0 \notin L^{d/\gamma}(\mathbf{R}^d)$, unless $V_0 = 0$.

In the remainder of this section we review results on the equation (3) using as a tool the scale of spaces $E^{\rho,m}$ which consist of functions from $C^m(\mathbf{R}^d)$ satisfying natural decay estimates at infinity, and their homogeneous counterparts $\dot{E}^{\rho,m}$ featuring estimates of the singularity at the origin (like symbols of classical pseudo-differential operators), see Cannone (1995) and Biler (1995, Theorem 4). More formally, for $\rho > 0$ and $m \in \mathbf{N}$, we define the following Banach spaces of functions on $\mathbf{R}^d$: $E^{\rho,m} = E^{\rho,m}(\mathbf{R}^d) =$

$$\{v \in C^m(\mathbf{R}^d) : |D^\beta v(x)| \leq C(1 + |x|)^{-\rho - |\beta|}, \ |\beta| \leq m\}, \qquad (7)$$

and $\dot{E}^{\rho,m} = \dot{E}^{\rho,m}(\mathbf{R}^d) =$

$$\{v \in C^m(\mathbf{R}^d \setminus \{0\}) : |D^\beta v(x)| \leq C|x|^{-\rho - |\beta|}, \ |\beta| \leq m\}, \qquad (8)$$

with the norms of v defined as the least constants in (7-8), respectively. Note that for $\rho < d$, $\dot{E}^{\rho,m} \subset L^1_{\text{loc}}$, since the singularity at the origin is integrable.

We look for solutions to (3) of the form $U = V_0 + \mathcal{N}(U) \in E^{\gamma,m}$ with $u_0 \in \dot{E}^{\gamma,m}$, $r > 1$, $\lfloor r \rfloor \geq m$ if r is not an integer, and $m + \gamma < d$, in order to avoid nonintegrable singularities in further analysis.

Certainly, the spaces $E^{\rho,m}$ are more natural tools to study (3) than Morrey or Besov spaces, but the estimates of the operator $\mathcal{N}$ in (4) (now without the use of Fourier transforms) are far more subtle than the frequency bound (6) which is particularly well adapted to the Littlewood-Paley decomposition of functions. The reason is that the fundamental solution $\mathcal{F}^{-1}(\exp(-t|\xi|^\alpha))$ of the evolution operator $\partial/\partial t + (-\Delta)^{\alpha/2}$ decays for $\alpha < 2$ much slower than for the heat equation with $\alpha = 2$.

We begin with formulation of the well-known asymptotic estimates for the kernel of the Lévy semigroup e^{tA} (see, e.g. Komatsu (1984)).

Lemma 3. *Denote by* $p_{\alpha,t}$ *the convolution kernel of the Lévy semigroup* e^{tA} *with* $A = -(-\Delta)^{\alpha/2}$ *in* $\mathbf{R}^d$, $0 < \alpha < 2$. *Then*

$$p_{\alpha,t}(z) = t^{-d/\alpha} p_{\alpha,1}(t^{-1/\alpha} z),$$

and

$$0 \leq p_{\alpha,1}(z) \leq C_{\alpha,d} \left(1 + |z|^{d+\alpha}\right)^{-1} \tag{9}$$

for some $C_{\alpha,d} > 0$. Moreover,

$$|\nabla p_{\alpha,1}(z)| \leq \tilde{C}_{\alpha,d} |z|^{d-1+\alpha} \left(1 + |z|^{d+\alpha}\right)^{-2} \tag{10}$$

for another constant $\tilde{C}_{\alpha,d} > 0$. In fact, $p_{\alpha,1}(z)(1 + |z|^{d+\alpha})$ is bounded from above and below by some positive constants, and there exists the limit of the above expression as $|z| \to \infty$.

Only L^q-estimates (for large q) of $p_{\alpha,1}$, $\nabla p_{\alpha,1}$ are needed in the proof of Theorem 2.

From the scaling properties of $p_{\alpha,t}$ and (9-10) we get immediately

$$p_{\alpha,t}(z) \leq C t^{-d/\alpha} \left(1 + |z|^{d+\alpha} t^{-d/\alpha-1}\right)^{-1},$$

and

$$|\nabla p_{\alpha,t}(z)| \leq C t^{-1-2d/\alpha} |z|^{d-1+\alpha} \left(1 + |z|^{d+\alpha} t^{-d/\alpha-1}\right)^{-2}.$$

Theorem 2. *Let $\alpha \in (1,2)$, $\gamma = (\alpha - 1)/(r - 1)$, $r > 1$, $m + \gamma < d$, and if $r \notin \mathbf{N}$, $m \leq \lfloor r \rfloor$. If $u_0 \in E^{\gamma,m}$ is homogeneous of degree $-\gamma$ and has a sufficiently small norm, then there exists a self-similar solution $t^{-\gamma/\alpha} U(xt^{-1/\alpha})$ with $U \in E^{\gamma,m}$, and $\mathcal{N}(U) \in E^{\rho,m}$, for all $\rho < \min(r\gamma, d)$. Such a solution is unique among those satisfying $\|U; E^{\gamma,m}\| \leq R$, with R from the proof of Lemma 2.*

Remark. In the case $\alpha = 2$, the result of Theorem 7.2 remains true, and the reasoning (in the style of Biler (1995, Theorem 4)) is even simpler.

8.4 Fractal nonlinear Markov processes and propagation of chaos

In this section we consider the one-dimensional fractal Burgers' equation

$$u_t = -\nu D^\alpha - u\nabla u, \qquad u(x,0) = u_0(x), \tag{1}$$

where $D^\alpha = (-\partial^2/\partial x^2)^{\alpha/2}$, $0 < \alpha \leq 2$, $\nabla = \nabla_x = \partial/\partial x$, $x \in \mathbf{R}$, $t \geq 0$, $u : \mathbf{R} \times \mathbf{R}^+ \to \mathbf{R}$, and apply the existence and uniqueness results of Section 8.1 to obtain existence of related McKean processes (anomalous nonlinear diffusions) driven by Lévy α-stable processes and a related propagation of chaos result. The reviewed material is taken from Funaki and Woyczynski (1997).

Remark 1. Theorem 8.1.1 establishes the existence of solutions which are L^∞ in variable t. Actually, this assertion can be strengthened and one can prove that these solutions $u \in C([0,T]; H^1(\mathbf{R}))$, as long as $\alpha \in (3/2, 2]$. This follows (since for any $\alpha > 1$ we have that $H^{\alpha/2}(S) \hookrightarrow C(S)$) from the fact that, for every $T > 0$, the sequence of Galerkin approximations $\{u^N\}$ is relatively compact in $C([0,T], H^{\alpha/2}(S))$, and therefore also in $C([0,T] \times S)$. Indeed, since

$$\|u^N(t_2) - u^N(t_1)\|_{H^\alpha}$$

$$\leq \int_{t_1}^{t_2} \|\partial_t u^N(t)\|_{H^{\alpha/2}} dt \leq (t_2 - t_1)^{1/2} \left(\int_0^T \|\partial_t u^N(t)\|_{H^{\alpha/2}}^2 dt \right)^{1/2},$$

the sequence $\{u^N(t)\}$ is equicontinuous in $H^{\alpha/2}$ in view of the inequality

$$\int_0^T \|\partial_t u(t)\|_{H^{\alpha/2}}^2 dt \leq C = C_T, \qquad \forall t \in [0,T]. \tag{2}$$

Now, by Rellich's theorem, the embedding $H^1 \hookrightarrow H^{\alpha/2}$ is compact, and $\sup_{N, 0 \leq t \leq T} \|u^N(t)\|_{H^1} < \infty$ in view of the inequality

$$\|u(t)\|_{H^1} \leq C = C_T, \qquad \forall t \in [0,T]. \tag{3}$$

This gives the compactness statement. Inequalities (2-3) are routine consequences of the standard Sobolev spaces embedding theorems (see, e.g. Adams (1975), Theorem 7.57 (a), p. 217, or Henry (1982)).

Construction of the McKean processes. Let $u = u(x,t)$ be a local in time weak solution of the fractal Burgers equation with viscosity $\nu = 1$

$$u_t = -D^\alpha u - u \nabla u, \tag{4}$$

where $x \in \mathbf{R}$, $t \in [0,T]$. We assume that the solution is bounded, that is

$$\sup_{x \in \mathbf{R}, t \in [0,T]} |u(x,t)| < \infty. \tag{5}$$

Indeed, (5) follows from Theorem 2.1 when $\alpha \in (1/2, 2]$ and, furthermore, $u(x,t)$ is continuous in (x,t) from Remark 1 when $\alpha \in (1,2]$; note that $H^\beta(\mathbf{R})$ is continuously embedded in $L^\infty(\mathbf{R}) \cap C(\mathbf{R})$ if $\beta > 1/2$.

Consider a solution $X(t)$ of the stochastic differential equation

$$dX(t) = dS(t) + \frac{1}{2} u(X(t), t) \, dt, \qquad t \in [0,T],$$

$$X(0) \sim u(x,0)\, dx, \qquad \text{in law}, \tag{6}$$

where $S(t)$ is a standard $(E \exp(i\xi S(t)) = \exp(-t|\xi|^\alpha))$ symmetric α-stable process with independent increments. In the following we always assume $\alpha \in (1,2)$. Then, since the coefficient $u(x,t)$ is bounded, the stochastic differential equation (6) has a unique weak solution (see, e.g., Komatsu (1985)). The measure-valued function of t

$$v(dx, t) := P\{X(t) \in dx\}, \qquad t \in [0, T]$$

satisfies the weak forward equation

$$\frac{d}{dt}\langle v(t), \varphi \rangle = \langle v(t), \mathcal{L}_{u(t)}\varphi \rangle, \qquad \forall \varphi \in \mathcal{S}(\mathbf{R})$$

$$v(0) = u(x,0)\, dx, \tag{7}$$

where $\mathcal{S}(\mathbf{R})$ denotes the Schwartz class of functions on $\mathbf{R}$ and

$$\mathcal{L}_u = -D^\alpha + \frac{1}{2}u(x)\nabla, \qquad u = u(x).$$

Theorem 1.   *Process $X(t)$ is the McKean process (nonlinear Markov process) corresponding to the fractal Burgers equation (4), namely, it satisifes condition*

$$P\{X(t) \in dx\} = u(x,t)\, dx. \tag{8}$$

PROOF. In view of results of Echeveria and Varadhan (see, e.g., Funaki (1984)) the following two statements are equivalent:

(a) The martingale problem for the operator $\mathcal{L}_{u(t)}$ is well posed, and

(b) The existence and uniqueness theorem holds for the corresponding weak forward equation (7).

In our case, (a) holds for the martingale problem associated with (6). However, $u(dx, t) := u(x,t)\, dx$ is also a solution of (7) since, by (4) or the definition of the qeak solution of the problem (8.1.2-3)

$$\frac{d}{dt}\langle u(t), \varphi \rangle = \langle u(t), -D^\alpha\varphi \rangle + \langle u(t)^2/2, \nabla\varphi \rangle = \langle u(t), \mathcal{L}_{u(t)}\varphi \rangle.$$

Now, the uniqueness for the (linear) equation (7) implies that

$$v(dx, t) = u(dx, t),$$

which, consequently, yields (8). ∎

Remark 2. Note that for $0 < \alpha < 2$ the fractional Laplacian has a representation

$$D^\alpha \varphi(x) = -K \int_{\mathbf{R}} \left\{ \varphi(x+y) - \varphi(x) - \nabla\varphi(x) \cdot \frac{y}{1+|y|^2} \right\} \frac{dy}{|y|^{1+\alpha}},$$

with some positive constant K, see, e.g., Dawson and Gorostiza (1990) and, in view of the Itô formula,

$$
\begin{aligned}
d\varphi(X_t) &= \left\{ -D^\alpha \varphi(X_t) + \frac{1}{2} u(t, X_t) \cdot \nabla\varphi(X_t) \right\} dt + M_t \\
&= \mathcal{L}_{u(t)} \varphi(X_t)\, dt + M_t,
\end{aligned}
$$

where $\mathcal{L}_u = -D^\alpha + u(x) \cdot \nabla/2$ and M_t is an S_t-martingale.

Remark 3. The process $S(t)$ has a representation

$$S(t) = \int_0^{t+} \int_{0<|y|<1} y\, \tilde{N}(ds\,dy) + \int_0^{t+} \int_{|y|\geq 1} y\, N(ds\,dy),$$

where $N(ds\,dy)$ is a Poisson point process with intensity $\hat{N}(ds\,dy) = ds\,\nu(dy)$, $\nu(dy) = K dy/|y|^{1+\alpha}$ is the Lévy measure and $\tilde{N}(ds\,dy) = N(ds\,dy) - \hat{N}(ds\,dy)$ (see, e.g., Ikeda and Watanabe (1981)).

Propagation of chaos for regularized fractal Burgers equation. The propagation of chaos result for the usual Burgers' equation was discussed in section 2.3. Here, we will review an analogous result for the fractal Burgers equation. The standing assumption is that $\alpha \in (1, 2)$, which permits us to operate with expectations of the corresponding α-stable processes. Let $\{S^i(t),\ i = 1, 2, \ldots, n\}$ be independent, symmetric, real-valued Lévy α-stable processes with the infinitesimal generator $\Delta_\alpha = -D^\alpha$, and let

$$\delta_\epsilon(x) := \frac{1}{\sqrt{2\pi\epsilon}} \exp[-x^2/2\epsilon], \qquad \epsilon > 0 \tag{9}$$

be a regularizing kernel. We consider a system of n interacting particles with positions

$$\{X^i(t)\}_{i=1,\ldots,n} \equiv \{X^{i,n,\epsilon}(t)\}_{i=1,\ldots,n} \tag{10}$$

and the corresponding measure-valued process (empirical distribution)

$$\bar{X}^n(t) \equiv \bar{X}^{n,\epsilon}(t) := \frac{1}{n} \sum_{i=1}^n \delta(X^{i,n,\epsilon}(t)) \tag{11}$$

with the dynamics provided by the system of regularized singular stochastic differential equations

$$dX^i(t) = dS^i(t) + \frac{1}{2n}\sum_{j \neq i}\delta_\epsilon(X^i(t) - X^j(t))\,dt, \quad i = 1,\ldots,n. \quad (12)$$

We assume that the initial positions $\{X^i(0)\}_{i=1,\ldots,n}$ satisfy the condition

$$\sup_n \sup_{\lambda \in \mathbf{R}} \frac{1}{1 + |\lambda|^a} E[|\langle \eta^n(0), \chi_\lambda\rangle|] < \infty,$$

for some $a \geq 0$, where $\chi_\lambda(x) = \exp(i\lambda x)$.

Theorem 2. *For each $\epsilon > 0$, the empirical process*

$$\bar{X}^{n,\epsilon}(t) \Longrightarrow u^\epsilon(x,t)\,dx, \quad \text{in prob.} \quad \text{as} \quad n \to \infty, \quad (13)$$

where $\Rightarrow$ denotes the weak convergence of measures, and the limit density $u^\epsilon \equiv u^\epsilon(x,t)$, $t > 0$, $x \in \mathbf{R}$, satisfies the regularized fractal Burgers equation

$$u_t^\epsilon = -D^\alpha u^\epsilon - \frac{1}{2}\nabla\big((\delta_\epsilon * u^\epsilon) \cdot u^\epsilon\big). \quad (14)$$

Moreover, for each $\epsilon > 0$, there exists a constant $C_\epsilon > 0$ such that, for any $\phi \in \mathcal{S}(\mathbf{R})$,

$$E\big[|\langle \bar{X}^{n,\epsilon}(t) - u^\epsilon(t), \phi\rangle|\big] \leq C_\epsilon n^{-1+1/\alpha}\int_{\mathbf{R}}(1 + |\lambda|^a)|\hat{\phi}|(d\lambda), \quad (15)$$

where $|\hat{\phi}|(d\lambda)$ denotes the total variation measure of $\hat{\phi}(d\lambda)$.

SKETCH OF THE PROOF. The assertion (13) essentially follows from (15); recall that $\alpha > 1$. Therefore, to prove Theorem 2, it suffices to demonstrate the speed of convergence result (15), that is, the boundedness of

$$\eta^n(t) \equiv \eta^{n,\epsilon}(t) := n^\beta\big(\bar{X}^{n,\epsilon}(t) - u^\epsilon(t)\big) \quad (16)$$

with $\beta = 1 - 1/\alpha$. The critical step is the following lemma which is a consequence of the Itô's formula and a number of estimates for the quantities appearing in it.

Lemma 1. *For each $\phi \in C_b^\infty(\mathbf{R})$,*

$$\langle \eta^n(t), \phi\rangle = \langle \eta^n(0), \phi\rangle + m^n(\phi; t) + \int_0^t b^n(\phi; s)\,ds + \int_0^t \langle \eta^n(s), -D^\alpha\phi\rangle\,ds,$$

where

$$m^n(\phi; t) = n^{\beta-1} \sum_{i=1}^{n} \int_0^{t+} \int_{\mathbf{R}} \{\phi(X^i(s-) + y) - \phi(X^i(s-))\} \tilde{N}^i(dsdy)$$

with $\tilde{N}^i = N^i - \hat{N}$, and $N^i = N^i(dsdy)$ being independent, Poisson random measures with identical intensity $\hat{N}(dsdy) = Kdsdy/|y|^{1+\alpha}$, and where

$$\begin{aligned}
b^n(\phi; t) &= \frac{1}{2} n^{-\beta} \langle \eta^n(dx, t)\eta^n(dy, t), \delta_\epsilon(x - y)\nabla\phi(x) \rangle \\
&+ \frac{1}{2} \langle u^\epsilon(dx, t)\eta^n(dy, t) + u^\epsilon(dy, t)\eta^n(dx, t), \delta_\epsilon(x - y)\nabla\phi(x) \rangle \\
&- \frac{1}{2} n^{\beta-1} \delta_\epsilon(0) \langle \bar{X}^n(t), \nabla\phi \rangle.
\end{aligned}$$

Complete proofs of the above results can be found in Funaki and Woyczynski (1998). For extensions to other nonlinear and nonlocal evoution problems, see Biler, Funaki and Woyczynski (1998b).

Bibliography

1. Adams R.A., *Sobolev Spaces*, Academic Press (1975)

2. Albeverio S., Molchanov S.A. and Surgailis D. (1994), Stratified structure of the Universe and Burgers equation; a probabilistic approach, *Prob. Theory Rel. Fields* 100, 457-484.

3. Ali A.H.A., Gardner G.A. and Gardner L.R.T. (1992), A collocation solution for Burgers' equation using cubic B-spine finite elements, *Computer Methods in Applied Mechanics and Engineering* 100, 325.

4. Andjel E.D., Bramson M.D. and Liggett T.M. (1988), Shocks in the asymmetric exclusion process, *Probab. Theory Rel. Fields* 78, 231-247.

5. Andjel E.D. and Kipnis C. (1984), Derivation of the hydrodynamical equation for the zero-range interaction process, *Ann. Prob.* 12, 325-334.

6. Arnold V.I. (1978), *Mathematical Methods of Classical Mechanics*, Springer.

7. Aurell E., Gurbatov S.N. and Wertgeim I.I. (1993), Self-preservation of large-scale structures in Burgers' turbulence, *Physics Letters A* 182, 109-113.

8. Avellaneda M. and E W. (1995), Statistical properties of shocks in Burgers turbulence, *Comm. Math. Phys.* 172,13-38.

9. Avellaneda M. (1995), Statistical properties of shocks in Burgers turbulence, II; Tail probabilities for velocities, shock-strengths and rarefaction intervals, *Comm. Math. Phys.* 169, 45-59.

10. Avrin J.D. (1987), The generalized Burgers' equation and the Navier-Stokes equation in $\mathbf{R}^n$ with singular initial data, *Proc. A.M.S.* 101, 29-40.

11. Barabasi A.L. and Stanley H.E. (1995), *Fractal Concepts in Surface Growth*, Cambridge Univ. Press.

12. Bardos C., Penel P., Frisch U., Sulem P.L. (1979), Modified dissipativity for a nonlinear evolution equation arising in turbulence, *Arch. Rat. Mech. Anal.* 71, 237-256.

13. Benassi A., Fouque J.P. (1987), Hydrodynamic limit for the asymmetric exclusion process, *Ann. Probab.* 15, 546-560.

14. Bernardeau F., Kofman L. (1995), Properties of the cosmological density distribution function, *Astrophysical J.* 443, 479-498.

15. Bertini L., Cancrini N., Jona-Lasinio G. (1994), The stochastic Burgers equation, *Comm. Math. Phys.* 165, 211-232.

16. Bertoin J. (1997), Large deviation estimates in Burgers turbulence with stable noise initial data, 11 pp., preprint.

17. Biagioni H. A., Oberguggenberger M. (1992), Generalized solutions to Burgers' equation, *J. Diff. Equations* 97, 263.

18. Biler P. (1984), Asymptotic behaviour in time of solutions to some equations generalizing the Korteweg-de Vries-Burgers equation, *Bull. Pol. Acad. Sci., Mathematics* 32, 275-282.

19. Biler P. (1995), The Cauchy problem and self-similar solutions for a nonlinear parabolic equation, *Studia Math.* 114, 181-205.

20. Biler P., Funaki T. and Woyczynski W.A. (1998a), Fractal Burgers equations, *J. Diff. Equations* 147, 1-38.

21. Biler P., Funaki T. and Woyczynski W.A. (1998b), Interacting jump Markov processes associated with nonlocal quadratic evolution problems, *Ann. Appl. Probability*, to appear.

22. Biler P. and Woyczynski W.A. (1998), Global and exploding solutions for nonlocal quadratic evolution problems, *SIAM J. Appl. Math.* 58(2), to appear.

23. Billingsley P. (1986), *Probability and Measure*, Wiley.

24. Bingham N.H., Goldie C.M., Teugels J.L. (1987), *Regular Variation*, Cambridge University Press.

25. Boghosian B. M., Levermore C. D. (1987), A cellular automaton for Burgers' equation, *Complex Systems* 1, 17-30.

26. Bona J. L., Dougalis V. A., Karakashian O. A. (1992), Computations of blown-up and decay for periodic solutions of the generalized Korteweg-deVries-Burgers equation, *Trans. of Appl. Numerical Mathematics* 10, 335.

27. Bossy M., Talay D. (1996), Convergence rate for the approximation of the limit law of weakly interacting particles: application to the Burgers equation, *Ann. Appl. Prob.* 6, 818-861.

28. Boyd J. P. (1992), The energy spectrum of fronts: time evolution of shocks in Burgers' equation, *J. Atmosph. Sci.* 49, 128-139.

29. Bramson M., Liggett T. M. (1986), *Lectures in Probability*, University of Nagoya Press.

30. Breuer H.P., Petruccione F. (1992), A stochastic formulation of Burgers' equation, *Physics Letters* 172, 49-55.

31. Breuer, P. and Major P. (1983), Central limit theorem for non-linear functionals of Gaussian fields, *Z. Wahr. verw. Geb.* 13, 425-441.

32. Brieger L., Bonomi E. (1992), A stochastic lattice gas for Burgers' equation: a practical study, *J. Stat. Physics* 69, 837.

33. Brio, M., Hunter J.K. (1992), Mach reflection for the two-dimensional Burgers equation, *Physica D, Nonlinear Phenomena* 60, 194.

34. Brissaud, A. (1973), et al., *Ann. Geophys.* 29, 539-45.

35. Bulinski, A.V. (1987), Limit theorems under weak dependence conditions, in *Probability Theory and Math. Stat., Proc. Fourth Vilnius Conf.* v.1, Utrecht VNP, pp. 307-327.

36. Bulinski A.V., Molchanov S.A. (1991), Asymptotically Gaussian solutions of the Burgers equation with random initial data, *Theory Prob. Appl.* 36, 217-235.

37. Bulinskii A. V. (1990) *CLT for families of integral functionals arising in solving multidimensional Burgers' equation*, in: Proc. 5th Vilnius Conf. Probab. Theory and Math. Statist., vol. 1, VSP-Mokslas, pp. 207-216.

38. Burgers J.M. (1939), On the application of statistical mechanics to the theory of turbulent fluid motion, *Proc. Roy. Neth.Acad.Sci.* 32, 414,643,818.

39. Burgers J.M. (1972), Statistical problems connected with asymptotic solutions of the one-dimensional nonlinear diffusion equation, in *Statistical Models and Turbulence*, M. Rosenblatt and C. van Atta, Eds., Springer Verlag, p.41.

40. Burgers J.M. (1974), *The Nonlinear Diffusion Equation*, Reidel.

41. Burkill H. (1964), A note on rearrangements of functions, *Amer. Math. Monthly.* 71, 887-888.

42. Calderoni P., Pulvirenti M. (1983), Propagation of chaos for Burgers equation, *Ann. Inst. H. Poincaré, Sec. A, Phys. Theor.* 29, 85-97.

43. Cannone M. (1995), *Ondelettes, paraproduits et Navier-Stokes*, Paris: Diderot Editeur; Arts et Sciences.

44. Carmona R. and Lacroix J. (1990), *Spectral Theory of Schrödinger Operators*, Birkhäuser-Boston.

45. Carmona R. and Molchanov S.A. (1994), *Parabolic Anderson Model and Intermittency*, Mem. Amer. Math. Soc. 518.

46. Chen H., Chen S., Kreichnan R.H. (1989), Probability distribution of a stochastically advected scalar field, *Phys. Rev. Letters* 63, 2657.

47. Chorin A.J. (1975), *Lectures on Turbulence Theory*, Publish or Perish, Inc.

48. Cole J. (1951), On a quasi-linear parabolic equation occuring in hydrodynamics, *Q. Appl. Math.* 9, 225-236.

49. Courant R., Hilbert D. (1953), *Methods of Mathematical Physics*, vol. 1, Interscience.

50. Csanady G.T. (1980), *Turbulent Diffusion in the Environment*, D. Reidel, Dordrecht.

51. Curro C., Donato A., Povzner A. Ya. (1992), Perturbation method for a generalized Burgers' equation, *Int. J. of Non-linear Mech.* 27, 149.

52. Cutler C. D. (1991), Some results on the behavior and estimation of the fractal dimensions of distributions on attractors, *J. Statist. Physics* 62, 651–708.

53. Da Prato G., Zabczyk J. 1992, *Stochastic Equations in Infinite Dimension*, Cambridge University Press.

54. Da Prato G. and Gątarek D. (1995), Stochastic Burgers' equation, *Stochastics and Stoch. rep.* 52, 29-41.

55. Davey A. (1972), The propagation of a weak nonlinear wave, *J. Fluid Mech.* 53, 769-781.

56. Davis R.E. (1982), On relating Eulerian and Lagrangian velocity statistics: single particles in homogeneous flows, *J. Fluid Mech.* 114,1-26.

57. Dawson D., Gorostiza L.G. (1990), Generalized solutions of a class of nucleatr space valued stochastic evolution equations, *Appl. Math. Optim.* 22, 241-264.

58. De Masi A., Presutti E. (1989), An introductory course on the collective behavior of particle systems, *CARR Reports in Math. Physics.*

59. De Masi A., Presutti E., Scacciatelli E. (1989), The weakly asymmetric simple exclusion process, *Ann. Inst. H. Poincaré, Sec. B, Prob. Stat.* 25, 1-38.

60. De Masi A., Janiro N., Pellegrinotti A., Presutti E. (1984), A survey of hydrodynamical behavior in many particle systems, in *Nonequilibrium Phenomena*, J.L. Lebowitz and E.W. Montroll, eds., Vol.11, North-Holland.

61. De Masi A. and Presutti E. (1991), *Mathematical Methods for Hydrodynamic Limits*, Lect. Notes in Math. 1501, Springer.

62. Dermoune A., Hamadene S., Ouknine Y. (1997), Limit theorem for the statistical solution of Burgers' equation, *Universites du Maine et d'Angers, Lab. de Statistique et Processus*, Preprint 97-1, 16 pp.

63. DiPerna R.J. (1983), Convergence of the viscosity method for isentropic gas dynamics, *Comm. Math. Phys.* 91, 1-30.

64. DiPerna R.J. (1983), Convergence of approximate solutions to conservation laws, *Arch. Rat. Math. Mech.* 82, 27-70.

65. Dobrushin R. L. (1979), Gaussian and their subordinated self-similar random generalized fields, *Ann. Probab.* 7, 1-28.

66. Dobrushin R. L. (1980), *Automodel generalized random fields and their renorm group*, in: Multicomponent Random Systems, R.L. Dobrushin and Ya. G. Sinai (eds), Dekker, pp. 153-198.

67. Dobrushin R. L. and Major P. (1979) *Non-central limit theorems for nonlinear functionals of Gaussian fields*, Z. Wahr. verw. Geb. 50, 27-52.

68. Donsker M.D. (1964), On function space integrals, in *Analysis in Function Space*, W.T. Martin and I. Segal, eds., MIT Press, Cambridge, Mass., pp.17-30.

69. E W. (1992), Convergence of spectral methods for Burgers' equation, *SIAM J. Num. Anal.* 29, 110.

70. E W., Rykov Yu.G. and Sinai Ya.G. (1996), General variational principles, global weak solutions and behavior with random initial data for systems of conservation laws arising in adhesion particle dynamics, *Comm. Math. Phys.* 177, 349-380.

71. Einstein A. (1905), On the movement of small particles suspended in a stationary liquid demanded by the molecular-kinetic theory of heat, *Ann. Physik* 17, 549-560.

72. Falconer K. J. (1990), *Fractal Geometry: Mathematical Foundation and Applications*, Wiley, New York.

73. Feller W. (1971), *An Introduction to Probability Theory and Its Applications*, Vol. 2, Wiley and Sons, Inc.

74. Ferrari P.A., Kipnis C , Saada E. (1991), Microscopic structure of travelling waves in the asymmetric simple exclusion process, *Ann. Probab.* 19, 226-244.

75. Ferrari P.A., Presutti E., Vares M.E. (1988), Nonequilibrium fluctuations for a zero range process, *Ann. Inst. H. Poincaré, Sec. B, Prob. Stat.* 24, 237-268.

76. Ferrari P.A. (1992), Shock fluctuation in asymmetric simple exclusion, *Probab. Theory Rel. Fields* 91, 81-101.

77. Forgacs G., Lipowsky P., Nieuwenhuizen Th. M. (1991), The behaviour of interfaces in ordered and disordered systems, in: *Phase Transitions and Critical Phenomena* 14, 139-363, Academic Press, New York.

78. Forsyth A.R. (1906), *Theory of Differential Equations, Part IV–Partial Differential Equations, Vol. VI*, Cambridge University Press.

79. Fournier J.-D., Frisch U. (1983), L'équation de Burgers déterministe et statistique, *J. Méc. Theor. Appl.* 2, 699-750.

80. Freidlin M.I., Wentzell A.D. (1984), *Random Perturbations of Dynamical Systems*, Springer-Verlag.

81. Frisch,U., Lessieur M., Brissaud A.J. (1974), Markovian random coupling model for turbulence, *J. Fluid Mech.* 65, 145-52.

82. Frisch U. (1996), *Turbulence*, Cambridge University Press.

83. Funaki T. (1984), A certain class of diffusion processes associated with nonlinear parabolic equations, *Z. Wahr. verw. Geb.* 67, 331-348.

84. Funaki T. (1991), Regularity properties for stochastic partial differential equations of parabolic type, *Osaka J. Math.*, 28, 495-516.

85. Funaki T., Surgailis D. and Woyczynski W.A. (1995), Gibbs-Cox random fields and Burgers turbulence, *Ann Appl. Prob.* 5, 461-492.

86. Funaki T. and Woyczynski W.A., Eds. (1996), *Nonlinear Stochastic PDE's: Hydrodynamic Limit and Burgers' Turbulence*, IMA Volume 77, Springer-Verlag.

87. Funaki T. and Woyczynski W.A. (1998), Interacting particle approximation for fractal Burgers' equation, in *Stochastic Processes and Related Topics. A Volume in Memory of Stamatis Cambanis*, Birkhäuser-Boston, 1-27.

88. Furutsu K. (1963), On the statistical theory of electromagnetic waves in a fluctuating media, *J. Res. NBS* D-67, 303.

89. Garbaczewski P. and Kondrat G. (1996), Burgers velocity fields and dynamical transport processes, *Phys. Rev. Lett.* 77, 2608-2611.

90. Garbaczewski P., Kondrat G. and Olkiewicz R. (1997), Burgers velocity fields and the electromagnetic forcing in diffusive (Markovian) matter transport, *Acta Phys. Pol.* B28, 1731-1746.

91. Garbaczewski P., Kondrat G. and Olkiewicz R. (1997), Burgers flows as Markovian diffusion processes, *Phys. Rev.* E55, 1401-1412.

92. Garbaczewski P., Kondrat G. and Olkiewicz R. (1997), Schrödinger interpolating dynamics and Burgers flows, *Chaos, Solitons and Fractals* 9, 29-41.

93. Gardiner C.W. (1985), *Handbook of Stochastic Methods for Physics, Chemistry and the Natural Sciences*, Springer-Verlag, Berlin.

94. Garner J., Presutti E. (1990), Shock fluctuations in particle systems, *Ann. Inst. H. Poincaré, Sec. A, Prob. Stat.* 53, 1-14.

95. Gärtner J., Molchanov S.A. (1990), Parabolic problems for the Anderson model, *Comm. Math. Phys.*, 132, 913-955.

96. Giga Y., Kambe T. (1988), Large time behavior of the vorticity of two-dimensional viscous flow and its application to vortex formation, *Comm. Math. Phys.* 117, 549-568.

97. Giraitis L. and D. Surgailis (1985) *CLT and other limit theorems for functionals of Gaussian processes*, Z. Wahr. verw. Geb. 70, 191-212.

98. Giraitis L., Molchanov S. A., and Surgailis D. (1992), *Long memory shot noises and limit theorems with application to Burgers' equation*, in: New Directions in Time Series Analysis, Part II , P. Caines, J. Geweke and M. Taqqu (eds), IMA Volumes in Mathematics and Its Applications, Springer.

99. Gossard E.E., Khuk U.K. (1978), *Waves in Atmosphere*, Nauka, Moscow.

100. Gotoh T. (1993), Inertial range spectrum of Burgers turbulence by mapping closure,*Phys. Fluids* A5, 445-457.

101. Gotoh T. and Kreichnan R. H. (1993), Statistics of decaying Burgers turbulence, *Phys. Fluids* A5, 445-457.

102. Grigoryev Yu.N. (1979), Diagram method of Prigogine-Balesky with respect to the simplest models of hydrodynamic turbulence, Preprint no 3, Inst. Teor. i Prikl. mech. SO AN USSR, Novosibirsk.

103. Gurbatov S., Malakhov A., and Saichev A. (1991), *Nonlinear Random Waves and Turbulence in Nondispersive Media; Waves, Rays and Particles*, Manchester University Press.

104. Gurbatov S., Saichev A. (1993), Inertial nonlinearity and chaotic motion of particle fluxes, *Chaos* 3, 333-358.

105. Gurbatov S.N., Saichev A.I., Shandarin S.F. (1984), Large-scale structure of the Universe within the framework of model equation for nonlinear diffusion, *Soviet Physics Doklady* 30, 321-324.

106. Gurbatov S.N., Saichev A.I., Shandarin S.F. (1989), The large scale structure of the Universe in the frame of the model equation of non-linear diffusion, *Mon. Not. R. Astro. Soc.* 236, 385-402.

107. Gurbatov S.N., Saichev A.I. (1981), Degeneracy of one-dimension acoustic turbulence at large Reynolds numbers, *Sov. Phys. JETP* 53, 347.

108. Gurbatov S., Saichev A. (1993), Inertial nonlinearity and chaotic motion of particle fluxes, *Chaos* 3 , 333-358.

109. Gurbatov S., Simdyankin S.I., Aurell E., Frisch U., Toth G. (1997), On the decay of Burgers turbulence, *J. Fluid Mech.* 344 , 339-374.

110. Gutkin E., Kac M. (1983), Propagation of chaos and the Burgers equation, *SIAM J. Appl. Math.* 43, 971-980.

111. Gwa Leh-Hun, Spohn H. (1992), Be the solution for the dynamical-scaling exponent of the noisy Burgers equation, *Physical Review., Statistical Physics* 46, 844.

112. Gyöngi I. (1997), On the stochastic Burgers equation, *University of Edinburgh*, preprint, 28 pp.

113. Halford W.D., Vlieg-Hulstman M. (1992), Korteweg-deVries-Burgers equation and the Painlevé property, J. Physics., A. Mathematical and General 25, 2375.

114. Handa K. (1993), A remark on shocks in inviscid Burgers' turbulence, in *Nonlinear Waves and Weak Turbulence*, N. Fitzmaurice et al., Eds., Birkhäuser-Boston.

115. Handa K. (1995), On a stochastic PDE related to Burgers' equation with noise, in *Nonlinear Stochastic PDE's: Hydrodynamic Limit and Burgers' Turbulence*, T. Funaki and W.A. Woyczynski, eds., IMA Volume 77, Springer-Verlag, pp. 147-156.

116. Henry D.B. (1982), How to remember the Sobolev inequalities, in *Differential Equations, Saõ Paulo 1981*, D.G. de Figueiredo and C.S. Hönig, Eds., Lecture Notes in Mathematics **957**, pp. 97-109, Berlin: Springer.

117. Holden H., Lindstrøm T., Øksendal B., Ubøe J., and Zhang. T-S.(1994), The Burgers equation with noisy force and the stochastic heat equation, *Comm. Partial Diff. Eq.* 19, 119-141.

118. Holden H., Øksendal B., Ubøe J., and Zhang. T-S.(1996), *Stochastic Partial Differential Equations. A Modeling, White Noise, Functional Approach*, Birkhäuser-Boston.

119. Hopf E. (1950), The partial differential equation $u_t + uu_x = \mu u_{xx}$, *Comm. Pure Appl. Math.* 3, 201-230.

120. Hu Y. and Woyczynski W.A. (1994), An extremal rearrangement property of statisitcal solutions of the Burgers equation, *Ann. Appl. Probab.* 4, 838-858.

121. Hu Y., Woyczynski W.A. (1995a), A maximum principle for unimodal moving average data for Burgers' equation, *Prob. and Math. Stat.* 15,153-171.

122. Hu Y., Woyczynski W.A. (1995b), Limit behavior of quadratic forms of moving averages and statistical solutions of the Burgers' equation, *J. Multivariate Analysis*, 52, 15-44.

123. Hu Y., Woyczynski W.A. (1995c), Large-scale structure of the Universe and asymptotics of Burgers' turbulence with heavy-tailed dependent data, in *Chaos: The Interplay Between Stochastic and Deterministic Behavior,* P. Garbaczewski et al. Eds., Springer's Lecture Notes in Physics 457.

124. Hu Y., Woyczynski W.A. (1996), Shock density in Burgers' turbulence, in *Nonlinear Stochastic PDE's: Burgers Turbulence and Hydrodynamic Limit,* T. Funaki and W.A.Woyczynski, Eds., IMA Volume 77, Springer-Verlag, pp. 211-226.

125. Hörmander L.(1983), *The Analysis of Linear Partial Differential Operators I,* Springer-Verlag, Berlin, Heidelberg, New York, Tokyo.

126. Ikeda N. and Watanabe S. (1981), *Stochastic Differential Equations and Diffusion Processes*, North-Holland-Kodansha.

127. Iskandar L., Mohsen A. (1992), Some numerical experiments on the splitting of Burgers' equation, *Num. Meth. for Partial Diff. Equations* 8, 267.

128. Ivanov A.V. and Leonenko N.N. (1989), *Statistical Analysis of Random Fields*, Kluwer, Dordrecht.

129. Janicki A., Surgailis D., Woyczynski W.A. (1995), Statistics and geometric thermodynamics of passive tracer densities in 2-D forced Burgers turbulence, preprint.

130. Janicki A. and A. Weron (1994), Simulation and Chaotic Behavior of α-stable Stochastic Processes, Marcel Dekker, N.Y.

131. Janicki A., Woyczynski W.A. (1997), Hausdorff dimension of regular points instochastic Burgers flows with Levy α-stable initial data, *J. Stat. Physics* 86, 277-299.

132. John F. (1990), *Nonlinear Wave Equations, Formation of Singularities,* Amer. Math. Soc., Providence.

133. Kallenberg O. (1986), *Random Measures*, 4th ed., Academic Press.

134. Kahng W-H. and Siegel A. (1970), The Cameron-Martin-Wiener method in turbulence and in Burgers' model: general formulae, and application to late decay, *J. Fluid mech.* 41, 593-618.

135. Karatzas I. and Shreve S.E. (1988), *Brownian Motion and Stochastic Calculus*, Springer-Verlag.

136. Kardar M., Parisi G., Zhang Y.C. (1986), Dynamical scaling of growing interfaces, *Phys. Rev. Lett.* 56, 889.

137. Keleti S., Reed Jr. X.B. (1987), Exact solutions to Burgers' equation at high Reynold's numbers: II. Their numerical evauation at large but finite Re_0, *Dev. Mech.* 14(a), 63.

138. Keleti S., Reed Jr. X.B. (1988), Spectral properties of exact random solutions to Burgers' equation for modified Thomas initial conditions, *Computat. Fluids* 16, 147.

139. Keleti S., Reed Jr. X.B. (1996), Evaluation of spectral behavior for large ensembles of exact solutions to Burgers' equation with Thomas initial conditions, in *Nonlinear Stochastic PDE's: Hydrodynamic Limit and Burgers' Turbulence*, T. Funaki and W.A. Woyczynski, Eds., IMA Volume 77, Springer-Verlag, pp. 193-236.

140. Khristov Kh. (1980), *Teoretichna i prilozhna mechanika (Sofia)* 11, 59.

141. Kida S. (1979), Asymptotic properties of Burgers turbulence, *J. Fluid Mech.* 93, 337-377.

142. Kifer Y. (1997), The Burgers equation with a random force and a general model for directed polymers in random environments, *Prob. Theory Rel. Fields*, 108, 29-65.

143. Kipnis C. (1986), Central limit theorems for infinite series of queues and applications to simple exclusions, *Ann. Prob.* 14, 397-408.

144. Kofman L., Bertschinger E., Gelb J.M., Nusser A., Dekel, A. (1994), Evolution of one-point distributions from Gaussian initial fluctuations, *Astrophys. J.* 420, 44-57.

145. Kofman L., Pogosyan D., Shandarin S., Mellott A. (1992), Coherent structures in the Universe and the adhesion model, *Astrophysical J.* 393, 437-449.

146. Kofman L., Raga A.C. (1992), Modeling structures of knots in jet flows with the Burgers equation, *Astrophys. J.* 390, 359-364.

147. Komatsu T. (1984), Pseudo-differential operators and Markov processes, *J. Math. Soc. Japan* 36, 387-418.

148. Komatsu T. (1984), On the martingale problem for generators of stable processes with perturbations, *Osaka J. Math.* 21, 113-132.

149. Korsunsky S. V. (1992), Nonlinear MHD waves in a dissipative plasma with Hall dispersion and the modified Burgers-Schrödinger equation, *J. Plasma Physics* 48, 237.

150. Kotani S. and Osada H. (1985), Propagation of chaos for Burgers' equation, *J. Math. Soc. Japan* 37, 275-294.

151. Kraichnan R.H. (1959), The structure of isotropic turbulence at very high Reynolds numbers, *J. Fluid Mech.* 5, 497-543.

152. Kraichnan R.H. (1968), Lagrangian-history statistical theory for Burgers' equation, *Physics of Fluids* 11, 265-277.

153. Kraichnan R.H. (1990), Models of intermittency in hydrodynamic turbulence, *Phys. Rev. Letters* 65, 575-578.

154. Krankel R.A., Pereira J.G., Manna M.A. (1992), Surface perturbations of a shallow viscous fluid heated from below and the $(2+1)$-dimensional Burgers equation, *Physical Review., Statistical Physics* 45, 838.

155. Kružkov J.N. (1970), First order quasilinear equations with several space variables, *Math. USSR-Sb.* 10, 217-243.

156. Kunita H. (1990), *Stochastic Flows and Stochastic Differential Equations* , Cambridge Univ. Press, Cambridge.

157. Kuramoto Y. (1985), *Chemical Oscillations, Waves and Turbulence,* Springer.

158. Kuwabara S. (1978), Statistical hydrodynamics for Burgers turbulence, *Mem. Fac. Eng. Nagoya Univ.* 30, 245-88.

159. Kwapien S. and Woyczynski W.A. (1992), *Random Series and Stochastic Integrals: Single and Multiple*, Birkhäuser-Boston.

160. Ladyženskaja O.A., Solonnikov V.A., Ural'ceva N.N. (1988), *Linear and Quasilinear Equations of Parabolic Type,* Providence, Amer. Math. Soc.

161. Landim C. (1991), Hydrodynamic equation for attractive particle systems on Z^d, *Annals of Prob.* 19, 1537-1558.

162. Lasseigne D. G., Olmstead W.E. (1990), Stability of a viscoelastic Burgers flow, *SIAM J. Appl. Math* 50, 352-360.

163. Lax P.D. (1973), *Hyperbolic Systems of Conservation Laws and the Mathematical Theory of Shock Waves*, SIAM.

164. Lax, P.D. (1991), The zero dispersion limit, a deterministic analogue of turbulence, *Comm. Pure Appl. Math.* 44, 1047-1056.

165. Lax P.D., Levermore C.D. (1983), The small dispersion limit of the Korteweg-de Vries equation, *Comm. Pure Appl. Math.* 36;I, 253-290; II, 571-593; III, 809-829.

166. Leadbetter M.R., Lindgren G., Rootzen H. (1983), *Extremes and Related Properties of Random Sequences and Processes*, Springer-Verlag.

167. Lee J. (1980), *J. Fluid Mech.* 101, 349-76.

168. Leonenko N.N. and Olenko A.Ya. (1991), Tauberian and Abelian theorems for correlation function of homogeneous isotropic random fields, *Ukrainian Math. J.* 43, 1652-1664.

169. Leonenko N.N., Deriev I.I. (1994), Limit theorems for the solutions of multidimensional Burgers equation with weak dependent random initial conditions, *Theory Prob. Math. Stat.* 51, 98-110.

170. Leonenko N.N., Orsingher E. and Parkhomenko V.N. (1995), Scaling limits of solutions of the Burgers equation with singular non-Gaussian data, *Random Operators and Stochastic Eq.* 3, 101-112.

171. Leonenko N.N., Orsingher E. and Parkhomenko V.N. (1996), On the rate of convergence to the normal law for solutions of the Burgers equation with singular initial data, *J. Stat. Phys.* 82, 915-930.

172. Leonenko N.N., Orsingher E. and Rybasov K.V. (1994), Limit distributions of solutions of multidimensional Burgers equation with random initial data, I and II, *Ukrainian Math. J.* 46, 870-877, and 1003-110.

173. Leonenko N.N., Zhanbing L. (1994), Non-Gaussian limit distributions for solutions of Burgers equation with strongly dependent random initial conditions, *Random Operators and Stoch. Eq.* 2, 79-86.

174. Leonenko N.N., Parkhomenko V.N., Woyczynski W.A. (1996), Spectral properties of the scaling limit solutions of the Burgers equation with singular data, *Random Operators and Stochastic Eq.* 4, 229-238.

175. Leonenko N.N., Woyczynski W.A. (1998), Exact parabolic asymptotics for singular n-D Burgers' random fields: Gaussian approximation, *Stoch. Proc. Appl.* 75, 1-25.

176. Leonenko N.N., Woyczynski W.A. (1998), Scaling limits of solutions of the heat equation for singular non-gaussian data, *J. Stat. Phys.* 91, 423-438.

177. Leonenko N.N., Woyczynski W.A. (1998), Parameter identification for stochastic Burgers' flows via parabolic rescaling, *Ann Stat.*, 55 pp., to appear.

178. Leonenko N.N., Woyczynski W.A. (1998), Statistical inference for long-memory random fields arising in Burgers' turbulence, *Proc. Umea Conf.*, 11 pp., to appear.

179. Leonenko N.N., Woyczynski W.A. (1998), Parameter estimation for random fields arising in three-dimensional Burgers' turbulence, Functional Analysis V, Dubrovnik 1997, Butkovic, H. Kraljevic and G. Peskir, Eds., Various Publications Series, Aarhus University, 11 pp., to appear.

180. Liggett T.M. (1985), *Interacting Particle Systems*, Springer.

181. Liggett T.M. (1997), Stochastic models of interacting particles, *Ann. Prob.* 25, 1-29.

182. Lighthill M. J. (1956), Viscosity effects in sound waves of finite amplitude. In *Surveys in Mechanics*, eds. G. K. Batchelor and R. M. Davies, pp. 250-351, Cambridge University Press.

183. Lions P.A. (1982), *Generalized Solutions of Hamilton-Jacobi Equation*, Pitman.

184. Liu Shi-da, Liu Shi-kuo (1992), KdV-Burgers equation modelling of turbulence, *Science in China* A35, 576-586.

185. Maddox S.J., Efstathiou G., Sutherland W.J. and Loveday J. (1990), *Monthly Notices Roy. Astron. Soc.* 242, 44p.

186. Majda A.J. (1993), Explicit inertial range renormalization theory in a model for turbulent diffusion, *J. Stat. Phys.* 73, 515-542.

187. Major P. (1981), *Multiple Wiener-Ito Integrals*, Springer's Lecture Notes in Math. **849**.

188. Mandelbrot B. (1982), *The Fractal Geometry of Nature*, W. H. Freeman and Co., San Francisco.

189. Mann J.A., Woyczynski W.A. (1997), Rough surfaces generated by nonlinear transport, Invited paper, *Symposium on Nonlinear Diffusion*, TMS International Meeting. Indianapolis.

190. Margolin L.G. (1992), Jones D.A., An approximate inertial manifold for computing Burgers' equation, *Physica D, Nonlinear phenomena* 60, 175.

191. McKean Jr. H.P. (1967), Propagation of chaos for a class of non-linear parabolic equations, in *Lecture Series in Diff. Eq.*, Catholic Univ, pp. 177-194.

192. Meecham W.C., Iyer P. and Clever W.C. (1975),, Burgers' model with a renormalized Wiener-Hermite representation, *Phys Fluids* 18, 1610-72.

193. Medina E., Hwa T., Kardar M. and Zhang Y.-C. (1989), Burgers equation with correlated noise: renormalization-group analysis and application to directed polymers and interface growth, *Phys. Rev. A* 39, 3053-3075.

194. Mellott A.L., Shandarin S.F., Weinberg D.H. (1994), A test of the adhesion approximation for gravitational clustering, *Astrophys. J.* 428, 28-34.

195. Molchan G.M. (1997), Burgers equation with self-similar Gaussian initial data: tail probabilities, *J. Stat. Phys.* 88, 1139-1150.

196. Molchanov S.A. (1994), Lectures in Random Media, in *Springer's Lecture Notes in Mathematics* **1581**, 242-406.

197. Molchanov S.A., Surgailis D., and Woyczynski W.A. (1995), Hyperbolic asymptotics in Burgers turbulence, *Comm. Math. Phys.* 168, 209-226.

198. Molchanov S.A., Surgailis D., Woyczynski W.A. (1995), Spectral methods in analysis of forced Burgers turbulence, CWRU preprint.

199. Molchanov S.A., Surgailis D., Woyczynski W.A. (1997), The large-scale structure of the Universe and quasi-Voronoi tessellation of shock fronts in forced inviscid Burgers' turbulence in R^d, *Ann. Appl. Prob.* 7, 200-228.

200. Molchanov S.A. and Woyczynski W.A., Eds. (1997a), *Stochastic Models in Geosystems*, IMA Volume 85, Springer-Verlag.

201. Mond M., Knorr G. (1980), *Phys. Fluids* 23,1306-10.

202. Musha T., Kosugi Y., Matsumoto G., Suzuki M. (1981), Modulation of the time relation of action potential impulses propagating along the axon, *IEEE Trans. Biomedical Eng.* BME-28, 616-623.

203. Møller J. (1994), *Lectures on Random Voronoi Tessellations*, Springer's Lecture Notes in Statistics.

204. Nakazawa H. (1980), Probabilistic aspects of equation of motion of forced Burgers and Navier-Stokes turbulence, *Prog. Theor. Phys.* 64, 1551-1564.

205. Naumkin P.I., Shishmarev I. A. (1992), On the decay of step for the Korteweg-de Vries-Burgers equation, *Functional Analysis and Its Applications* 26, 148.

206. Newell G.F. (1993), A simplified theory of kinematic waves in highway traffic, Part I: General theory, *Transportation Res.* 27B, 281-287.

207. Novikov E.A. (1964), Functionals and the random-force method in turbulence theory, *Sov. Phys. JETP* 20 (5), 1290.

208. Oelschläger, K.A. (1985), Law of large numbers for moderately interacting diffusion processes, *Z. Wahr. verw. Geb.* 69, 279-322.

209. Okabe A., Boots B., Sugihara K., (1992), *Spatial Tessellations*, Wiley and Sons.

210. Oleinik O. (1957), Discontinuous solutions of nonlinear differential equations, *Russ. Math. Survey*, Amer. Math. Transl. Series 2, 26, 95-172.

211. Oleinik O.A. (1985), On the Cauchy problem for nonlinear equations in a class of discontinuous functions, *Doklady AN USSR* 95, 451-454.

212. Orszag S.A., Bissonette L.R. (1967), *Phys. Fluids* 10, 2603-13.

213. Parker A. (1992), On the periodic solution of the Burgers equation: a unified approach,*Proc. Royal Soc. London* 438, 113.

214. Peebles P.J.E. (1980), *The Large Scale Structure of the Universe*, Princeton University Press.

215. Ravishankar K. (1992), Fluctuations from the hydrodynamical limit for the symmetric simple exclusion in Z^2, *Stoch. Proc. Appl.* 42, 31-37.

216. Ravishankar K. (1992), Interface fluctuations in the two-dimensional weakly asymmetric simple exclusion process, *Stoch. Proc. . Appl.* 43, 223-247.

217. Reid W.H. (1957), On the transfer of energy in Burgers' model of turbulence, *Appl. Sci. Res.* A6, 85.

218. Revuz D. and Yor M. (1994), *Continuous Martingales and Brownian Motion*, Springer.

219. Rosenblatt M. (1981), Limit theorems for Fourier transforms of functionals of Gaussian sequences, *Z. Wahr. verw. Geb.* 55, 123-132.

220. Rosenblatt M. (1968), Remarks on the Burgers equation, *J. Math.Phys.* 9, 1129-1136.

221. Rosenblatt M. (1978), Energy transfer for the Burgers' equation, *Phys. Fluids* 21 , 1694-1697.

222. Rosenblatt M. (1987), Scale renormalization and random solutions of the Burgers equation, *J. Appl. Prob.* 24, 328-338.

223. Rudenko O.V., Soluyan S.I. (1977), *Theoretical Foundation of Nonlinear Acoustics*, Plenum Cons. B, New York.

224. Ruelle D. (1969), *Statistical Mechanics: Rigorous Results*, Benjamin, New York.

225. Ryan R. (1996), Large deviation analysis of Gaussian fields and the statistics of Burgers' turbulence, *Ph. D. Dissertation*, Courant Institute, N.Y.U, 128 pp.

226. Ryan R. (1998), Large deviation analysis of Burgers' turbulence with white-noise initial data, *Comm. Pure Appl. Math.*, 51, 47-75.

227. Ryan R. (1998), The statistics of Burgers turbulence initialized with fractional Brownian noise data, *Comm. Math. Phys.*, 31 pp.,to appear.

228. Sahni V., Sathyaprakash B.S., Shandarin S.F. (1994), The evolution of voids in the adhesion approximation, *Astrophys. J.* 431, 20-40.

229. Saichev A.I., Woyczynski W.A. (1996), *Distributions in the Physical and Engineering Sciences*: Volume 1: *Distributional and Fractal Calculus, Integral Transforms, Wavelets*; Volume 2: *Partial Differential Equations and Random Fields*, Birkhauser-Boston,to appear.

230. Saichev A.I., Woyczynski W.A. (1996a), Model description of passive tracer density fields in the framework of Burgers' and other related model equations, in *Nonlinear Stochastic PDE's: Burgers Turbulence and Hydrodynamic Limit*, IMA Volume 77, Springer-Verlag, pp. 167-192.

231. Saichev A.I., Woyczynski W.A. (1996b), Density fields in Burgers and KdV-Burgers turbulence, *SIAM J. Appl. Math.*, 56, 1008-1038.

232. Saichev A.I., Woyczynski W.A (1996c), Probability distributions of passive tracers in randomly moving media, in *Stochastic Models in Geosystems*, S.A. Molchanov and W.A. Woyczynski, Eds., IMA Volume 85, Springer-Verlag, pp. 359-400.

233. Saichev A.I., Woyczynski W.A. (1997a), Evolution of Burgers' turbulence in presence of external forces, *J. Fluid Mech.*, 331, 313-343.

234. Saichev A.I., Woyczynski W.A. (1997b), Advection of passive and reactive tracers in multidimensional Burgers' velocity field, *Physica D*, 100, 119-141.

235. Saichev A.I., G.M. Zaslavsky (1997), Fractional kinetic equations: solutions and applications, *Chaos*, 7, 753-764.

236. Samorodnitsky G., Taqqu M. S. (1994), *Stable non–Gaussian Random Processes: Stochastic Models with Infinite Variance*. Chapman & Hall, London.

237. Satsuma J. (1987), Exact solutions of Burgers' equation with reaction terms,in *Topics in Soliton Theory and Exactly Solvable Nonlinear Equations*, M. Ablowitz, B. Fuchssteiner, M. Kruskal, Eds., World Scientific, 255-262.

238. Schilder M. (1966), Some asymptotic formulas for Wiener integrals, *Trans. Amer. Math. Soc.* 20, 63-85.

239. Schult R.L., Wyld H.W. (1992), Using wavelets to solve the Burgers equation: A comparative study, *Physical Review, Statistical Physics* 46, 7953.

240. Seppäläinen T. (1996), A microscopic model for the Burgers equation and longest increasing subsequences, *Electronic J. Prob.* 1, Paper 5, 1-51.

241. Seppäläinen T. (1997), Coupling the totally asymmetric simple exclusion process with a moving interface, *Iowa State U, Appl. Math. Report* AM97-35, 34 pp.

242. Shandarin S.F. and Zeldovich Ya.B. (1989), Turbulence, intermittency, structures in a self-gravitating medium: the large scale structure of the Universe, *Rev. Modern Phys.* 61, 185-220.

243. She Z.-S., Aurell E., and Frisch U. (1992), The inviscid Burgers equation with initial data of Brownian type, *Comm. Math. Phys.* 14, 623-641.

244. Shih Y.-C., Reed Jr. X B (1985), Solution to the piecewise linear continuous random initial value problem for Burgers'equation, *Phys. Fluids* 28, 2088.

245. Shlesinger M.F., Zaslavsky G.M., Frisch U., Eds. (1995), *Lévy Flights and Related Topics in Physics*, Lect. Notes in Phys. **450**, Springer.

246. Sinai Ya. G. (1992), Statistics of shocks in solutions of inviscid Burgers equation, *Commun. Math. Phys.* 148, 601–621.

247. Sinai Ya.G. (1992), Two results concerning asymptotic behavior of solutions of the Burgers equation with force, *J. Stat. Phys.* 64, 1-12.

248. Sinai Ya. G. (1996), Burgers system driven by a periodic stochastic flow, Princeton University preprint, *Itô's Stochastic Calculus and Probability Theory*, N. Ikeda et al., Eds., Springer-Verlag, 1997, 347-353.

249. Smith R., Walton I., (1992), A Burgers concentrations dispersion equation,*J. Fluid Mech.* 239, 65.

250. Smoller J. (1994), *Shock Waves and Reaction-Diffusion Equations*, Springer-Verlag, Berlin, Heidelberg, New York, Tokyo.

251. Smoluchowski M. (1906), *Ann. Physik* 21, 756.

252. Spohn H. (1991), *Dynamics of Systems with Many Particles*, Springer-Verlag.

253. Stroock D. (1975), Diffusion processes associated with Lévy generators, *Z. Wahr. verw. Gebiete* 32, 209-244.

254. Stroock D.W. and Varadhan S.R.S. (1979), *Multidimensional Diffusion Processes*, Springer-Verlag.

255. Sugimoto N. (1989), "Generalized" Burgers equations and fractional calculus, in *Nonlinear Wave Motion*, A. Jeffrey, Ed., pp. 162-179, Harlow: Longman Scientific.

256. Sugimoto N. (1991), Burgers equation with a fractional derivative; hereditary effects on nonlinear acoustic waves, *J. Fluid Mech.* 225, 631-653.

257. Sugimoto N. (1992), Propagation of nonlinear acoustic waves in a tunnel with an array of Helmholtz resonators, *J. Fluid Mech.* 244, 55-78.

258. Sugimoto N., Kakutani T. (1985), Generalized Burgers equation for nonlinear viscoelastic waves, *Wave Motion* 7, 447-458.

259. Sun J. (1993), Tail probabilities of the maxima of gaussian random fields, *Ann. Prob.* 21, 34-71.

260. Surgailis D. (1981), On infinitely divisible self-similar random fields, *Z. Wahr. verw. Geb.* 58, 453-477.

261. Surgailis D. (1996), Intermediate asymptotics of statistical solutions of Burgers' equation, in *Nonlinear Stochastic PDE's: Hydrodynamic Limit and Burgers' Turbulence*, T. Funaki and W.A. Woyczynski, Eds., IMA Volume 77, Springer-Verlag, pp. 137-146.

262. Surgailis D. (1997), Asymptotics of solutions of Burgers' equation with random piecewise constant data, in *Stochastic Models in Geosystems*, S.A. Molchanov and W.A. Woyczynski, Eds., IMA Volume 85, Springer-Verlag, pp. 427-442.

263. Surgailis D., Woyczynski W.A. (1993), Long range prediction and scaling limit for statistical solutions of the Burgers' equation, in *Nonlinear Waves and Weak Turbulence*, Birkhäuser-Boston, 313-338.

264. Surgailis D. and Woyczynski W. A. (1994a), Burgers' equation with nonlocal shot noise data, *J. Appl. Prob.* 31A, 351-362.

265. Surgailis D. and Woyczynski W. A. (1994b), Scaling limits of solutions of Burgers' equation with singular Gaussian initial data, in *Chaos Expansions, Multiple Wiener–Itô Integrals and Their Applications*, C. Houdré and V. Perez–Abreu, Eds.,CRC Press, Boca Raton, 145-162.

266. Surgailis D., W.A. Woyczynski (1994c), Burgers' topology on random point measures, *Probability on Banach Spaces, 9*, Birkhäuser-Boston, pp. 209-221.

267. Sznitman A.S. (1987), A propagation of chaos result for Burgers' equation, in *Hydrodynamic Behavior and Interacting Particle Systems*, Springer-Verlag, New York.

268. Sznitman A. (1988), A propagation of chaos result for Burgers equation, *Prob. Th. Rel. Fields* 71, 581-613.

269. Sznitman A.S.(1991), Topics in propagation of chaos, *Lecture Notes in Math. 1464*, Springer-Verlag, New York.

270. Taqqu M. S. (1979), Convergence of integrated processes of arbitrary Hermite rank, *Z. Wahr. verw. Geb.* 50, 53-83.

271. Tatsumi T, Kida S. (1972), Statistical mechanics of the Burgers model of turbulence, *J. Fluid Mech.* 55, 659-675.

272. Tatsumi T. (1980), *Adv. Appl. Mech* 20, 39-133.

273. Taylor M. (1992), Analysis on Morrey spaces and applications to Navier-Stokes and other evolution equations, *Comm. PDE* **17**, 1407-1456.

274. Temam R. (1983), *Navier-Stokes Equations and Nonlinear Functional Analysis*, SIAM.

275. Triebel H. (1983 and 1992), *Theory of Function Spaces, I and II*, Monographs Math. **78** and **84**, Basel: Birkhäuser.

276. Tseng P.-L., Reed Jr. X.B. (1986), Some new results for weak shocks, *Bull. Amer. Phys. Soc.* 31, 1737.

277. Tseng P.-L., Reed Jr. X.B.(1987), Exact solutions to Burgers' equation at high Reynold's numbers: I. Inviscid solutions, *Dev. Mech.* 14(a), 57.

278. Van de Weygaert R. (1991), *Voids and the Geometry of Large Scale Structure*, Ph.D. Thesis, Leiden University.

279. Vanaja V., Sachdev P.L. (1992), Asymptotic solutions of a generalized Burgers equation, *Quart. Appl. Math.* 50, 627.

280. Varadhan S.R.S. (1993), Entropy methods in hydrodynamical scaling, in *Nonequilibrium Problems in Many-Particle Systems*, LNM 1551, Springer-verlag, 112-145.

281. Varadhan S.R.S. (1997), The complex story of simple exlusion, in *Itô's Stochastic Calculus and Probability Theory*, N. Ikeda et al., Eds., Springer-verlag, 385-399.

282. Vergassola M., Dubrulle B.D., Frisch. U., and Noullez A. (1994), Burgers equation, devil's staircases and mass distribution for the large-scale structure, *Astr. and Astrophys.* 289, 325-356.

283. Vishik M. J., Fursikov A. V. (1988), *Mathematical Problems of Statistical Hydrodynamics*, Kluwer.

284. Walsh J. B.(1986), *An introduction to stochastic partial differential equations*, in: Lect. Notes in Math., 1180, pp. 265-439, Springer-Verlag, Berlin.

285. Walton J.J. (1970), *Phys. Fluids* 13, 1634-5.

286. Weinberg D., Gunn J. (1990), Large-scale structure and the adhesion approximation, *Mon. Not. R. Astro.Soc.* 247, 260-286.

287. Weissler F.B. (1980), Local existence and nonexistence for semilinear parabolic equations in L^p, *Indiana Univ. Math. J.* **29**, 79-102.

288. White L. W. (1993), A study of uniqueness for the initialization problem for Burgers' equation, *J. Math. Anal.. Appl.* 172, 412.

289. Wick W.D. (1985), A dynamical phase transition in an infinite particle system, *J. Stat. Phys.* 38, 1005-1025.

290. Williams B.G., Heavens A.F., Peacock J.A. (1991), Shandarin S.F., Exact hierarchical clustering in one dimension, *Mon. Not .R. Astro. Soc.* 250, 458-476.

291. Winter H. (1938), *Asymptotic Distributions and Infinite Convplutions*, Edwards Brothers, Ann Arbor, Michigan.

292. Whitham G.B. (1974), *Linear and Nonlinear Waves*, Wiley, New York.

293. Wolansky G. (1992), Stationary and quasi-stationary shock waves for non-spatially homogeneous Burger's equation in the limit of small dissipation, *Indiana U. Math. J.* 41, 43.

294. Woyczynski W.A. (1993), Stochastic Burgers' flows, in *Nonlinear Waves and Weak Turbulence*, N. Fitzmaurice et al., eds., Birkhäuser-Boston, pp.279-311.

295. Woyczynski W.A. (1997), Computing with Brownian and Lévy α-stable path integrals, in *9th 'Aha Huliko'a Hawaiian Winter Workshop on Methods of Theoretical Physics in Oceanography*, University of Hawaii, 1997, pp. 1-12.

296. Wyller J., Wellander N., Larson F. (1992), Burgers equation as a model for the IP phenomenon, *Geophysical Prospecting* 40, 325.

297. Yakushkin I.G. (1981), Description of turbulence in the Burgers model, *Radiophysics and Quantum Electronics* 24, 41-48.

298. Yang H.Q., Przekwas A.J. (1992), A comparative study of advanced shock-capturing schemes applied to Burgers' equation, *J. Comp. Phys.* 102, 139.

299. Zambianchi E., Griffa A. (1994), Effects of finite scales of turbulence on dispersions estimates, *J. Marine Res.* 52, 129-148.

300. Zaslavsky G.M. (1994), Fractional kinetic equations for Hamiltonian chaos, *Physica D* 76, 110-122.

301. Zaslavsky, G.M., Abdullaev S.S. (1995), Scaling properties and anomalous transport of particles inside the stochastic layer, *Phys. Rev. E* 51(5), 3901-3910.

302. Zel'dovich Ya. B. (1970), Gravitational instability: An approximate theory for large density perturbations. *Astro. Astrophys.* 5, 84.

303. Zeldovich Ya.B., Einasto J, Shandarin S.F. (1982), Giant voids in the Universe, *Nature* 300, 407-413.

304. Zeldovich Ya.B. (1970), Gravitational instability; an approximate theory for large density perturbations, *Astro. Astrophys.* 5, 84.

305. Zheng W., Reed Jr. X.B. (1992), Some observations of bispectral behavior of large ensembles of exact solutions to the Burgers' equation for random initial conditions, *Physics of Fluids A* 4, 845-848.

306. Zheng W. (1995), Conditional propagation of chaos and a class of quasi-linear PDE's, *Ann. Prob.* 23, 1389-1413.

Index

Springer
und
Umwelt

Als internationaler wissenschaftlicher
Verlag sind wir uns unserer besonderen
Verpflichtung der Umwelt gegenüber
bewußt und beziehen umweltorientierte
Grundsätze in Unternehmens-
entscheidungen mit ein. Von unseren
Geschäftspartnern (Druckereien,
Papierfabriken, Verpackungsherstellern
usw.) verlangen wir, daß sie sowohl
beim Herstellungsprozess selbst als
auch beim Einsatz der zur Verwendung
kommenden Materialien ökologische
Gesichtspunkte berücksichtigen.
Das für dieses Buch verwendete Papier
ist aus chlorfrei bzw. chlorarm
hergestelltem Zellstoff gefertigt und im
pH-Wert neutral.

Springer

Lecture Notes in Mathematics

For information about Vols. 1–1504
please contact your bookseller or Springer-Verlag

Vol. 1547: P. Harmand, D. Werner, W. Werner, M-ideals in Banach Spaces and Banach Algebras. VIII, 387 pages. 1993.

Vol. 1548: T. Urabe, Dynkin Graphs and Quadrilateral Singularities. VI, 233 pages. 1993.

Vol. 1549: G. Vainikko, Multidimensional Weakly Singular Integral Equations. XI, 159 pages. 1993.

Vol. 1550: A. A. Gonchar, E. B. Saff (Eds.), Methods of Approximation Theory in Complex Analysis and Mathematical Physics IV, 222 pages, 1993.

Vol. 1551: L. Arkeryd, P. L. Lions, P.A. Markowich, S.R. S. Varadhan. Nonequilibrium Problems in Many-Particle Systems. Montecatini, 1992. Editors: C. Cercignani, M. Pulvirenti. VII, 158 pages 1993.

Vol. 1552: J. Hilgert, K.-H. Neeb, Lie Semigroups and their Applications. XII, 315 pages. 1993.

Vol. 1553: J.-L- Colliot-Thélène, J. Kato, P. Vojta. Arithmetic Algebraic Geometry. Trento, 1991. Editor: E. Ballico. VII, 223 pages. 1993.

Vol. 1554: A. K. Lenstra, H. W. Lenstra, Jr. (Eds.), The Development of the Number Field Sieve. VIII, 131 pages. 1993.

Vol. 1555: O. Liess, Conical Refraction and Higher Microlocalization. X, 389 pages. 1993.

Vol. 1556: S. B. Kuksin, Nearly Integrable Infinite-Dimensional Hamiltonian Systems. XXVII, 101 pages. 1993.

Vol. 1557: J. Azéma, P. A. Meyer, M. Yor (Eds.), Séminaire de Probabilités XXVII. VI, 327 pages. 1993.

Vol. 1558: T. J. Bridges, J. E. Furter, Singularity Theory and Equivariant Symplectic Maps. VI, 226 pages. 1993.

Vol. 1559: V. G. Sprindžuk, Classical Diophantine Equations. XII, 228 pages. 1993.

Vol. 1560: T. Bartsch, Topological Methods for Variational Problems with Symmetries. X, 152 pages. 1993.

Vol. 1561: I. S. Molchanov, Limit Theorems for Unions of Random Closed Sets. X, 157 pages. 1993.

Vol. 1562: G. Harder, Eisensteinkohomologie und die Konstruktion gemischter Motive. XX, 184 pages. 1993.

Vol. 1563: E. Fabes, M. Fukushima, L. Gross, C. Kenig, M. Röckner, D. W. Stroock, Dirichlet Forms. Varenna, 1992. Editors: G. Dell'Antonio, U. Mosco. VII, 245 pages. 1993.

Vol. 1564: J. Jorgenson, S. Lang, Basic Analysis of Regularized Series and Products. IX, 122 pages. 1993.

Vol. 1565: L. Boutet de Monvel, C. De Concini, C. Procesi, P. Schapira. M. Vergne. D-modules, Representation Theory, and Quantum Groups. Venezia, 1992. Editors: G. Zampieri, A. D'Agnolo. VII, 217 pages. 1993.

Vol. 1566: B. Edixhoven, J.-H. Evertse (Eds.), Diophantine Approximation and Abelian Varieties. XIII, 127 pages. 1993.

Vol. 1567: R. L. Dobrushin, S. Kusuoka, Statistical Mechanics and Fractals. VII, 98 pages. 1993.

Vol. 1568: F. Weisz, Martingale Hardy Spaces and their Application in Fourier Analysis. VIII, 217 pages. 1994.

Vol. 1569: V. Totik, Weighted Approximation with Varying Weight. VI, 117 pages. 1994.

Vol. 1570: R. deLaubenfels, Existence Families, Functional Calculi and Evolution Equations. XV, 234 pages. 1994.

Vol. 1571: S. Yu. Pilyugin, The Space of Dynamical Systems with the C^0-Topology. X, 188 pages. 1994.

Vol. 1572: L. Göttsche, Hilbert Schemes of Zero-Dimensional Subschemes of Smooth Varieties. IX, 196 pages. 1994.

Vol. 1573: V. P. Havin, N. K. Nikolski (Eds.). Linear and Complex Analysis – Problem Book 3 – Part I. XXII, 489 pages. 1994.

Vol. 1574: V. P. Havin, N. K. Nikolski (Eds.), Linear and Complex Analysis – Problem Book 3 – Part II. XXII, 507 pages. 1994.

Vol. 1575: M. Mitrea, Clifford Wavelets, Singular Integrals, and Hardy Spaces. XI, 116 pages. 1994.

Vol. 1576: K. Kitahara, Spaces of Approximating Functions with Haar-Like Conditions. X, 110 pages. 1994.

Vol. 1577: N. Obata, White Noise Calculus and Fock Space. X, 183 pages. 1994.

Vol. 1578: J. Bernstein, V. Lunts, Equivariant Sheaves and Functors. V, 139 pages. 1994.

Vol. 1579: N. Kazamaki, Continuous Exponential Martingales and BMO. VII, 91 pages. 1994.

Vol. 1580: M. Milman, Extrapolation and Optimal Decompositions with Applications to Analysis. XI, 161 pages. 1994.

Vol. 1581: D. Bakry, R. D. Gill, S. A. Molchanov, Lectures on Probability Theory. Editor: P. Bernard. VIII, 420 pages. 1994.

Vol. 1582: W. Balser. From Divergent Power Series to Analytic Functions. X, 108 pages. 1994.

Vol. 1583: J. Azéma, P. A. Meyer. M. Yor (Eds.), Séminaire de Probabilités XXVIII. VI. 334 pages. 1994.

Vol. 1584: M. Brokate. N. Kenmochi. I. Müller. J. F. Rodriguez, C. Verdi, Phase Transitions and Hysteresis. Montecatini Terme. 1993. Editor: A. Visintin. VII. 291 pages. 1994.

Vol. 1585: G. Frey (Ed.), On Artin's Conjecture for Odd 2-dimensional Representations. VIII. 148 pages. 1994.

Vol. 1586: R. Nillsen, Difference Spaces and Invariant Linear Forms. XII. 186 pages. 1994.

Vol. 1587: N. Xi, Representations of Affine Hecke Algebras. VIII, 137 pages. 1994.

Vol. 1588: C. Scheiderer, Real and Étale Cohomology. XXIV, 273 pages. 1994.

Vol. 1589: J. Bellissard, M. Degli Esposti. G. Forni, S. Graffi, S. Isola, J. N. Mather, Transition to Chaos in Classical and Quantum Mechanics. Montecatini Terme. 1991. Editor: 2S. Graffi. VII. 192 pages. 1994.

Vol. 1590: P. M. Soardi, Potential Theory on Infinite Networks. VIII, 187 pages. 1994.

Vol. 1591: M. Abate, G. Patrizio. Finsler Metrics – A Global Approach. IX, 180 pages. 1994.

Vol. 1592: K. W. Breitung, Asymptotic Approximations for Probability Integrals. IX, 146 pages. 1994.

Vol. 1593: J. Jorgenson & S. Lang. D. Goldfeld, Explicit Formulas for Regularized Products and Series. VIII, 154 pages. 1994.

Vol. 1594: M. Green, J. Murre, C. Voisin. Algebraic Cycles and Hodge Theory. Torino, 1993. Editors: A. Albano, F. Bardelli. VII, 275 pages. 1994.

Vol. 1595: R.D.M. Accola, Topics in the Theory of Riemann Surfaces. IX, 105 pages. 1994.

Vol. 1647: D. Dias, P. Le Barz, Configuration Spaces over Hilbert Schemes and Applications. VII, 143 pages. 1996.

Vol. 1648: R. Dobrushin, P. Groeneboom, M. Ledoux, Lectures on Probability Theory and Statistics. Editor: P. Bernard. VIII, 300 pages. 1996.

Vol. 1649: S. Kumar, G. Laumon, U. Stuhler, Vector Bundles on Curves – New Directions. Cetraro, 1995. Editor: M. S. Narasimhan. VII, 193 pages. 1997.

Vol. 1650: J. Wildeshaus, Realizations of Polylogarithms. XI, 343 pages. 1997.

Vol. 1651: M. Drmota, R. F. Tichy, Sequences, Discrepancies and Applications. XIII, 503 pages. 1997.

Vol. 1652: S. Todorcevic, Topics in Topology. VIII, 153 pages. 1997.

Vol. 1653: R. Benedetti, C. Petronio, Branched Standard Spines of 3-manifolds. VIII, 132 pages. 1997.

Vol. 1654: R. W. Ghrist, P. J. Holmes, M. C. Sullivan, Knots and Links in Three-Dimensional Flows. X, 208 pages. 1997.

Vol. 1655: J. Azéma, M. Emery, M. Yor (Eds.), Séminaire de Probabilités XXXI. VIII, 329 pages. 1997.

Vol. 1656: B. Biais, T. Björk, J. Cvitanic, N. El Karoui, E. Jouini, J. C. Rochet, Financial Mathematics. Bressanone, 1996. Editor: W. J. Runggaldier. VII, 316 pages. 1997.

Vol. 1657: H. Reimann, The semi-simple zeta function of quaternionic Shimura varieties. IX, 143 pages. 1997.

Vol. 1658: A. Pumarino, J. A. Rodríguez, Coexistence and Persistence of Strange Attractors. VIII, 195 pages. 1997.

Vol. 1659: V. Kozlov, V. Maz'ya, Theory of a Higher-Order Sturm-Liouville Equation. XI, 140 pages. 1997.

Vol. 1660: M. Bardi, M. G. Crandall, L. C. Evans, H. M. Soner, P. E. Souganidis, Viscosity Solutions and Applications. Montecatini Terme, 1995. Editors: I. Capuzzo Dolcetta, P. L. Lions. IX, 259 pages. 1997.

Vol. 1661: A. Tralle, J. Oprea, Symplectic Manifolds with no Kähler Structure. VIII, 207 pages. 1997.

Vol. 1662: J. W. Rutter, Spaces of Homotopy Self-Equivalences – A Survey. IX, 170 pages. 1997.

Vol. 1663: Y. E. Karpeshina; Perturbation Theory for the Schrödinger Operator with a Periodic Potential. VII, 352 pages. 1997.

Vol. 1664: M. Väth. Ideal Spaces. V, 146 pages. 1997.

Vol. 1665: E. Giné, G. R. Grimmett, L. Saloff-Coste, Lectures on Probability Theory and Statistics 1996. Editor: P. Bernard. X, 424 pages, 1997.

Vol. 1666: M. van der Put, M. F. Singer, Galois Theory of Difference Equations. VII, 179 pages. 1997.

Vol. 1667: J. M. F. Castillo, M. González, Three-space Problems in Banach Space Theory. XII, 267 pages. 1997.

Vol. 1668: D. B. Dix. Large-Time Behavior of Solutions of Linear Dispersive Equations. XIV, 203 pages. 1997.

Vol. 1669: U. Kaiser, Link Theory in Manifolds. XIV, 167 pages. 1997.

Vol. 1670: J. W. Neuberger, Sobolev Gradients and Differential Equations. VIII, 150 pages. 1997.

Vol. 1671: S. Bouc. Green Functors and G-sets. VII, 342 pages. 1997.

Vol. 1672: S. Mandal, Projective Modules and Complete Intersections. VIII. 114 pages. 1997.

Vol. 1673: F. D. Grosshans, Algebraic Homogeneous Spaces and Invariant Theory. VI, 148 pages. 1997.

Vol. 1674: G. Klaas, C. R. Leedham-Green, W. Plesken, Linear Pro-p-Groups of Finite Width. VIII, 115 pages. 1997.

Vol. 1675: J. E. Yukich, Probability Theory of Classical Euclidean Optimization Problems. X, 152 pages. 1998.

Vol. 1676: P. Cembranos, J. Mendoza, Banach Spaces of Vector-Valued Functions. VIII, 118 pages. 1997.

Vol. 1677: N. Proskurin, Cubic Metaplectic Forms and Theta Functions. VIII, 196 pages. 1998.

Vol. 1678: O. Krupková, The Geometry of Ordinary Variational Equations. X, 251 pages. 1997.

Vol. 1679: K.-G. Grosse-Erdmann. The Blocking Technique. Weighted Mean Operators and Hardy's Inequality. IX, 114 pages. 1998.

Vol. 1680: K.-Z. Li, F. Oort, Moduli of Supersingular Abelian Varieties. V, 116 pages. 1998.

Vol. 1681: G. J. Wirsching, The Dynamical System Generated by the 3n+1 Function. VII. 158 pages. 1998.

Vol. 1682: H.-D. Alber, Materials with Memory. X, 166 pages. 1998.

Vol. 1683: A. Pomp, The Boundary-Domain Integral Method for Elliptic Systems. XVI. 163 pages. 1998.

Vol. 1684: C. A. Berenstein, P. F. Ebenfelt, S. G. Gindikin, S. Helgason, A. E. Tumanov, Integral Geometry, Radon Transforms and Complex Analysis. Firenze. 1996. Editors: E. Casadio Tarabusi, M. A. Picardello, G. Zampieri. VII, 160 pages. 1998.

Vol. 1685: S. König. A. Zimmermann. Derived Equivalences for Group Rings. X. 146 pages. 1998.

Vol. 1686: J. Azéma, M. Émery. M. Ledoux, M. Yor (Eds.). Séminaire de Probabilités XXXII. VI. 440 pages. 1998.

Vol. 1687: F. Bornemann, Homogenization in Time of Singularly Perturbed Mechanical Systems. XII. 156 pages. 1998.

Vol. 1688: S. Assing, W. Schmidt. Continuous Strong Markov Processes in Dimension One. XII, 137 page. 1998.

Vol. 1689: W. Fulton, P. Pragacz. Schubert Varieties and Degeneracy Loci. XI, 148 pages. 1998.

Vol. 1690: M. T. Barlow. D. Nualart. Lectures on Probability Theory and Statistics. Editor: P. Bernard. VIII, 237 pages. 1998.

Vol. 1691: R. Bezrukavnikov. M. Finkelberg. V. Schechtman, Factorizable Sheaves and Quantum Groups. X. 282 pages. 1998.

Vol. 1692: T. M. W. Eyre, Quantum Stochastic Calculus and Representations of Lie Superalgebras. IX, 138 pages. 1998.

Vol. 1694: A. Braides, Approximation of Free-Discontinuity Problems. XI, 149 pages. 1998.

Vol. 1695: D. J. Hartfiel. Markov Set-Chains. VIII, 131 pages. 1998.

Vol. 1696: E. Bouscaren (Ed.): Model Theory and Algebraic Geometry. XV, 211 pages. 1998.

Vol. 1697: B. Cockburn. C. Johnson. C.-W. Shu. E. Tadmor, Advanced Numerical Approximation of Nonlinear Hyperbolic Equations. Cetraro. Italy, 1997. Editor: A. Quarteroni. VII, 390 pages. 1998.

Vol. 1698: M. Bhattacharjee, D. Macpherson, R. G. Möller. P. Neumann, Notes on Infinite Permutation Groups. XI, 202 pages. 1998.

Vol. 1700: W. A. Woyczyński. Burgers-KPZ Turbulence. XI, 318 pages. 1998.